Boundary Value Problems for Linear Partial Differential Equations

Boundary value problems play a significant role in modeling systems characterized by established conditions at their boundaries. On the other hand, initial value problems hold paramount importance in comprehending dynamic processes and foreseeing future behaviors. The fusion of these two types of problems yields profound insights into the intricacies of the conduct exhibited by many physical and mathematical systems regulated by linear partial differential equations.

Boundary Value Problems for Linear Partial Differential Equations provides students with the opportunity to understand and exercise the benefits of this fusion, equipping them with realistic, practical tools to study solvable linear models of electromagnetism, fluid dynamics, geophysics, optics, thermodynamics and specifically, quantum mechanics. Emphasis is devoted to motivating the use of these methods by means of concrete examples taken from physical models.

Features

- No prerequisites apart from knowledge of differential and integral calculus and ordinary differential equations.
- Provides students with practical tools and applications
- Contains numerous examples and exercises to help readers understand the concepts discussed in the book.

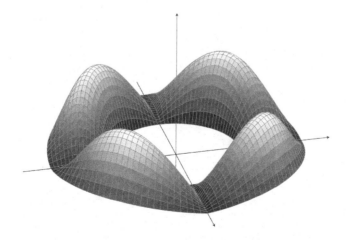

Boundary Value Problems for Linear Partial Differential Equations

Manuel Mañas
Luis Martínez Alonso

Universidad Complutense de Madrid

CRC Press
Taylor & Francis Group
Boca Raton London New York

CRC Press is an imprint of the
Taylor & Francis Group, an **informa** business

First edition published 2024
by CRC Press
2385 NW Executive Center Drive, Suite 320, Boca Raton FL 33431

and by CRC Press
4 Park Square, Milton Park, Abingdon, Oxon, OX14 4RN

CRC Press is an imprint of Taylor & Francis Group, LLC

ISBN: 978-1-032-66252-7 (hbk)
ISBN: 978-1-032-66450-7 (pbk)
ISBN: 978-1-032-66451-4 (ebk)

DOI: 10.1201/9781032664514

Typeset in LM Roman
by KnowledgeWorks Global Ltd.

Publisher's note: This book has been prepared from camera-ready copy provided by the authors.

To Montse, Clara and David

To Esther, Esther and Luis

Table of Contents

Preface

THE subject matter of this book, "Boundary and/or Initial Value Problems for Linear Partial Differential Equations," has been taught for several years as a semester course at the undergraduate level Physics curriculum of the Physics Faculty at Complutense University of Madrid. The book has evolved from a set of lecture notes that we have been preparing for the last 20 years. Our aim is to provide students with realistic practical tools to study solvable linear models of electromagnetism, fluid dynamics, geophysics, optics, thermodynamics and specifically, quantum mechanics.

Boundary value problems involve the quest for a solution to a linear partial differential equation that adheres to specified conditions set at the boundaries of the defined domain. These conditions come into play when constraints or prerequisites pertain to specific points or surfaces within the domain. Notably, boundary value problems prove to be especially applicable when circumstances entail constraints at boundary interfaces.

Boundary value problems play a significant role in modeling systems characterized by established conditions at their boundaries. On the other hand, initial value problems hold paramount importance in comprehending dynamic processes and foreseeing future behaviors. The fusion of these two types of problems yields profound insights into the intricacies of the conduct exhibited by many physical and mathematical systems regulated by linear partial differential equations.

Separation of variables, Green functions, eigenfunction series expansions, Fourier series, and transforms are included as some of the most important analytic tools involving an extensive description of the special functions of mathematical physics. Emphasis is devoted to motivating the use of these methods by means of concrete examples taken from physical models.

Prerequisites: The book only assumes knowledge of differential and integral calculus and ordinary differential equations. However, in order to substantially expand the scope of the applications, we provide elements of Hilbert space theory, differential operators as linear operators on Hilbert spaces, as well as their spectral theory.

Outline of Chapters: The basic notions of the theory of partial differential equations (PDEs) are presented in the Introduction. It also introduces five of the main second-order PDEs of mathematical physics: the Laplace equation, the Poisson equation, the wave equation, the heat equation, and the Schrödinger equation. Local solvability and the Cauchy–Kovalevskaya theorem are presented. The notion of characteristic and of general solution are also discussed. In Chapter 2, linear boundary and/or initial value problems for linear PDEs are defined and formulated in terms of linear differential operators. The chapter also contains an introduction to eigenvalue problems of linear differential operators, which play a unifying fundamental role along the methods provided in the book.

Chapters 3, 6, and 8 describe the two main methods presented in our discussion: the separation of variables method and the eigenfunction expansion method, and how they apply when boundary and/or initial conditions arise. They are illustrated with many examples based on the equations of Mathematical Physics expressed in the standard systems of curvilinear coordinates.

Mathematical tools such as spectral theory of symmetric operators in Hilbert spaces, Green functions, Sturm–Liouville operators, Fourier series, Fourier transforms, and special functions are supplied when they are required by the applications in Chapters 4, 5, and 7.

With the reader in mind, we have included many examples as well as exercises illustrating and motivating the ideas we develop. In particular, we have included 96 exercises, and presented the solutions for 61 of them. We have also included 140 figures.

Historical Matters: While we do not specialize in the field of mathematics history, we are eager to share some essential information about the pioneers behind the theories discussed in this book. In addition to the historical comments scattered throughout the text, we have included 25 concise biographies, which, following the model in [67], we refer to as 'encapsulated biographies,' at the section entitled *Remarkable Lives and Achievements* at the end of Chapters 1, 3, 4, 5, and 7. We also highly recommend two outstanding references written by esteemed professionals in the field of mathematics history: [11] and [33]. Additionally, the book [60] contains a collection of interesting historical facts. For a great resource on-line we refer the interest reader to MacTutor at: https://mathshistory.st-andrews.ac.uk.

Margin Marks: We have used symbols in the margin to indicate examples (✎) or observations (☞or ☜). Additionally, we have followed the Bourbaki tradition (as employed by Donald Knuth in *The TeXbook*) and denoted sections or subsections containing advanced material that can be skipped during the initial reading with the symbol ☿. We have also included comments on further reading, indicated by the symbol 📓, which guides the reader to the recommended bibliography. Historical comments are indicated by 🖋.

Acknowledgments: We would like to extend our heartfelt gratitude to the students who, over the years, have shared, corrected, and improved these notes. Their dedication and commitment have been invaluable in shaping this book.

We also want to express our appreciation to our colleagues who have contributed to enhancing this book, especially Professors Ricardo Brito, Juan M. R. Parrondo, and Ángel Rivas. Their insights and expertise have played a significant role in making this work more comprehensive and insightful. Thanks to all of them for their unwavering support and contributions.

We extend our sincere gratitude to Professors Francisco Marcellán and Juan Luis Vázquez, whose diligent reviews of the manuscript significantly enhanced the readability of this book. Any misprints, typos, or errors remain the sole responsibility of the authors.

LMA would like to express his sincere gratitude to his teacher, Professor Alberto Galindo. He has been a permanent source of inspiration in shaping his understanding of mathematics and mathematical physics and fostering his passion for learning.

MM also dedicates this book with deep gratitude and love to his parents, Luis Mañas (1931-2018) and Charo Baena (1935-1997), whose unwavering support, patience, and wisdom have been his guiding light and formed the heart of all.

Last, but certainly not least, MM wants to express his gratitude to his wife Montse and to his two children, Clara and David, and LMA would like to thank his wife Esther and his two children Esther and Luis. They have supported us and patiently shared in our research, teaching, and the endless journey of writing this book.

Manuel Mañas
Luis Martínez Alonso
November, 2023
Complutense University
Madrid

About the authors

Who is Manuel Mañas? Born in Madrid, Spain, in 1964, Manuel Mañas obtained his PhD in Physics in 1991 from the Complutense University of Madrid (UCM). Currently holding the position of full professor in Theoretical Physics at UCM, he has also served as a faculty member at the Instituto de Ciencias Matemáticas. Mañas has demonstrated his commitment to the academic sphere by being an active member of the Government Board of the Royal Spanish Mathematical Society.

He has served as a visiting professor at several prestigious international institutions, including Johannes Kepler Linz University (Austria), University of Aveiro (Portugal), University of California at Berkeley (USA), China University of Mining and Technology (Beijing, China), Université Catholique de Louvain (Belgium), SISSA (Trieste, Italy), La Sapienza (Italy), and Leeds University (UK).

His dedication to teaching is evident in his comprehensive curriculum. His instructional portfolio spans a diverse array of topics in mathematical physics. He specializes in teaching courses related to partial differential equations, with a particular focus on their applications in physics, encompassing boundary value problems. Moreover, his teaching extends to the fundamental principles of complex analysis in one variable and their practical applications. Furthermore, Mañas has shared his expertise through postgraduate courses, imparting knowledge on a wide range of advanced topics. These include functional analysis, Lie group theory, integrable systems, and orthogonal polynomials.

With 130 scientific publications to his name, his research primarily focuses on mathematical physics and applied mathematics. His expertise spans various areas, with notable contributions in the theory of integrable systems, the field of orthogonal polynomials, and the theory of special functions. Of late, he achieved a significant breakthrough by  establishing a spectral theorem similar to Favard's for bounded banded operators, effectively extending the well-known spectral Favard theorem for Jacobi matrices. His research also encompasses investigations into random walks and Markov chains beyond birth and death processes. His contributions also include foundational work in discovering the Lax pair for the Krichever–Novikov equation, creating Darboux transformations for the nonlinear Schrödinger equation, and identifying new solutions for supersymmetric integrable systems.

Mañas has demonstrated effective leadership and administrative skills throughout his career. He has held several key administrative roles, such as Vice Chancellor of Innovation, Vice Dean, and Coordinator of the Doctoral Program in Physics.

For more details see https://www.ucm.es/manuel_manas.

Who is Luis Martínez Alonso? Born in 1949 in Tolosa, Spain, Luis Martínez-Alonso earned his PhD in Physics in 1975 from the Complutense University of Madrid (UCM). Over more than 40 years as a full professor at UCM, he imparted mathematical methods to physicists. His teaching portfolio encompassed a wide range of mathematical physics subjects, from fundamental courses in linear algebra and calculus to specialized classes in functional analysis, partial differential equations, and Lie group theory.

Throughout his career, he has profoundly influenced several generations of physicists and earned recognition as an esteemed educator. Additionally, he served as the Director of the Theoretical Physics department at UCM during various periods. He currently holds the title of emeritus professor at UCM.

Martínez Alonso has also held visiting professorships at several prominent institutions, including the Universities of Montpellier and Paris VI in France, as well as Rome I and III in Italy, Lecce (Italy), and the Newton Institute at the University of Cambridge (UK).

He boasts authorship and co-authorship of over 120 research articles spanning topics such as symmetries, conservation laws, integrable nonlinear differential equations, matrix models, and mathematical aspects of cosmology. His notable research achievements include proving the converse of the Noether theorem, providing a counterexample to the Gel'fand–Kirillov conjecture on the enveloping algebras of Lie groups, formulating the soliton-radiation interactions in nonlinear integrable models (both continuous and discrete), establishing hierarchies of nonlinear integrable models associated with Schrödinger spectral problems featuring energy-dependent potentials, and contributing to the development of the four-dimensional integrable model known as the Martínez-Alonso–Shabat model.

Currently, his research interests center on inflationary quantum cosmology and the foundations of quantum physics.

1. Introduction

Contents

THIS book starts with a brief introduction to partial differential equations (PDEs). We recall some elementary definitions of complex number arithmetic and complex functions. We will also present some of the most paradigmatic linear PDEs of mathematical physics, such as the wave, heat, Laplace, Poisson, and Schrödinger equations.

We will also discuss changes of variables in a PDE, Cauchy problems and the existence of local solutions for PDEs in normal form, and the notion of a general or complete solution and the hodograph method as well.

§1.1. Partial Differential Equations

BASIC ideas and concepts for the study of linear PDEs are introduced here, such as complex numbers, partial derivatives and their notations, dependent and independent variables, and the order of a PDE. Additionally, we will introduce the most relevant PDEs of mathematical physics that we will discuss throughout this course, including the Poisson equation, the wave equation, the Schrödinger equation, and the heat equation. Finally, we will delve into changes of variables.

In the book we will often use integer and complex numbers; here we fix the notation

$$\mathbb{Z} := \{\ldots, -2, -1, 0, 1, 2, \ldots\},$$
$$\mathbb{N}_0 := \{0, 1, 2, 3, \ldots\},$$
$$\mathbb{N} := \{1, 2, 3, \ldots\}.$$

We consider functions that take complex values and depend on a certain number of real variables $(t, x, \ldots)$. We write

$$u = u(t, x, \ldots) = u_1(t, x, \ldots) + iu_2(t, x, \ldots)$$

to denote a function that depends on the real variables $(t, x, \ldots)$ and takes complex values with real and imaginary parts given by $\operatorname{Re} u = u_1 = u_1(t, x, \ldots)$ and $\operatorname{Im} u = u_2 = u_2(t, x, \ldots)$, respectively.

As complex numbers, the function u can be conjugated as $\bar{u} = u_1 - iu_2$ and has a modulus and an argument:

$$|u| = \sqrt{u_1^2 + u_2^2} = \sqrt{u\bar{u}}$$

and

$$\arg u = \arctan\left(\frac{u_2}{u_1}\right).$$

Hence, if we denote

$$r = |u|, \quad \theta = \arg u,$$

we can write

$$u_1 = r\cos\theta, \quad u_2 = r\sin\theta.$$

The complex exponential function Recall that in the arithmetic of complex numbers, for any given real numbers a and b, exponentials with complex exponents are defined using the Euler formulas

$$e^{ib} = \cos b + i\sin b, \quad e^{a+ib} = e^a e^{ib} = e^a(\cos b + i\sin b),$$

for $a, b \in \mathbb{R}$.

Euler's formula allows us to express functions with complex values in the form:

$$u = r\cos\theta + ir\sin\theta = re^{i\theta},$$

where

$$r = |u|, \quad \theta = \arg u.$$

This expression is exceedingly useful for applications in wave phenomena and quantum physics.

Note that $\arg u$ is a multivalued function, as for any possible argument, adding integer multiples of $2\pi i$ yields another possible argument. Consequently, we have the flexibility to select a determination, which confines the argument to a semi-closed interval with a length of 2π. For the principal argument, $\operatorname{Arg}(u) \in [-\pi, \pi)$.

Examples:

(1) Let $u(x, y) = xy + ie^{x^2+y^2}$. In this case $u_1(x, y) = xy$ and $u_2(x, y) = e^{x^2+y^2}$. Consequently, $\bar{u} = xy - ie^{x^2+y^2}$ and $|u| = \sqrt{(xy)^2 + e^{2(x^2+y^2)}}$.

(2) Let $u(x, y) = e^{xy+i(x^2+y^2)}$. Using Euler formulas $u = e^{xy}(\cos(x^2 + y^2) + i\sin(x^2+y^2))$, hence $u_1 = e^{xy}\cos(x^2+y^2)$ and $u_2 = e^{xy}\sin(x^2 + y^2)$, with modulus and argument given by $|u| = e^{xy}$ and $\arg u = x^2 + y^2$.

Sometimes, we will also make use of the multivalued **complex logarithm** defined as:

$$\log z = \log |z| + \mathrm{i}(\arg z + 2n\pi), \quad n \in \mathbb{Z},$$

where $\log |z|$ represents the natural logarithm (also known as the Napierian logarithm) of the positive real number $|z|$, and $\arg z$ can be any of the possible arguments of z. If we use the principal argument the logaritm is denoted by $\mathrm{Log}(z) = \log |z| + \mathrm{i}\,\mathrm{Arg}(z)$.

Examples:

   (1) Solve the equation

$$e^u = 1.$$

     Taking $\arg 1 = 0$ we get

$$u = \log 1 = 2n\pi\mathrm{i},$$

     for $n \in \mathbb{Z}$.

   (2) Solve the equation $e^u = -2\mathrm{i}$. Taking $\arg(-2\mathrm{i}) = -\pi/2$ we get

$$u = \log(-2\mathrm{i}) = \log 2 + \mathrm{i}\left(-\frac{\pi}{2} + 2n\pi\right),$$

     for $n \in \mathbb{Z}$.

§1.1.1. PDEs

> **Extended Notation for Partial Derivatives**
>
> On many occasions, the partial derivatives of u will be denoted as shown in the following examples:
>
> $$u_t = \frac{\partial u}{\partial t}, \quad u_x = \frac{\partial u}{\partial x}, \quad u_{xx} = \frac{\partial^2 u}{\partial x^2}, \quad u_{xy} = \frac{\partial^2 u}{\partial x \partial y}.$$
>
> We will always assume that the functions we handle admit derivatives to the order required by the operations that we perform. The most convenient situation arises when we assume that all derivatives of all orders exist. In that case, according to the Schwarz–Clairaut theorem, the result of multiple derivations is independent of the order in which we take the individual derivations.

As an example, we will have

$$u_{xxyxzy} = u_{xxxyyz} = u_{zxyxyx}.$$

The basic rule to derive and integrate functions u with complex values is to think of the imaginary unit i as a constant. So, for example

$$u_t = \frac{\partial u}{\partial t} = \frac{\partial u_1}{\partial t} + i\frac{\partial u_2}{\partial t},$$

$$u_x = \frac{\partial u}{\partial x} = \frac{\partial u_1}{\partial x} + i\frac{\partial u_2}{\partial x},$$

$$u_{xx} = \frac{\partial^2 u}{\partial x^2} = \frac{\partial^2 u_1}{\partial x^2} + i\frac{\partial^2 u_2}{\partial x^2},$$

$$u_{xy} = \frac{\partial^2 u}{\partial x \partial y} = \frac{\partial^2 u_1}{\partial x \partial y} + i\frac{\partial^2 u_2}{\partial x \partial y},$$

$$\int u \, dx = \int (u_1 + i u_2) \, dx = \int u_1 \, dx + i \int u_2 \, dx.$$

Specifically, the exponential function, denoted as e^u, applied to a function u with complex values obeys the same differentiation rule as if u were a real-valued function. For instance,

$$\frac{\partial e^u}{\partial x} = e^u u_x.$$

To prove it, just proceed as follows:

$$\frac{\partial e^u}{\partial x} = \frac{\partial}{\partial x}\left(e^{u_1} e^{i u_2}\right) = \frac{\partial}{\partial x}\left(e^{u_1} \cos u_2 + i e^{u_1} \sin u_2\right)$$

$$= \frac{\partial}{\partial x}\left(e^{u_1} \cos u_2\right) + i\frac{\partial}{\partial x}\left(e^{u_1} \sin u_2\right)$$

$$= e^{u_1} \frac{\partial u_1}{\partial x} \cos u_2 - e^{u_1} \sin u_2 \frac{\partial u_2}{\partial x} + i\left(e^{u_1} \frac{\partial u_1}{\partial x} \sin u_2 + e^{u_1} \cos u_2 \frac{\partial u_2}{\partial x}\right)$$

$$= \left(e^{u_1} \cos u_2 + i e^{u_1} \sin u_2\right)\left(\frac{\partial u_1}{\partial x} + i\frac{\partial u_2}{\partial x}\right) = e^u u_x.$$

Note the use of the chain rule.

Examples:

(1) Given $u(x, y) = xy + i e^{x^2+y^2}$ we have

$$u_x = y + 2ix e^{x^2+y^2},$$

$$u_{xx} = i(2 + 4x^2)e^{x^2+y^2}.$$

(2) For $u(x, y) = e^{xy+i(x^2+y^2)}$ the first-order derivatives are

$$u_x = (y + 2ix)e^{xy}(\cos(x^2 + y^2) + i\sin(x^2 + y^2)),$$

$$u_y = (x + 2iy)e^{xy}(\cos(x^2 + y^2) + i\sin(x^2 + y^2)).$$

(3) The indefinite integral of the exponential function $u(x) = e^{(a+ib)x}$, $a + ib \neq 0$, with exponent depending linearly on x, is done as if

$c = a + ib$ was a real number. That is to say,

$$\int e^{(a+ib)x}\mathrm{d}x = \frac{e^{(a+ib)x}}{a+ib}.$$

In this manner, one can compute the indefinite integral as follows

$$\int_0^\pi e^{(1+i)x}\mathrm{d}x = \left[\frac{e^{(1+i)x}}{1+i}\right]_0^\pi$$

$$= \frac{1}{1+i}\left(e^{(1+i)\pi} - e^0\right)$$

$$= -\frac{1}{1+i}(e^\pi + 1)$$

$$= \frac{1+e^\pi}{2}(-1+i).$$

Let's consider complex functions

$$u = u(x) = u_1(x) + iu_2(x)$$

that depend on a vector

$$x = (x_0, x_1, \ldots, x_{n-1})$$

in $\mathbb{R}^n$. Frequently, but not always, the variable x_0 will be identified with a time variable t.

Condensed Notation for Partial Derivatives

For derivatives, we will use the notation:

$$D^\alpha u := \frac{\partial^{|\alpha|}u}{\partial x_0^{\alpha_0} \partial x_1^{\alpha_1} \cdots \partial x_{n-1}^{\alpha_{n-1}}}, \quad |\alpha| = \alpha_0 + \alpha_1 + \cdots + \alpha_{n-1},$$

where vector indices are given by:

$$\alpha = (\alpha_0, \alpha_1, \cdots, \alpha_{n-1}) \in \mathbb{N}_0^n,$$

with n non-negative integer components. It is important to observe that $|\alpha|$ represents the order of the derivative $D^\alpha u$. By definition, if $\alpha = (0, 0, \ldots, 0)$, then $D^\alpha u \equiv u$.

Definition

The relationship between the two types of notations is easy to establish. For example, if $(x_0, x_1, x_2, x_3) = (t, x, y, z)$, then $u_{xxzyz} = D^\alpha u$ where $\alpha = (0, 2, 1, 2)$. When we have a single independent variable x, we will use the notation $D^n u := \frac{\mathrm{d}^n u}{\mathrm{d}x^n}$.

To define the concept of a PDE, it is convenient to use condensed notation.

> ## What is a PDE?
>
> A PDE is a relation of the form:
>
> $$(1.1) \qquad\qquad F(x, D^\alpha u) = 0,$$
>
> where F is a function depending on
>
> $$x = (x_0, x_1, \ldots, x_{n-1}),$$
>
> for $n > 1$, and on a finite number of derivatives $D^\alpha u$. The nomenclature is as follows:
>
> (1) The variables x_i, $i = \{0, \ldots, n-1\}$, are named as the **independent variables** of the PDE.
> (2) The unknown function u in the PDE is called the **dependent variable**.
> (3) If N is the maximum order of derivatives $D^\alpha u$ upon which the function F depends, then N is, by definition, the **order** of the PDE.
>
> Definition

The most frequent situation is that the PDE is defined by a function F that is a polynomial in the variables $D^\alpha u$. However, there are physical situations in which more general equations appear. For example, the so-called sine-Gordon equation,

$$u_{tt} - u_{xx} = \sin u.$$

In particular, if F is a polynomial of degree one in the variables $D^\alpha u$, it is said that the PDE is a **linear PDE**. In this case, the PDE takes the form:

$$(1.2) \qquad \boxed{\sum_\alpha{}' a_\alpha(x) D^\alpha u - f(x) = 0,}$$

where $\sum_\alpha{}'$ means that the sum extends to a finite set of multi-indices α with $|\alpha| \geq 0$. The functions $a_\alpha(x)$ and $f(x)$ are assumed to be given.

Normally, when we deal with a PDE as in (1.2), we rearrange the equation by moving $f(x)$ to the right-hand side (RHS) and write:

$$\sum_\alpha{}' a_\alpha(x) D^\alpha u = f(x).$$

We refer to $f(x)$ as the **inhomogeneous term** of the equation. If the function $f(x) \equiv 0$, then we will say that the linear PDE is **homogeneous**. Otherwise, if $f \neq 0$, then we say that the linear PDE is **inhomogeneous**.

In general, nonlinear PDEs are much more difficult to treat than the linear ones. We will not study them in this course. To consider a specific PDE, the extended notation is more convenient.

Examples:

(1) The PDE

$$u_x + e^{x+y} u_y - u = x^2 y^2,$$

is linear and has order 1. It is inhomogeneous with an independent term given by $x^2 y^2$.

(2) The PDE

$$u_{xx} u + u_y + xy = 0$$

is nonlinear due to the appearance of the term $u_{xx} u$, and it has order 2.

§1.1.2.   Linear PDEs in Physics

Physics is replete with linear and nonlinear PDEs. In both electromagnetism and quantum mechanics, the basic equations are linear. However, in other areas, such as continuous media dynamics or general relativity, the fundamental equations are nonlinear.

There are five examples of linear second-order PDEs to which we will devote particular attention. These main versions, with three spatial variables (x, y, z), are as follows:

Laplace Equation: $u_{xx} + u_{yy} + u_{zz} = 0$

The solutions of the Laplace equation are so important in mathematics that they have a special name, harmonic functions. Harmonic functions are very special functions with excellent properties for the analysis of differential equations. The real and imaginary parts of an analytic function in the complex plane are harmonic functions.

Poisson Equation: $u_{xx} + u_{yy} + u_{zz} = f$

Here $f = f(x, y, z)$ is a given function, like a mass density or an electrical charge distribution. Poisson equation appears often in electrostatics and geophysics. If $f \equiv 0$, the PDE is the Laplace equation.

Wave Equation: $u_{tt} = c^2(u_{xx} + u_{yy} + u_{zz})$

c a positive real number that represents the propagation speed of the wave. It appears in the description of waves in linear media, like the vibration of a string in a violin or of the sound in a room.

Schrödinger Equation:

$$i\hbar u_t = -\frac{\hbar^2}{2m}\left(u_{xx} + u_{yy} + u_{zz}\right) + V(x, y, z)u$$

Which describes the quantum dynamics of a particle of mass m in a field of forces with potential function $V = V(x, y, z)$. The symbol $\hbar$ represents the normalized Planck constant. Notice the presence of the imaginary number i in the coefficient of u_t. This fact is the main reason why, in this book, we consider functions with complex values.

Heat Equation: $u_t = a^2\left(u_{xx} + u_{yy} + u_{zz}\right)$

Which is relevant in thermal diffusion and fluid diffusion processes in general. The symbol a^2 represents the diffusion coefficient.

To write the above equations in abbreviated form, it is convenient to use the notation of the Laplacian operator:

$$\Delta u := u_{xx} + u_{yy} + u_{zz},$$

which is a fundamental example of a concept known as **differential operator**, to which we will devote significant attention in this course. In terms of the Laplacian, the previous equations are expressed as follows:

(1) Laplace equation:

$$\Delta u = 0.$$

(2) Poisson equation:

$$\Delta u = f.$$

(3) Wave equation:

$$u_{tt} = c^2 \Delta u.$$

(4) Schrödinger equation:

$$i\hbar u_t = -\frac{\hbar^2}{2m}\Delta u + Vu.$$

(5) Heat equation:

$$u_t = a^2 \Delta u.$$

Sometimes, we will consider simplified versions of the previous equations in which u does not depend on some of the variables x, y, and z. Thus, a version in $1 + 2$ dimensions of wave, Schrödinger, or heat equations is a PDE in which we assume that u depends on (t, x, y) only.

§1.1.3. Change of Variables

Given a PDE (1.1), one of the most frequent manipulations that we must do is determine the form it acquires when we perform a change of independent variables $x \mapsto x' = x'(x)$ with transformation equations:

$$x_i' = x_i'(x_0, x_1, \ldots, x_{n-1}), \quad i \in \{0, 1, \ldots, n - 1\},$$

which we always assume to be invertible, i.e., $x' \mapsto x = x(x')$, where the inverse transformation is given by the equations

$$x_i = x_i(x_0', x_1', \ldots, x_{n-1}'), \quad i \in \{0, 1, \ldots, n - 1\}.$$

For the sake of simplicity, we will not use a new function symbol for the compound function $u(x(x'))$, which we will simply denote as $u(x')$.

The form that (1.1) takes in the new variables is determined by substituting the variables x in $F(x, D^\alpha u)$ with $x(x')$, and the derivatives with respect to x $(D^\alpha u)$ with their expressions in terms of derivatives with respect to x'. For the latter, we must use the chain rule. The expressions derived for the first and second orders are as follows:

$$\frac{\partial u}{\partial x_j} = \sum_k \frac{\partial u}{\partial x_k'} \frac{\partial x_k'}{\partial x_j},$$

$$\frac{\partial^2 u}{\partial x_i \partial x_j} = \frac{\partial}{\partial x_i}\left(\sum_k \frac{\partial x_k'}{\partial x_j} \frac{\partial u}{\partial x_k'} \right)$$

$$= \sum_k \frac{\partial^2 x_k'}{\partial x_i \partial x_j} \frac{\partial u}{\partial x_k'} + \sum_l \sum_k \frac{\partial x_l'}{\partial x_i} \frac{\partial x_k'}{\partial x_j} \frac{\partial^2 u}{\partial x_l' \partial x_k'}$$

Sometimes, a change of variables can convert a PDE into a simpler one. A typical example is the $1 + 1$ dimensional wave equation.

Examples:

(i) Consider the wave equation, $u_{tt} - u_{xx} = 0$, and let us perform the change of variables $y_1 = t + x$ and $y_2 = t - x$, with the inverse change given by $t = \frac{1}{2}(y_1 + y_2)$ and $x = \frac{1}{2}(y_1 - y_2)$, respectively. We immediately obtain:

$$u_t = u_{y_1}\frac{\partial y_1}{\partial t} + u_{y_2}\frac{\partial y_2}{\partial t} = u_{y_1} + u_{y_2},$$

$$u_x = u_{y_1}\frac{\partial y_1}{\partial x} + u_{y_2}\frac{\partial y_2}{\partial x} = u_{y_1} - u_{y_2},$$

$$u_{tt} = \left(\frac{\partial}{\partial y_1} + \frac{\partial}{\partial y_2} \right)(u_{y_1} + u_{y_2}) = u_{y_1 y_1} + u_{y_2 y_2} + 2u_{y_1 y_2},$$

$$u_{xx} = \left(\frac{\partial}{\partial y_1} - \frac{\partial}{\partial y_2}\right)(u_{y_1} - u_{y_2}) = u_{y_1 y_1} + u_{y_2 y_2} - 2u_{y_1 y_2}.$$

As a consequence, the PDE is transformed into:

$$4u_{y_1 y_2} = 0.$$

In this new form, the PDE can be integrated immediately, and the solution is:

$$u = f(y_1) + g(y_2) = f(x + t) + g(t - x),$$

with f and g being arbitrary functions of one variable only.

(ii) Let us write the Laplace equation

$$u_{xx} + u_{yy} = 0,$$

in polar coordinates:

$$\boxed{\begin{array}{ll} x = r\cos\theta, & y = r\sin\theta, \\[2mm] r = \sqrt{x^2 + y^2}, & \theta = \arctan\dfrac{y}{x}. \end{array}}$$

Applying the chain rule, we get:

$$u_x = \frac{x}{\sqrt{x^2 + y^2}} u_r - \frac{y}{x^2 + y^2} u_\theta = \cos\theta\, u_r - \frac{\sin\theta}{r} u_\theta,$$

$$u_y = \frac{y}{\sqrt{x^2 + y^2}} u_r + \frac{x}{x^2 + y^2} u_\theta = \sin\theta\, u_r + \frac{\cos\theta}{r} u_\theta,$$

$$u_{xx} = \left(\cos\theta\frac{\partial}{\partial r} - \frac{\sin\theta}{r}\frac{\partial}{\partial\theta}\right)\left(\cos\theta\, u_r - \frac{\sin\theta}{r} u_\theta\right)$$

$$= \cos^2\theta\, u_{rr} + 2\frac{\cos\theta\sin\theta}{r^2} u_\theta - 2\frac{\cos\theta\sin\theta}{r} u_{\theta r} + \frac{\sin^2\theta}{r} u_r + \frac{\sin^2\theta}{r^2} u_{\theta\theta},$$

$$u_{yy} = \left(\sin\theta\frac{\partial}{\partial r} + \frac{\cos\theta}{r}\frac{\partial}{\partial\theta}\right)\left(\sin\theta\, u_r + \frac{\cos\theta}{r} u_\theta\right)$$

$$= \sin^2\theta\, u_{rr} - 2\frac{\cos\theta\sin\theta}{r^2} u_\theta + 2\frac{\cos\theta\sin\theta}{r} u_{\theta r} + \frac{\cos^2\theta}{r} u_r + \frac{\cos^2\theta}{r^2} u_{\theta\theta}.$$

Therefore, the two-dimensional Laplace equation in polar coordinates is

$$\boxed{u_{rr} + \frac{1}{r} u_r + \frac{1}{r^2} u_{\theta\theta} = 0.}$$

§1.2.   Boundary and Initial Conditions

W E will now discuss the boundary conditions and/or initial conditions that typically accompany the PDEs that need to be solved. First, we introduce the concepts of domain and boundary, and then we delve into the idea of boundary value problems (BVPs) for PDEs. In particular, we will cover the Dirichlet and Neumann problems, where the concept of normal derivative to the surface of the boundary becomes essential. We will also explore periodic boundary conditions and clarify what is meant by an initial condition. Lastly, we briefly discuss the type of solutions that will be considered: the smooth functions.

§1.2.1. Domains and Boundaries

Domains and Boundaries for PDEs

In the context of PDEs, we consider an unknown function $u = u(x)$ defined over a given set Ω in $\mathbb{R}^n$. We assume that Ω satisfies the following two conditions:

(1) Ω is an open set. This means that for every point $a \in \Omega$, there exists an $r > 0$ such that any point $x \in \mathbb{R}^n$ with a distance to a less than r ($d(x,a) < r$) belongs to Ω.

(2) Ω is connected. This implies that it is not possible to find two nonempty open sets Ω_1 and Ω_2, such that $\Omega_1 \cap \Omega_2 = \varnothing$ and $\Omega_1 \cup \Omega_2 = \Omega$.

In such cases, we refer to Ω as a **domain** in $\mathbb{R}^n$. The **boundary** $\partial\Omega$ of Ω is the set of points $a \in \mathbb{R}^n$ such that for any $r > 0$, there exist points x inside Ω and points x outside Ω with $d(x,a) < r$. The closure of Ω is the union set given by:

$$\overline{\Omega} = \Omega \cup \partial\Omega.$$

Definition

Property (1) ensures that Ω does not share any points with its boundary $\partial\Omega$. Property (2) prevents Ω from being divided into two separate sectors. In the case of 1D ($n = 1$), the domain Ω is necessarily an open interval of the real line $\Omega = (a,b) = \{x \in \mathbb{R} : a < x < b\}$, and if the interval is finite, $\partial\Omega = \{a,b\}$.

Next, we present diagrams of a domain and sets that do not qualify as domains since they violate either property (1) or property (2).

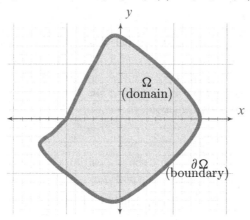

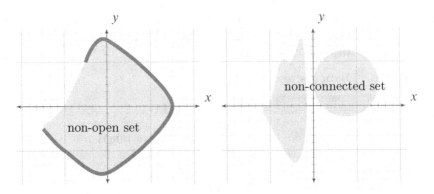

§1.2.2. Boundary Conditions

We will now introduce fundamental concepts related to boundary conditions for PDEs. Beginning with illustrative examples of the general approach, we will subsequently delve into significant instances, including Dirichlet, Neumann, and periodic boundary conditions.

Boundary Conditions for PDEs

Given a PDE defined in a domain $\Omega \subset \mathbb{R}^n$:

$$F(x, D^\alpha u) = 0, \quad x \in \Omega,$$

we are often required not only to find a function $u = u(x)$ that satisfies the PDE at all points of Ω, but also that satisfies certain conditions

$$f_i(x, D^\alpha u) = 0, \quad x \in S_i, \quad i \in \{1, \dots m\},$$

where S_i denotes parts of the boundary $\partial\Omega$ of Ω, and f_i are functions that depend on the variables x_i, as well as on a finite number of derivatives $D^\alpha u$ with $|\alpha| \geq 0$. Such conditions are known as **boundary conditions**. In this context, we will only consider linear boundary conditions, in which the functions f_i are degree-one polynomials in the variables $D^\alpha u$. That is, for $i \in \{1, \dots m\}$, they take the form

$$\sum_{\alpha}{}' b_{i,\alpha}(x) D^\alpha u - g_i(x) = 0, \quad x \in S_i,$$

or equivalently $\sum_{\alpha}' b_{i,\alpha} D^\alpha u|_{S_i} = g_i$. A problem consisting of solving a PDE over a domain Ω along with a set of boundary conditions is called a **boundary value problem**.

Definition

A PDE does not always admit the imposition of certain boundary conditions. As an example, consider the equation:

$$u_{xy} = 0$$

in the square domain $\Omega = \{(x, y) \in (0, 1) \times (0, 1)\}$. In this case, the boundary is the union of the four sides of the square:

$$\partial \Omega = S_1 \cup S_2 \cup S_3 \cup S_4,$$

where:

$$S_1 := \{(x, 0) \in \mathbb{R}^2 : 0 < x \leq 1\},$$
$$S_2 := \{(1, y) : 0 < y \leq 1\},$$
$$S_3 := \{(x, 1) \in \mathbb{R}^2 : 0 \leq x < 1\},$$
$$S_4 := \{(0, y) : 0 \leq y < 1\}.$$

Now, let's consider the following boundary conditions:

$$u|_{S_1} = f_1(x),$$
$$u|_{S_2} = f_2(y),$$
$$u|_{S_3} = f_3(x),$$
$$u|_{S_4} = f_4(y).$$

As the PDE, $u_{xy} = 0$, is satisfied, the function u_x does not depend on y, which implies $u_x|_{S_1} = u_x|_{S_3}$. Therefore, we must have $f_1'(x) = f_3'(x)$ and, similarly, we find $f_2'(y) = f_4'(y)$.

Hence, for the BVP to have a solution, it is necessary to impose additional conditions on the boundary data.

Dirichlet and Neumann Boundary Conditions

In problems involving a domain Ω in three-dimensional space $\mathbb{R}^3$, we use the following nomenclature for the simplest boundary conditions on a surface S contained within $\partial \Omega$:

(1) **Dirichlet condition:**

$$\boxed{u|_S = g.}$$

(2) **Neumann condition:**

$$\boxed{\left.\frac{\partial u}{\partial \boldsymbol{n}}\right|_S = g.}$$

Here, the normal derivative to the surface is given by

$$\frac{\partial u}{\partial \boldsymbol{n}} := \boldsymbol{n} \cdot \nabla u = n_1 u_x + n_2 u_y + n_3 u_z,$$

where

$$\boldsymbol{n} = (n_1, n_2, n_3)$$

represents a vector field of unit normals to the surface S.

Definition

Sometimes, we will consider $\mathbb{R}^2$ versions of the previously mentioned boundary conditions. In such cases, Ω will be a region in the plane, Γ will be a curve contained in $\partial \Omega$.

In any case, whenever we work with a boundary condition on a part S of the boundary of $\Omega \subset \mathbb{R}^n$, with $n \geq 2$, we will assume that S can be described by an implicit equation

$$f_S(x) = 0,$$

such that

$$\nabla f_S(x) \neq \mathbf{0}$$

for $x \in S$. We can define a unit normal vector field as follows

$$n(x) := \frac{\nabla f_S(x)}{\|\nabla f_S(x)\|}.$$

Examples: The following examples illustrate the normal vector fields and the corresponding directional derivative operations along the normal vectors.

(1) In this first example, consider Ω to be a disk in $\mathbb{R}^2$ centered at the origin with radius r , and its boundary is a circle $\Gamma = \Gamma(\Omega)$ of radius r. We can observe the following:

$$f_\Gamma(x, y) = x^2 + y^2 - r^2,$$

$$n = \left.\frac{(2x, 2y)}{\|(2x, 2y)\|}\right|_\Gamma = \frac{1}{r}(x, y),$$

$$\left.\frac{\partial u}{\partial n}\right|_\Gamma = \frac{1}{r}\left(x u_x + y u_y\right).$$

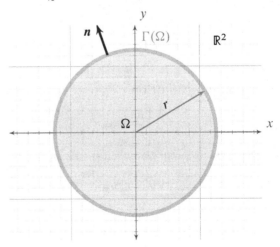

Domain, boundary and normal vector

(2) Secondly, we consider cylindrical coordinates.

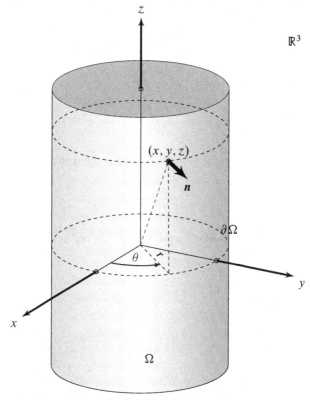

For that aim, let Ω be the inside of the infinite cylinder with axis OZ and a radius of r in $\mathbb{R}^3$, with $S = \partial\Omega$ being the cylindrical surface.

The points $(x, y, z) \in \mathbb{R}^3$ that satisfy the equation

$$f_S(x, y, z) = x^2 + y^2 - r^2,$$

form the boundary. The unit normal vector field to the surface is given by

$$\boldsymbol{n} = \left.\frac{(2x, 2y, 0)}{\|(2x, 2y, 0)\|}\right|_S = \frac{1}{r}(x, y, 0),$$

and the normal derivative over the surface is computed using the expression

$$\left.\frac{\partial u}{\partial \boldsymbol{n}}\right|_S = \frac{1}{r}\left(xu_x + yu_y\right).$$

(3) Now, we consider spherical coordinates. For that aim, let us consider Ω as the interior of the sphere centered at

$$a = (a_1, a_2, a_3) \in \mathbb{R}^3$$

and radius r, with

$$f_S(x, y, z) = (x - a_1)^2 + (y - a_2)^2 + (z - a_3)^2 - r^2,$$

and unit normal vetor field and normal derivatives given by

$$n = \frac{1}{r}(x - a_1, y - a_2, z - a_3),$$

$$\left.\frac{\partial u}{\partial n}\right|_S = \frac{1}{r}\left((x - a_1)u_x + (y - a_2)u_y + (z - a_3)u_z\right).$$

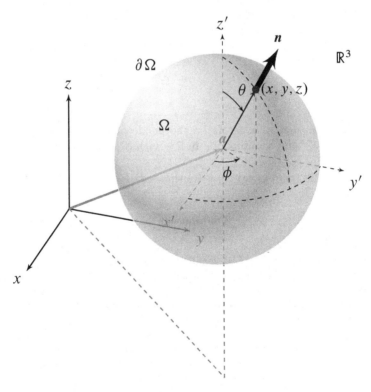

There are other types of boundary conditions associated with appropriate pairs of hypersurfaces on the boundary of Ω.

Periodic Conditions

Let us suppose that
$$S_1, S_2 \subset \partial\Omega \subset \mathbb{R}^n,$$
and there exists a bijective map between them:
$$\sigma : \quad S_1 \quad \longrightarrow \quad S_2$$
$$x \quad \longmapsto \quad \sigma(x).$$
Boundary conditions involving both S_1 and S_2 take the form:
$$\sideset{}{'}\sum_{\alpha} a_\alpha D^\alpha u(x) = \sideset{}{'}\sum_{\beta} b_\beta D^\beta u(\sigma(x)), \quad x \in S_1,$$

where $a_\alpha, b_\beta \in \mathbb{C}$ are constants.
In particular, periodic boundary conditions are expressed as equations of the form:
$$D^\alpha u(x) = D^\alpha u(\sigma(x)),$$

for $x \in S_1$.

Definition

Example: When $n = 1$, for $a \in \mathbb{R}$, we may consider
$$S_1 = \{0\} \subset \mathbb{R}$$
and
$$\sigma : \quad S_1 \quad \longrightarrow \quad S_2$$
$$x \quad \longmapsto \quad x + a.$$
The boundary conditions could be expressed as
$$u|_{x=0} = u|_{x=a},$$
$$u_x|_{x=0} = u_x|_{x=a}.$$

§1.2.3. Initial Conditions

In physical problems that involve the analysis of a system's evolution, both boundary and initial conditions typically coexist. These conditions provide essential information for determining a unique solution to the partial differential equation. While boundary conditions are imposed on the boundaries of the domain Ω, ensuring the behavior of the solution at those edges, initial conditions, on the other hand, dictate the system's state at a specific initial time or position, laying the foundation for the time evolution of the system. By combining these conditions, a well-defined solution to the PDE can be obtained that accurately describes the system's behavior over time and within the defined domain.

Initial Conditions

Another type of conditions that is usually required for the solutions of a PDE is the so-called *initial conditions* for one of the independent variables, denoted as t or x_0. These conditions typically take the form:

$$
\begin{cases}
u|_{t=t_0} = \Phi_0, \\
\dfrac{\partial u}{\partial t}\bigg|_{t=t_0} = \Phi_1, \\
\quad\vdots \\
\dfrac{\partial^{r-1} u}{\partial t^{r-1}}\bigg|_{t=t_0} = \Phi_{r-1},
\end{cases}
$$

where $r \geq 0$ and Φ_i, $i \in \{0, \ldots, r-1\}$, are functions that depend on the remaining independent variables. It is important to note that, in general, the initial conditions are not boundary conditions, as there are situations where the set determined by the equation $t = t_0$ can lie *inside* Ω.

Definition

Examples: The following examples illustrate two generic situations with different boundary and initial conditions: Consider the 1+1-dimensional heat equation:

$$u_t = u_{xx},$$

defined over the domain: $\Omega = \{(t, x) \in \mathbb{R}^2 : t > 0,\ 1 < x < 2\}$. This problem involves both boundary and initial conditions.

We can impose the following conditions:

$$
\begin{cases}
u|_{t=0} = (x-1)(x-2), \\
u|_{x=1} = 0, \\
u|_{x=2} = 0.
\end{cases}
$$

In this case, the initial condition also serves as a boundary condition. See figure below.

Next, let us consider the Schrödinger equation in 1+1 dimensions:

$$iu_t = -u_{xx},$$

defined over the domain: $\Omega = \{(t, x) \in \mathbb{R}^2 : -\infty < t < \infty,\ -1 < x < 1\}$.

We can impose the following conditions,

$$
\begin{cases}
u|_{t=0} = e^{-x^2} \sin \pi x, \\
u|_{x=-1} = 0, \\
u|_{x=1} = 0.
\end{cases}
$$

In this case, the initial condition is not a boundary condition. See the figure below.

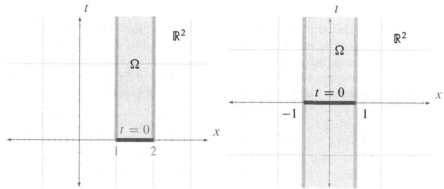

§1.2.4. Smooth Functions

When seeking solutions to a BVP, various types of functional spaces are considered to investigate the existence of such solutions. One common space used for boundary value problems is $C^{\infty}(\overline{\Omega})$, which consists of functions that are infinitely differentiable or "smooth" within the closure $\overline{\Omega}$ of Ω.

Smooth Functions

By definition $u \in C^{\infty}(\overline{\Omega})$ is a smooth function (sometimes we also use differentiable functions) if and only if all its derivatives $D^{\alpha}u(x)$ exist for all orders at every point x of some open set $\Omega_0 \supset \overline{\Omega}$.

Definition

Properties of Smooth Functions

(1) If $u, v \in C^{\infty}(\overline{\Omega})$, then the functions

$$\lambda u(x) + \mu v(x), \quad \lambda, \mu \in \mathbb{C},$$

$$u(x) \cdot v(x),$$

$$\frac{u(x)}{v(x)}, \text{(if } v(x) \neq 0 \text{ for all } x \in \overline{\Omega}),$$

also belong to $C^{\infty}(\overline{\Omega})$.

(2) If u is in $C^{\infty}(\overline{\Omega})$, then all its derivatives $D^{\alpha}u(x)$ also belong to $C^{\infty}(\overline{\Omega})$.

§1.3. Local Solvability

 HEN confronted with a partial differential equation defined over a domain $\Omega \subset \mathbb{R}^n$, the initial question that naturally arises pertains to the existence of solutions, denoted as $u = u(x)$.

Drawing parallels from our experience with ordinary differential equations, it's intuitive that the expected outcomes will largely be of a local nature. In simpler terms, these outcomes ensure the existence of solutions within open neighborhoods of each point within Ω.

An essential factor to contemplate revolves around the type of solutions we are pursuing, as well as the specific

$$F = F(x, D^\alpha u)$$

itself, which defines the given PDE.

In this context, our attention will now be directed towards a specific class of problems wherein we can derive a significant result regarding the existence of analytic solutions. It is pertinent, at this stage, to delve into the concept of an analytic function.

Analytic Functions

The space $A(\Omega)$ of analytic functions in an open set Ω is comprised of functions $f = f(x)$ such that for every point $a \in \Omega$, there exists a radius $r > 0$ and a multiple power series expansion of f given by

$$f(x) = \sum_{\alpha \in \mathbb{N}_0^n} c_\alpha (x - a)^\alpha,$$

$$(x - a)^\alpha := (x_0 - a_0)^{\alpha_0} (x_1 - a_1)^{\alpha_1} \cdots (x_n - a_n)^{\alpha_{n-1}},$$

which converges within the open ball

$$B(a, r) = \{x \in \mathbb{R}^{n+1} : |x - a| < r\}.$$

Definition

Every analytic function in Ω is also differentiable in Ω. Furthermore, its power series expansions around any $a \in \Omega$ coincide with its Taylor expansions. In other words,

$$f(x) = \sum_{\alpha \in \mathbb{N}_0^n} c_\alpha (x - a)^\alpha,$$

$$c_\alpha = \frac{1}{\alpha!} D^\alpha f(a).$$

The properties of analytic functions that are of primary interest to us now are as follows:

Properties of Analytic Functions

(1) If $f, g \in A(\Omega)$, then the functions

$$\lambda f(x) + \mu g(x), \qquad f(x) \cdot g(x), \qquad \frac{f(x)}{g(x)},$$

where in the last expression $g(x) \neq 0$, also belong to $A(\Omega)$ for all $\lambda, \mu \in \mathbb{C}$.

(2) The composition of two analytic maps

$$x \mapsto y = f(x) \mapsto z = g(y) = g(f(x)),$$

is also an analytic map. Here, we have

$$f : \Omega_1 \longrightarrow \Omega_2, \qquad g : \Omega_2 \longrightarrow \mathbb{C}.$$

We now introduce the notion of a PDE in normal or Kovalevskaya form.

PDE in Kovalevskaya Form

Consider a PDE with independent variables

$$x = (x_0 = t, \boldsymbol{x}), \quad \boldsymbol{x} := (x_1, \ldots, x_{n-1}).$$

We say that the PDE possesses normal form (or Kovalevskaya form) of order $r > 0$ with respect to the variable t if it can be written as:

(1.3) $$\frac{\partial^r u}{\partial t^r} = G\left(x, D^\alpha u\right), \quad r > 0,$$

where G is a function that depends polynomially on a finite number of derivatives but must be independent of the following: $\dfrac{\partial^r u}{\partial t^r}$ and $D^\alpha u$, with $|\alpha| > r$.

Definition

Observations:

(1) According to the previous definition, the PDE (1.3) is of order r.

(2) To analyze whether a PDE possesses normal form with respect to one of its independent variables t, the first step is to isolate the highest-order derivative with respect to t.

   Then, one should verify that in the right-hand side of the equation, no derivatives of strictly higher order appear.

Examples:

(1) The PDE

$$\frac{\partial u}{\partial x} - \frac{\partial u}{\partial y} = \log(xy)$$

possesses normal form with respect to either of its variables x or y, as it can be written as

$$\frac{\partial u}{\partial x} = \frac{\partial u}{\partial y} + \log(xy),$$

$$\frac{\partial u}{\partial y} = \frac{\partial u}{\partial x} - \log(xy)$$

and in both cases, the respective normality condition is satisfied. However, the function $\log(xy)$ is only analytic when $xy > 0$, thus the PDE is analytically normal in x and y within the domains

$$\Omega_1 = \left\{ (x, y) \in \mathbb{R}^2 : x > 0, y > 0 \right\},$$
$$\Omega_2 = \left\{ (x, y) \in \mathbb{R}^2 : x < 0, y < 0 \right\}.$$

(2) The heat equation in 1+1 dimensions

$$\frac{\partial u}{\partial t} - \frac{\partial^2 u}{\partial x^2} = 0,$$

is clearly in normal form with respect to the variable x, but not with respect to t, as when isolating u_t

$$\frac{\partial u}{\partial t} = \frac{\partial^2 u}{\partial x^2},$$

there remains a derivative of a higher order than that of u_t in the right-hand side.

(3) The Korteweg–de Vries equation

$$\frac{\partial^3 u}{\partial x^3} = -u \frac{\partial u}{\partial x} - \frac{\partial u}{\partial t},$$

is clearly normal with respect to x.

(4) The Poisson and Laplace equations in three dimensions are normal with respect to all their three independent variables

$$x, y, z.$$

(5) The wave equation in $1 + 3$ dimensions is normal with respect to all its four independent variables

$$t, x, y, z.$$

(6) The Schrödinger and heat equations in $1 + 3$ dimensions are not normal with respect to t, but they are with respect to x, y, z.

Cauchy Problem

Let's consider the PDE

(1.4)
$$\frac{\partial^r u}{\partial t^r} = G\left(x, D^\alpha u\right), \quad x \in \Omega \subset \mathbb{R}^n,$$

normal with respect to a variable t and order r, defined over a domain

$$\Omega = I \times \Lambda,$$

where I is an open interval in $\mathbb{R}$ and Λ is an open set in $\mathbb{R}^{n-1}$. A Cauchy problem with initial values consists of determining a solution $u = u(x)$ of this PDE that satisfies the r initial conditions

(1.5)
$$\begin{cases} u\left(t_0, x\right) = \Phi_0(x), \\ \dfrac{\partial u}{\partial t}\left(t_0, x\right) = \Phi_1(x), \\ \qquad \vdots \\ \dfrac{\partial^{r-1} u}{\partial t^{r-1}}\left(t_0, x\right) = \Phi_{r-1}(x), \end{cases}$$

where $t_0 \in I$ and $\Phi_i = \Phi_i(x)$, $i \in \{0, \ldots, r-1\}$, are given functions, referred to as the initial values of the problem.

Definition

Example: For the nonlinear Schrödinger equation, the Cauchy problem with initial conditions is formulated as

$$\begin{cases} \dfrac{\partial^2 u}{\partial x^2} = -\mathrm{i}\dfrac{\partial u}{\partial t} + u^3, \\ \quad u(t,0) = f(t), \\ \dfrac{\partial u}{\partial x}(t,0) = g(t). \end{cases}$$

One of the general results in the theory of PDEs applicable to both linear and nonlinear cases, is the following theorem due to Cauchy and Kovalevskaya, which we will refer to as the CK theorem.

The CK theorem is a fundamental result in the theory of PDEs, specifically focusing on the existence and uniqueness of solutions for certain types of initial value problems involving first-order PDEs. Named after the mathematicians Augustin-Louis Cauchy and Sofia Kovalevskaya, this theorem provides conditions under which solutions to these PDEs can be guaranteed to exist and be unique.

Cauchy–Kovalevskaya theorem emerged as a result of contributions from both Augustin-Louis Cauchy and Sofia Kovalevskaya, with Kovalevskaya's work in the late 19th century being particularly instrumental in the development of the theorem. Their combined efforts significantly advanced the understanding of the behavior of solutions to partial differential equations.

Cauchy–Kovalevskaya Theorem (1842, 1874)

Consider a Cauchy problem with initial values for a normal PDE (1.4), and let $x_0 = (t_0, \boldsymbol{x}_0) \in \Omega \subset \mathbb{R}^n$ be a point in its domain such that:

(1) **Analyticity condition of the PDE:** As a function of x, the function $G(x, D^\alpha u)$ in (1.4) is analytic at x_0.

(2) **Analyticity condition of the initial values:** The initial values $\Phi_i(\boldsymbol{x})$, $i \in \{0, \dots, r-1\}$, in (1.5), are analytic functions at $\boldsymbol{x}_0$.

Then, there exists a function $u = u(x)$ defined over an open set $\Omega_0 \subset \Omega$ containing x_0, such that:

(1) The function $u = u(x)$ satisfies the PDE in Ω_0 and the initial conditions at every point $(t_0, \boldsymbol{x})$, $\boldsymbol{x} \in \Omega_0$.

(2) The function $u = u(x)$ is the unique analytic function in Ω_0 that satisfies these properties.

Theorem

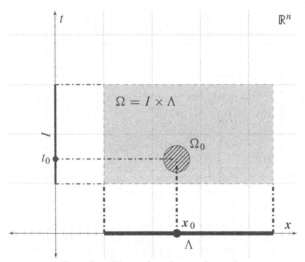

Cauchy–Kovaslevskaya domains

☞ **Observations:**

(1) The theorem guarantees the local existence and uniqueness of an analytic solution to a Cauchy problem with initial data, provided that the PDE is normal and both the PDE and the initial data depend analytically on the independent variables.

(2) There is a clear analogy between this result and the basic existence theorems in the theory of ordinary differential equations. We will say that a PDE is analytically normal if it is normal and satisfies

condition (i) of analyticity in the CK theorem. Then, under appropriate conditions, the local solution of an ODE of order r depends on r arbitrary constants.

The local analytic solution of an analytically normal PDE of order r depends on r arbitrary analytic functions.

(3) The proof of the theorem, which we will not provide, is based on generating a solution in the form of a multiple Taylor series expansion around x_0

$$u(t, x) = \sum_{\alpha \in \mathbb{N}_0^n} \frac{1}{\alpha!} D^\alpha u\,(x_0)\,(x - x_0)^\alpha.$$

The unknowns to be determined are the derivatives $D^\alpha u\,(x_0)$. To do this, note that by differentiating with respect to the variables x_i, $i \in \{1, 2, \ldots, n-1\}$, the initial conditions (1.5), we can find all derivatives of the form

$$\frac{\partial^{|\alpha|} u}{\partial t^{\alpha_0} \partial x_1^{\alpha_1} \cdots \partial x_{n-1}^{\alpha_{n-1}}}\,(t_0, x) = \frac{\partial^{|\alpha|-\alpha_0} \Phi_{\alpha_0}}{\partial x_1^{\alpha_1} \cdots \partial x_{n-1}^{\alpha_{n-1}}}\,(x),$$

with $\alpha_0 \le r - 1$. To calculate derivatives with $\alpha_0 \ge r$, we must use the previous derivatives as well as the PDE (1.5) and its derivatives. Thanks to the normal form of the PDE, it is possible to uniquely generate all the derivatives $D^\alpha u\,(x_0)$ in this way. This process demonstrates the uniqueness of the analytic solution of the Cauchy problem (1.4) and (1.5). The proof of the convergence of the series, i.e., that we are actually constructing a function, is more delicate.

(4)

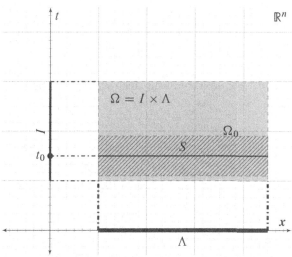

Cauchy–Kovalevskaya domains

The condition $t = t_0$ ($t_0 \in I$) determines a piece of a hyperplane in $S \subset \Omega$. If the function $G(x, D^\alpha u)$ is analytic in S and the initial data $\Phi_i(x)$ are analytic functions in Λ, then the theorem ensures the existence of a local analytic solution around each point of S. As a consequence, it can be shown that there exists a unique analytic solution of the Cauchy problem in an open set Λ containing S. See picture above.

To understand the scope of this important result, let's consider the following examples.

Examples:

(1) Consider the following problem

$$\begin{cases} u_t = u_x, & (t, x) \in \mathbb{R}^2, \\ u(0, x) = e^x. \end{cases}$$

Clearly, the PDE is normal with respect to t, and the conditions of analytic dependence are satisfied throughout $\mathbb{R}^2$. Let's seek a local solution around the point $(t, x) = (0, 0)$:

(1.6)
$$u(t, x) = \sum_{n,m=0}^{\infty} \frac{1}{n! m!} \frac{\partial^{n+m} u}{\partial t^n \partial x^m}(0, 0) t^n x^m.$$

Differentiating the initial condition with respect to x, we find that

(1.7)
$$\frac{\partial^m u}{\partial x^m}(0, 0) = \left. \frac{\partial^m e^x}{\partial x^m} \right|_{x=0} = e^0 = 1.$$

Moreover, differentiating the PDE with respect to t

$$\frac{\partial^{n+1} u}{\partial t^{n+1}} = \frac{\partial^{n+1} u}{\partial t^n \partial x},$$

for $n \in \mathbb{N}_0$, and iterating this result, we obtain

$$\frac{\partial^{n+1} u}{\partial t^{n+1}} = \frac{\partial^{n+1} u}{\partial t^{n-1} \partial x^2} = \cdots = \frac{\partial^{n+1} u}{\partial x^{n+1}}.$$

Differentiating this last relation with respect to x yields

$$\frac{\partial^{n+m+1} u}{\partial t^{n+1} \partial x^m} = \frac{\partial^{n+m+1} u}{\partial x^{n+m+1}},$$

for $n, m \in \mathbb{N}_0$.

Thus, for $n, m \in \mathbb{N}_0$, we find:

(1.8)
$$\frac{\partial^{n+m+1} u}{\partial t^{n+1} \partial x^m}(0, 0) = \frac{\partial^{n+m+1} u}{\partial x^{n+m+1}}(0, 0), \qquad \left. \frac{\partial^{n+m+1} e^x}{\partial x^{n+m+1}} \right|_{x=0} = 1.$$

Using (1.7) and (1.8) in (1.6), we obtain

$$u(t,x) = \sum_{n,m=0}^{\infty} \frac{1}{n!m!} t^n x^m = \sum_{n=0}^{\infty} \frac{1}{n!} t^n \sum_{m=0}^{\infty} \frac{1}{m!} x^m = e^t e^x.$$

It can be directly verified that we have constructed a solution to the problem. The general solution of the equation $u_t = u_x$ has the form $f(t+x)$ where f is any differentiable function. If we further require $u(0,x) = e^x$, then we must have $f(x) = e^x$ and thus $u(t,x) = e^{t+x}$ as we obtained.

(2) **Absence of normality:** Let's now consider the Cauchy problem:

$$\begin{cases} u_t = u_{xx}, & (t,x) \in \mathbb{R} \times (-1,1), \\ u(0,x) = \dfrac{1}{1-x}. \end{cases}$$

In this case, although the conditions of analyticity in the CK theorem are satisfied throughout the domain, the PDE is not normal with respect to t. Therefore, we cannot guarantee the conclusions of the CK theorem. In fact, we are going to demonstrate that they are not fulfilled. To do this, we will assume that there exists a local analytic solution at $(t,x) = (0,0)$

$$u(t,x) = \sum_{n,m=0}^{\infty} \frac{1}{n!m!} \frac{\partial^{n+m} u}{\partial t^n \partial x^m}(0,0) t^n x^m.$$

The series must converge for every (t,x) in an open set of the type $(-r_0, r_0) \times (-r_1, r_1)$. For any x_0 such that $|x_0| < r_1$, the function $u(t, x_0)$ will be an analytic function of t around $t = 0$, and thus it will have a power series expansion

$$u(t, x_0) = \sum_{n=0}^{\infty} a_n(x_0) t^n,$$

where

$$a_n(x_0) = \frac{1}{n!} \frac{\partial^n u}{\partial t^n}(0, x_0).$$

However, by differentiating the PDE

$$\frac{\partial^n u}{\partial t^n} = \frac{\partial^{n+1} u}{\partial t^{n-1} \partial x^2},$$

for $n \in \mathbb{N}_0$, and iterating this result, we obtain

$$\frac{\partial^n u}{\partial t^n} = \frac{\partial^{n+4} u}{\partial t^{n-2} \partial x^4} = \cdots = \frac{\partial^{2n} u}{\partial x^{2n}}.$$

On the other hand, by differentiating the initial condition with respect to x, we have

$$\frac{\partial^m u}{\partial x^m}(0, x_0) = \left. \frac{\partial^m (1-x)^{-1}}{\partial x^m} \right|_{x=x_0} = \frac{m!}{(1-x_0)^{m+1}}.$$

Thus,

$$a_n(x_0) = \frac{(2n)!}{n!\,(1-x_0)^{2n+1}}.$$

However, this result implies that the series (1.10) has a radius of convergence $r = 0$, since applying the well-known formula:

$$\frac{1}{r} = \lim_{n\to\infty} \left|\frac{a_{n+1}}{a_n}\right| = \lim_{n\to\infty} \frac{(2n+2)(2n+1)}{(n+1)\,(1-x_0)^2} = \infty.$$

Hence, it is not possible to find an analytic solution around $(0,0)$.

Even though there is no local analytic solution at $(0,0)$ for this Cauchy problem, it can be shown that there does exist a solution of class C^∞.

(3) **Absence of Condition (i) of Analyticity of the PDE** If Condition (i) of analyticity is not satisfied, we can encounter severe problems with the existence of local solutions. It's not just that we might lack analytic solutions, but perhaps the PDE has no admissible local solutions. (By "admissible," we mean solutions whose degree of differentiability is greater than or equal to the order of the PDE.)The construction of such examples is sophisticated; we will only mention the one due to H. Lewy (1957), which involves the following linear first-order PDE:

$$u_x + iu_y - 2i(x+iy)u_t = f(t).$$

It turns out that if $f = f(t)$ is a continuous function with real values, depending solely on t, and is not analytic at $t = 0$, then there is no C^1 local solution at $(t,x,y) = (0,0,0)$. This example illustrates the depth of the local existence problem for solutions of PDEs in the case of non-analyticity.

Among the most general results that can be used in such a context is the following important result:

Ehrenpreis–Malgrange Theorem (1955)

Given a linear PDE of the form:

$$\sum_{\alpha}{}' c_\alpha D^\alpha u(x) = f(x),$$

where the c_α are constants and f is a smooth function at $x_0 \in \mathbb{R}^n$, then there exists a smooth local solution at x_0.

Theorem

Historical Notes: In 1842, Cauchy proved an existence and uniqueness theorem for quasilinear first-order PDEs with real analytic data. For a set of real analytic matrix maps A_j from a neighborhood of $(0,0)$ in $\mathbb{R}^n \times \mathbb{R}^m$ into $\mathbb{R}^{m \times m}$, and a real analytic map f into $\mathbb{R}^m$, he showed that there exists a unique real analytic solution u in a neighborhood of the origin for the following system:

$$\begin{cases} \dfrac{\partial u}{\partial x_n} = \displaystyle\sum_{j=1}^{n-1} A^j(x, u)\dfrac{\partial u}{\partial x_j} + f(x, u), \\ u(x_1, \ldots, x_{n-1}, 0) = 0. \end{cases}$$

More than 30 years later, in 1874, Kovalevskaya independently proved a general nonlinear version of Cauchy's theorem in her PhD thesis, which is the second-order version, with

$$F = F\left(x, u, Du, D^2u\right) = 0,$$

of what we now refer to as the Cauchy–Kovalevskaya theorem.

Around 1955, Leon Ehrenpreis and Bernard Malgrange independently established the local solvability of any linear PDE with constant coefficients, as mentioned earlier. At that time, it was commonly believed that any linear operator L should be locally solvable, as it became well-known in the field of PDEs. This belief persisted until the example mentioned by Hans Lewy in 1957.

In 1963, Louis Nirenberg and François Trèves proposed a condition on the principal symbol of the N-th order PDE:

$$\sigma_L(\xi) := \sum_{|\alpha|=N} a_\alpha(x)\xi^\alpha,$$

$$\xi^\alpha := \prod_{i=1}^{n} \xi_i^{\alpha_i}.$$

This condition should be necessary and sufficient for local solvability. They introduced the concept of bi-characteristics of H, which are curves

$$(x(t), \xi(t))$$

solving the Hamilton equations:

$$\frac{\mathrm{d}x}{\mathrm{d}t} = \nabla_\xi H(x, \xi),$$

$$\frac{\mathrm{d}\xi}{\mathrm{d}t} = -\nabla_x H(x, \xi)$$

These curves have the important property that the function

$$H(x(t), \xi(t))$$

remains constant on them. When this constant is zero, we have null bi-characteristics of H.

The condition found by Nirenberg and Trèves is as follows:

On every null bi-characteristic of the real part of the symbol $\operatorname{Re}\sigma_L$, the imaginary part of the symbol $\operatorname{Im}\sigma_L$ has a constant sign.

The necessity and sufficiency of this condition for local solvability were established for an increasing number of cases in successive works by Nirenberg and Trèves themselves, followed by Beals and Fefferman, and Hörmander.

This historical note draws its information from Chapter X in Volume Two of [38] and from the informative paper [62].

§1.4. Characteristics

HARACTERISTIC hypersurfaces play a pivotal role in the study of partial differential equations, particularly within the realm of hyperbolic PDEs. They offer profound insights into the solution behavior and the intricate propagation of the initial data throughout the solution domain.

Hyperbolic PDEs constitute a class of equations characterized by their wave-like nature. They encompass phenomena such as the wave equation, fluid dynamics equations, and linearized approximations of nonlinear equations. These equations capture scenarios where disturbances advance at finite velocities.

Previously, we considered the question of local solution existence for a PDE. Now, we want to explore the same issue but for Cauchy problems with initial values over a subset S of $\mathbb{R}^n$ determined by an implicit equation:

$$(1.9) \qquad h\left(x_0, x_1, \ldots, x_{n-1}\right) = 0.$$

Such subsets S are referred to as hypersurfaces in $\mathbb{R}^n$. We can construct a field of unit normal vectors on S using the expression:

$$n(x) := \frac{\nabla h(x)}{\|\nabla h(x)\|}.$$

In the cases of $n = 2$ (the plane) and $n = 3$ (space), S will be a curve denoted as Γ and a surface, respectively.

The significance of characteristic hypersurfaces is that the Cauchy problem with initial values on them will generally be incompatible. This is because according to the given definition, the value of $F\left(x, D^\alpha u\right)$ on S is determined by the initial conditions, and therefore the existence of a solution to the PDE is only possible if that value is zero.

Note that characteristic hypersurfaces denoted as S serve to transmit values from a designated hypersurface S'. This holds true particularly within the framework of considering a Cauchy problem for hyperbolic systems. The values situated at the intersection of S and S' play a pivotal role in dictating the values along the hypersurface S'. It's worth mentioning that, as we'll delve into later, characteristics are the loci where discontinuities have the potential to propagate.

> ## Characteristic Hypersurfaces
>
> Given a PDE of order r over a domain Ω in $\mathbb{R}^n$,
>
> (1.10) $F\left(x, D^\alpha u\right) = 0,$ $x \in \Omega.$
>
> Given a hypersurface $S \subset \bar{\Omega}$ described by the implicit equation (1.9), a Cauchy problem with initial values on S is a system of equations determined by the previous PDE (1.10) and conditions
>
> (1.11)
> $$\begin{cases} u\big|_S = \Phi_0(x), \\ \dfrac{\partial u}{\partial \boldsymbol{n}}\bigg|_S = \Phi_1(x), \\ \quad \vdots \\ \dfrac{\partial^{r-1} u}{\partial \boldsymbol{n}^{r-1}}\bigg|_S = \Phi_{r-1}(x), \end{cases}$$
>
> where the initial data $\Phi_i(x)$, $i \in \{0, \dots, r-1\}$, are functions defined on S. We will say that S is a characteristic hypersurface of the PDE (1.10) if the initial conditions (1.11) determine the value of the function $F\left(x, D^\alpha u\right)$ on S.
>
> Definition

Examples:

(1) Let's consider the following Cauchy problem

$$\begin{cases} \dfrac{\partial u}{\partial t} = \dfrac{\partial^2 u}{\partial x^2}, & (t, x) \in \mathbb{R}^2, \\ \begin{cases} u|_{t=0} = \Phi_0(x), \\ \dfrac{\partial u}{\partial t}\bigg|_{t=0} = \Phi_1(x). \end{cases} \end{cases}$$

Note that the PDE is of second order and the two initial conditions are associated with the curve Γ determined by the equation $t = 0$. On Γ, the function $F = u_t - u_{xx}$ is determined by the initial conditions, as follows:

$$(u_t - u_{xx})|_\Gamma = \Phi_1(x) - \Phi_{0,xx}(x).$$

However, the PDE demands that $F = 0$ on Γ. Therefore, in general, the Cauchy problem is incompatible, as it will only have a solution for certain initial data. Those that satisfy

$$\Phi_1(x) - \Phi_{0,xx}(x) = 0.$$

(2) Let's study the following Cauchy problem

(1.12)
$$\begin{cases} A\left(\xi_1, \xi_2\right) u_{\xi_1} + B\left(\xi_1, \xi_2\right) u_{\xi_2} + G\left(\xi_1, \xi_2, u\right) = 0, \\ u\left(0, \xi_2\right) = \Phi\left(\xi_2\right). \end{cases}$$

Now the independent variables are ξ_1, ξ_2, and the curve Γ is determined by $\xi_1 = 0$. Observe that on Γ, all partial derivatives with respect to ξ_2 of the solutions are entirely determined by Φ and its derivatives:

$$\frac{\partial^n u}{\partial \xi_2^n}(0, \xi_2) = \Phi^{(n)}(\xi_2).$$

However, calculating the derivatives with respect to ξ_1 of u on Γ cannot be done using the initial condition. Therefore, the left-hand side of the PDE won't be completely determined on Γ, unless it happens that $A(0, \xi_2) = 0$. In such a case, for a solution of the Cauchy problem to exist, it must satisfy

$$B(0, \xi_2)\,\Phi'(\xi_2) + G(0, \xi_2, \Phi) = 0$$

which clearly is a constraint involving the coefficients B, G defining the PDE, and the function Φ. The curve $\Gamma : \xi_1 = 0$ is an example of what is understood by a characteristic curve, a concept that will be studied next.

§1.4.1. First-Order PDEs

Let's now consider the general Cauchy problem for a first-order PDE in $\Omega \subset \mathbb{R}^2$ of the form:

$$(1.13) \qquad \begin{cases} a(x,y)u_x + b(x,y)u_y + G(x,y,u) = 0, \\ u|_\Gamma = \Phi(x,y), \end{cases}$$

where the functions a and b take real values, and Γ is a curve in Ω defined by the equation:

$$\Gamma : \quad y = y(x).$$

Our goal is to determine the curves Γ that are characteristics. To do this, we perform a change of variables $(x, y) \mapsto (\xi_1, \xi_2)$

$$\xi_1 = \xi_1(x, y), \quad \xi_2 = \xi_2(x, y),$$

to bring our problem into the form (1.12). Thus, we need to impose the condition

$$\xi_1(x, y) = y - y(x).$$

This condition makes the equation of the curve Γ in the new variables become

$$\Gamma : \quad \xi_1 = 0.$$

As for the PDE, it takes the form

$$a\left(u_{\xi_1}\frac{\partial \xi_1}{\partial x} + u_{\xi_2}\frac{\partial \xi_2}{\partial x}\right) + b\left(u_{\xi_1}\frac{\partial \xi_1}{\partial y} + u_{\xi_2}\frac{\partial \xi_2}{\partial y}\right) + G(\xi_1, \xi_2, u) = 0$$

or

(1.14) $\left(a\xi_{1,x} + b\xi_{1,y}\right)u_{\xi_1} + \left(a\xi_{2,x} + b\xi_{2,y}\right)u_{\xi_2} + G\left(\xi_1, \xi_2, u\right) = 0.$

The initial condition is written as

(1.15) $u\left(0, \xi_2\right) = \Phi\left(0, \xi_2\right),$

where now we assume that all functions $a, b, u, G, \dots$ are expressed in terms of the variables (ξ_1, ξ_2). Differentiating (1.15) yields

$$u_{\xi_2}\left(0, \xi_2\right) = \Phi_{\xi_2}\left(0, \xi_2\right).$$

Therefore, the function F of the PDE (1.14)

$$F = \left(a\xi_{1,x} + b\xi_{1,y}\right)u_{\xi_1} + \left(a\xi_{2,x} + b\xi_{2,y}\right)u_{\xi_2} + G\left(\xi_1, \xi_2, u\right),$$

is determined by the initial condition on Γ except for the term involving u_{ξ_1}. However, this term is absent when

(1.16) $\left.\left(a\xi_{1,x} + b\xi_{1,y}\right)\right|_{\Gamma} = 0.$

Thus, this condition characterizes the characteristic curves. Since the curve Γ satisfies:

$$\xi_1\left(x, y(x)\right) = 0,$$

then differentiating this equation with respect to x

$$\xi_{1,x}(x, y(x)) + \xi_{1,y}(x, y(x))y'(x) = \left.\left(\xi_{1,x} + \xi_{1,y}y'\right)\right|_{\Gamma} = 0.$$

Thus, the condition (1.16) becomes

$$\left.\left((b - ay')\,\xi_{1,y}\right)\right|_{\Gamma} = 0.$$

This gives us the following ordinary differential equation that determines the characteristic curves:

(1.17) $$\boxed{y'(x) = \frac{b(x, y)}{a(x, y)}.}$$

Examples:

(1) Consider the PDE

$$u_x + u_y + xyu^2 = 0.$$

In this case:

$$a = 1, \quad b = 1,$$

Equation (1.17) for the characteristics is

$$y'(x) = 1.$$

Hence, the characteristics are straight lines

$$y = x + c.$$

(2) For the PDE

$$xu_x + yu_y + u = 0$$

we have

$$a = x, \quad b = y,$$

thus Equation (1.17) for the characteristics is

$$y'(x) = \frac{y}{x},$$

which, when integrated, leads to the following characteristic curves

$$y = cx.$$

(3) The PDE

$$\sin y \, u_x + \cos x \, u_y + u^3 = 0$$

has

$$a = \sin y, \quad b = \cos x,$$

Equation (1.17) for the characteristics is

$$y'(x) = \frac{\cos x}{\sin y}.$$

Thus, the characteristics are the curves

$$y = \arccos(-\sin x + c).$$

§1.4.2. Characteristics and Discontinuities I

Let's discuss, following G. B. Whitham in his authoritative book [76], the curves $\gamma(x, y) = 0$ where the PDE can have discontinuities.

Then, we are looking for solutions of the form

$$u(x, y) = v(x, y) + H(\gamma)w(x, y)$$

with v and w smooth functions and θ the Heaviside or step function

$$H(\gamma) = \begin{cases} 1, & \gamma \geq 0, \\ 0, & \gamma < 0. \end{cases}$$

Clearly, this presents an evident contradiction, as the underlying assumption of the PDE is that the function in question is not just continuous but also differentiable. To transcend this limitation, we transition from the realm of *classical* solution spaces and venture into the realm of *generalized functions* or *distributions*, which allows us to contemplate *weak* solutions. This involves progressing from the conventional space of solutions to a more encompassing perspective.

To embark on this journey, we employ a suitable space of test functions, such as the linear space comprised of smooth functions with compact support, symbolized by $\mathcal{D}$, or the Schwartz space $\mathcal{S}$, which we will introduce in Chapter

5. By considering the topological dual of these spaces, we access the domain of linear functionals.

Prominent among these linear functionals is the renowned Dirac delta, denoted as $\delta(x - a)$. It is defined as the linear functional that assigns to each function f its value at a. Within this realm of generalized functions, we establish a notion of derivatives, and intriguingly, it is readily established that all these distributions admit derivatives of any order. For instance, the unit step distribution's derivative corresponds to the Dirac delta functional $\delta(x)$ at $a = 0$. Furthermore, all the derivatives $\delta^{(n)}(x - a)$ are well-defined up to any order n, i.e., $\langle \delta^{(n)}(x - a), f \rangle = (-1)^n f^{(n)}(a)$.

The Dirac distribution can be visualized as an abstraction of mass or charge concentrated precisely at the point a, which presents a significantly distinct aspect compared to conventional functions. As we delve into higher-order derivatives, the distribution exhibits increasingly conspicuous singular behavior.

Further reading: For a beginner-friendly introduction to distribution theory, we highly recommend Laurent Schwartz's book [65]. It offers a thorough and accessible exploration of this complex subject. If you're looking for a more in-depth study, consider referring to [30, 66] and [73], which delve deeper into the intricacies of distribution theory.

Let's introduce the expression $u(x, y) = v(x, y) + H(\gamma)w(x, y)$ into the PDE in (1.13). This substitution yields

$$(a\gamma_x + b\gamma_y)\delta(x)w + a(v_x + \theta w_y) + b(v_y + \theta v_y) + G(x, y, u) = 0.$$

Given the varying degrees of singularity present within the terms of this PDE, the initial step involves eliminating those terms with higher singular distributions. In this case, we target the coefficients corresponding to the Dirac functional. Consequently, we search for a curve that satisfies the condition

$$a\gamma_x + b\gamma_y = 0.$$

Remarkably, this equation aligns precisely with the characteristic curves derived previously (1.16).

§1.4.3. Second-Order PDEs

Let's consider now the Cauchy problem for a second-order PDE in $\Omega \subset \mathbb{R}^2$ of the form:

$$(1.18) \quad \begin{cases} a(x, y)u_{xx} + 2b(x, y)u_{xy} + c(x, y)u_{yy} + G\big(x, y, u, u_x, u_y\big) = 0, \\ \quad u|_\Gamma = \Phi_0(x, y), \\ \quad \dfrac{\partial u}{\partial \boldsymbol{n}}\bigg|_\Gamma = \Phi_1(x, y), \end{cases}$$

where the functions a, b, and c take real values, and Γ is a curve in Ω defined by the equation:

$$(1.19) \qquad\qquad \Gamma : \quad y = y(x).$$

Our goal once again is to determine the curves Γ that are characteristics. To do this, we also perform a change of variables $(x, y) \mapsto (\xi_1, \xi_2)$

$$\xi_1 = \xi_1(x, y), \quad \xi_2 = \xi_2(x, y),$$

with the condition

$$\xi_1(x, y) = y - y(x),$$

which makes the equation of the curve Γ in the new variables become

$$\Gamma: \quad \xi_1 = 0.$$

Let's see what derivatives we can determine from the initial conditions. The first of these conditions takes the form

$$u(0, \xi_2) = \Phi_0(0, \xi_2),$$

and through differentiation, it's clear that we obtain all derivatives with respect to ξ_2 on Γ:

$$u_{\xi_2}(0, \xi_2) = \Phi_{0,\xi_2}(0, \xi_2), \quad u_{\xi_2\xi_2}(0, \xi_2) = \Phi_{0,\xi_2\xi_2}(0, \xi_2), \dots$$

Regarding the second initial condition, we have first that as a consequence of (1.19), the normal vector field is

$$\boldsymbol{n} = \frac{1}{\sqrt{\xi_{1,x}^2 + \xi_{1,y}^2}} \left(\xi_{1,x}, \xi_{1,y}\right)$$

and the corresponding derivative is

$$\frac{\partial u}{\partial \boldsymbol{n}} = \frac{1}{\sqrt{\xi_{1,x}^2 + \xi_{1,y}^2}} \left(\xi_{1,x} u_x + \xi_{1,y} u_y\right).$$

Since through the chain rule we can express the derivatives u_x and u_y in terms of u_{ξ_1} and u_{ξ_2}, we have an expression of the form

$$\left.\frac{\partial u}{\partial \boldsymbol{n}}\right|_{\Gamma} = \lambda(\xi_2) u_{\xi_1}(0, \xi_2) + \mu(\xi_2) u_{\xi_2}(0, \xi_2),$$

where we know the functions λ and μ. Thus, the second initial condition is written as

$$\lambda(\xi_2) u_{\xi_1}(0, \xi_2) + \mu(\xi_2) u_{\xi_2}(0, \xi_2) = \Phi_1(0, \xi_2),$$

which is in the form

$$u_{\xi_1}(0, \xi_2) = \tilde{\Phi}_1(\xi_2).$$

Through differentiation, we obtain

$$u_{\xi_1\xi_2}(0, \xi_2) = \tilde{\Phi}_{1,\xi_2}(\xi_2).$$

In conclusion, the initial conditions allow us to determine on Γ the function u and all its derivatives with respect to ξ_i up to the second order, except for

$$u_{\xi_1\xi_1}(0, \xi_2).$$

By writing the PDE (1.18) on Γ in the variables ξ_i, the only term that won't be determined by the initial conditions is the one containing

$$u_{\xi_1 \xi_1}(0, \xi_2).$$

To find its coefficient, we can see that

$$
\begin{aligned}
u_{xx} &= \left(\xi_{1,x} \frac{\partial}{\partial \xi_1} + \xi_{2,x} \frac{\partial}{\partial \xi_2} \right) \left(\xi_{1,x} \frac{\partial u}{\partial \xi_1} + \xi_{2,x} \frac{\partial u}{\partial \xi_2} \right) \\
&= \xi_{1,x}^2 u_{\xi_1 \xi_1} + \cdots, \\
u_{yy} &= \left(\xi_{1,y} \frac{\partial}{\partial \xi_1} + \xi_{2,y} \frac{\partial}{\partial \xi_2} \right) \left(\xi_{1,y} \frac{\partial u}{\partial \xi_1} + \xi_{2,y} \frac{\partial u}{\partial \xi_2} \right) \\
&= \xi_{1,y}^2 u_{\xi_1 \xi_1} + \cdots, \\
u_{xy} &= \left(\xi_{1,x} \frac{\partial}{\partial \xi_1} + \xi_{2,x} \frac{\partial}{\partial \xi_2} \right) \left(\xi_{1,y} \frac{\partial u}{\partial \xi_1} + \xi_{2,y} \frac{\partial u}{\partial \xi_2} \right) \\
&= \xi_{1,x} \xi_{1,y} u_{\xi_1 \xi_1} + \cdots
\end{aligned}
$$

From this, it follows that the term containing $u_{\xi_1 \xi_1}(0, \xi_2)$ in (1.18) is

$$\left(a \xi_{1,x}^2 + 2b \xi_{1,x} \xi_{1,y} + c \xi_{1,y}^2 \right) u_{\xi_1 \xi_1}.$$

Therefore, the characteristics are determined by the condition

$$\boxed{\left(a \xi_{1,x}^2 + 2b \xi_{1,x} \xi_{1,y} + c \xi_{1,y}^2 \right)\big|_\Gamma = 0.}$$

Alternatively, when $a \neq 0$

$$\xi_{1,x}(x, y(x)) + \frac{1}{a} \left(b \pm \sqrt{b^2 - ac} \right) \xi_{1,y}(x, y(x)) = 0.$$

As $\xi_1(x, y(x)) = 0$, differentiating with respect to x yields

$$\xi_{1,x}(x, y(x)) + \xi_{1,y}(x, y(x)) y'(x) = 0,$$

which provides us with the following pair of ordinary differential equations for the characteristics

(1.20)
$$y' = \frac{1}{a} \left(b \pm \sqrt{b^2 - ac} \right).$$

Each of these equations can determine a family of characteristics.

Furthermore, this allows us to establish the following classification of second-order PDEs (1.18) into subsets $\Lambda \subseteq \Omega$ where the sign of the function $b^2 - ac$ doesn't change.

Characteristic Curves

- **Elliptic** in Λ if

$$b^2 - ac < 0, \quad (x, y) \in \Lambda.$$

They do not have characteristic curves in Λ since (1.20) has no meaning.
- **Hyperbolic** in Λ if

$$b^2 - ac > 0, \quad (x, y) \in \Lambda.$$

They have two families of characteristic curves in Λ determined by (1.20).
- **Parabolic** in Λ if

$$b^2 - ac = 0, \quad (x, y) \in \Lambda.$$

They have only one family of characteristic curves in Λ, given by the solutions of $y' = \frac{b}{a}$.

Definition

Examples:

(1) Laplace's equation:

$$u_{xx} + u_{yy} = 0$$

verifies

$$a = 1, \quad b = 0, \quad c = 1,$$

thus $b - ac = -1 < 0$. It's elliptic throughout the plane. It doesn't have characteristics.

(2) Wave equation:

$$u_{tt} - u_{xx} = 0$$

verifies

$$a = 1, \quad b = 0, \quad c = -1,$$

thus $b - ac = 1 > 0$. It's hyperbolic throughout the plane. Its characteristics are described by the equations

$$x'(t) = \pm 1,$$

which correspond to the two families of lines

$$x = \pm t + k.$$

(3) Heat equation:

$$u_t - u_{xx} = 0,$$

verifies

$$a = -1, \quad b = 0, \quad c = 0,$$

thus $b - ac = 0$. It's parabolic throughout the plane. Its characteristics satisfy

$$t'(x) = 0.$$

Therefore, they are the family of lines $t = k$.

(4) Tricomi's equation:

$$y u_{xx} + u_{yy} = 0$$

has

$$a = y, \quad b = 0, \quad c = 1.$$

Thus,

$$b^2 - ac = -y.$$

Therefore, the equation is

$$\begin{cases} \text{elliptic for} & y > 0, \\ \text{parabolic for} & y = 0, \\ \text{hyperbolic for} & y < 0. \end{cases}$$

In the hyperbolic case ($y < 0$), the differential equation of the characteristics is

$$y' = \pm \sqrt{-\frac{1}{y}}$$

hence, the characteristic curves are

$$y(x) = -\left(\mp \frac{3}{2}x + c\right)^{2/3}$$

with $c \in \mathbb{R}$ an arbitrary constant.

§1.4.4. Characteristics and Discontinuities II

Let's delve into the case at hand, as guided by G. B. Whitham [76], and explore the curves $\gamma(x, y) = 0$ where our second-order PDE may exhibit discontinuities.

As we previously discussed for the first-order case, we are in search of solutions following the form:

$$u(x, y) = v(x, y) + H(\gamma)w(x, y).$$

Applying this to the PDE in (1.18), we arrive at the expression:

$$(a\gamma_x^2 + 2b\gamma_x\gamma_y + c\gamma_y^2)w\delta'(x) + \text{terms of lower order} = 0.$$

From this, we can deduce:

$$a\gamma_x^2 + 2b\gamma_x\gamma_y + c\gamma_y^2 = 0.$$

This realization leads us to the conclusion that discontinuities could potentially occur exclusively along the characteristic curves.

§1.5. General Solutions

ENERAL solutions for an ordinary differential equation or a partial differential equation arise when attempting to solve the equation by applying integration operations. In principle, for an ODE of order N,

$$F\left(x, u, Du, \ldots, D^N u\right) = 0$$

indefinite integrations are required, and each integration introduces an arbitrary integration constant. As a result, the obtained solution will be an expression with N arbitrary constants, known as the **general solution** of the ODE

$$u = u\left(x, c_1, \ldots, c_N\right).$$

These N integration constants are arbitrary and directly related to the initial conditions.

For example, the differential equation

$$D^2 u - u = 0$$

has the general solution

$$u = c_1 \exp(x) + c_2 \exp(-x).$$

However, in reality, this process can be challenging. First, obtaining explicit solutions through indefinite integration operations is only possible for simple cases. Even in situations where it is possible, there is no guarantee that a single expression with N arbitrary constants can represent all the solutions of an ODE of order N.

The following examples illustrate this situation.

Examples:

(1) The equation

$$\frac{d^2 u}{dx^2}\left(\frac{du}{dx} - u\right) = 0,$$

has order 2. However, we cannot characterize all of its solutions through a single expression with two arbitrary constants since u is a solution if and only if at least one of the two factors is canceled: $\frac{d^2 u}{dx^2}$ or $\left(\frac{du}{dx} - u\right)$. In other words, all its solutions are given by two expressions:

$$u = c_1 x + c_2 \text{ or } u = c_3\, e^x,$$

where (c_1, c_2, c_3) are arbitrary constants. Note that the expression $c_1\, x + c_2$ is a general solution, but not the expression $c_3\, e^x$ although it is a solution to the equation.

(2) The equation

$$\frac{du}{dx} = \sqrt{10 - u},$$

has order 1. All its solutions are given by

$$u = 10 - \frac{(x - c_1)^2}{4},$$

which is a general solution, and the constant solution

$$u = 10,$$

not included in the previous general solution.

In the realm of PDEs, the complexity deepens further. In an N-th order PDE with n independent variables, the process of indefinite integration in relation to one of the independent variables introduces not an arbitrary constant, but rather an arbitrary function of the remaining $(N - 1)$ variables. If this integration procedure can be executed comprehensively, then it necessitates N indefinite integrations. Each integration introduces an arbitrary function involving $(n - 1)$ variables. Consequently, the outcome of the derived solution becomes a formulation containing N arbitrary functions tied to $(n - 1)$ variables each. This comprehensive expression is termed the **general solution** of the PDE. Despite the limitations inherent in this concept, its application holds value in evaluating the extent to which the methods employed in obtaining PDE solutions accurately characterize a robust solution space.

When dealing with a PDE denoted as

$$F(x, D^\alpha u) = 0, \quad x \in \Omega \subset \mathbb{R}^n,$$

of order N, a general solution is one that relies on N arbitrary functions of $(n-1)$ variables. These variables themselves are contingent on the independent variables $x_0, \ldots, x_{n-1}$.

We have already seen that the general solution of the wave equation

$$u_{tt} - u_{xx} = 0$$

is

$$u(t, x) = f(t + x) + g(t - x).$$

The order is $N = 2$, and two arbitrary functions of one variable ($n - 1 = 2 - 1 = 1$) will appear —note that these variables are functions of x and t.

We must note that the arbitrary functions that appear in the construction methods of general solutions of PDEs do not always depend on the independent variables of the PDE. For example, for the PDE:

$$u_x + u_y = 0,$$

with $r = 1$ and $n = 2$, we can proceed as follows to build a general solution. First, we observe that:

$$u = e^{i\lambda(x-y)}$$

is a solution for any possible value $\lambda \in \mathbb{R}$. Also, as the PDE is linear, we conclude that:

$$u = \int_{\mathbb{R}} c(\lambda) e^{i\lambda(x-y)} d\lambda,$$

with $c(\lambda)$ an arbitrary function of a variable, is a general solution of the PDE. These types of general solutions are the ones that will appear later.

Just like in the case of ODEs, the general solution of a PDE might not contain all solutions. For example, if a PDE factors as

$$F(x, D^\alpha u) = F_1(x, D^\alpha u) F_2(x, D^\alpha u) = 0,$$

where both F_1 and F_2 are of order N, the general solution of $F_1(x, D^\alpha u) = 0$ is a general solution of $F(x, D^\alpha) = 0$. However, in this general solution, we do not encounter the solutions of $F_2(x, D^\alpha u) = 0$ which are also solutions of $F(x, D^\alpha u) = 0$.

§1.5.1. First-Order PDEs

Now, we will study a method for finding **general (or complete) solutions** of first-order equations of the form

$$F\left(x_0, \ldots, x_{n-1}, u, \frac{\partial u}{\partial x_1}, \ldots, \frac{\partial u}{\partial x_{n-1}}\right) = 0.$$

In physics, these equations often arise in various circumstances. For example, in theoretical mechanics, we encounter the **Hamilton–Jacobi equation**

$$\frac{\partial S}{\partial t} + H\left(q_1, \ldots, q_{n-1}, \frac{\partial S}{\partial q_1}, \ldots, \frac{\partial S}{\partial q_{n-1}}\right) = 0,$$

and in geometric optics, we find the **eikonal equation**:

$$\left(\frac{\partial S}{\partial x_1}\right)^2 + \cdots + \left(\frac{\partial S}{\partial x_{n-1}}\right)^2 = 1.$$

We are going to deal with the enveloping solution of a multiparameter family of solutions. For this purpose, we introduce

---

Multiparameter Solution

Given $a = (a_0, \ldots, a_{m-1}) \in \mathbb{R}^m$, we say that $u(x, a)$ is a multiparameter solution of the PDE

$$F(x, D^\alpha u) = 0$$

if it is a solution for all values of a.

Definition

There is a technical aspect about these multiparameter families: Do they really depend on m parameters? In other words, is it possible to introduce

$a_i = a_i(b_0, \ldots, b_{m-2})$, so that the solution will only depend on $m-1$ parameters? This question is resolved by requiring that the matrix

$$\begin{bmatrix} u_{a_0} & u_{x_0 a_0} & \cdots\cdots\cdots u_{x_{n-1} a_0} \\ \vdots & \vdots & \vdots \\ \vdots & \vdots & \vdots \\ u_{a_{m-1}} & u_{x_0 a_{m-1}} & \cdots\cdots u_{x_{n-1} a_{m-1}} \end{bmatrix}$$

has rank m. This way, we ensure that the solution truly depends on m parameters. When we have a solution that genuinely depends on n parameters, we call it a **complete integral**.

Examples:

(1) **Clairaut's equation:** This is a relevant PDE in differential geometry:

$$x \cdot \nabla u + f(\nabla u) = u.$$

Given $a \in \mathbb{R}^n$, we have the following complete integral:

$$u(x,a) = a \cdot x + f(a).$$

(2) **Eikonal equation:** In geometric optics, the following equation appears:

$$|\nabla u| = 1.$$

For $a \in \mathbb{R}^n$ and $b \in \mathbb{R}$, a complete integral is:

$$u(x,a,b) = a \cdot x + b, \text{ as long as } |a| = 1.$$

(3) **Hamilton–Jacobi equation:** In mechanics, we encounter the equation (here we denote the action S as u):

$$u_t + H(\nabla u) = 0, \quad \nabla u = \left(u_{x_1}, \ldots, u_{x_{n-1}}\right),$$

where the Hamiltonian H depends only on momentum. A complete solution is:

$$u(x,t,a,b) = a \cdot x - H(a)t + b, \quad a \in \mathbb{R}^{n-1}, \quad b \in \mathbb{R}.$$

Given a multiparameter family of solutions $u(x,a)$ of a first-order PDE, we can construct a solution from it, not initially included in the family, as described below:

Enveloping Solution

If $u(x,a)$ is a multiparameter solution of the PDE

$$F\left(x,u,\frac{\partial u}{\partial x_0},\dots,\frac{\partial u}{\partial x_{n-1}}\right)=0$$

and $a(x)$ is the function determined by the conditions

$$\frac{\partial u(x,a)}{\partial a_i}=0,\quad i\in\{0,\dots,m-1\},$$

then

$$\tilde u(x):=u(x,a(x))$$

is a new solution of the PDE, which we call the enveloping solution.

Theorem

This result follows from the following calculation:

$$\frac{\partial \tilde u}{\partial x_i}(x)=\frac{\partial u}{\partial x_i}(x,a(x))+\sum_{j=0}^{m-1}\frac{\partial u}{\partial a_j}(x,a(x))\frac{\partial a_j}{\partial x_i}(x)$$

$$=\frac{\partial u}{\partial x_i}(x,a(x)),$$

and then

$$F\left(x,\tilde u,\frac{\partial \tilde u}{\partial x_0},\dots,\frac{\partial \tilde u}{\partial x_{n-1}}\right)=F\left(x,u(x,a(x)),\frac{\partial u}{\partial x_0}(x,a(x)),\dots,\frac{\partial u}{\partial x_{n-1}}(x,a(x))\right)$$

$$=0.$$

The enveloping solution constructed from a multiparameter solution of a first-order PDE allows us to sometimes find the general solution:

Complete Solution

Let $u(x,a)$ be a complete integral of a first-order PDE

$$F\left(x,u,\frac{\partial u}{\partial x_0},\dots,\frac{\partial u}{\partial x_{n-1}}\right)=0.$$

A general (or complete) solution of the PDE is obtained by substituting

$$a_0=f(a_1,\dots,a_{n-1})$$

and finding the enveloping solution $\tilde u(x)$ of

$$u(x,f(a_1,\dots,a_{n-1}),a_1,\dots,a_{n-1}).$$

Theorem

A general solution for a first-order PDE will depend on a function of $(n-1)$ variables. But the constructed enveloping solution depends exactly on an arbitrary function of $(n-1)$ functions: $a_1(x), \ldots, a_{n-1}(x)$. Therefore, the theorem holds true.

Examples:

(1) For instance, we can consider the PDE

$$u_x = u_y^m.$$

A one-parameter solution is

$$u(x, y, a) = a^m x - ay + f(a),$$

whose enveloping solution is given by

$$\tilde{u}(x, y) = u(x, y, a(x, y)),$$

where $a(x, y)$ satisfies $m a^{m-1} x + y + f'(a) = 0$.

(2) The eikonal equation in the plane:

$$u_x^2 + u_y^2 = 1$$

has the complete solution

$$u(x, y, a, b) = x \cos a + y \sin a + b,$$

let's put $b = h(a)$. The general solution is given by

$$x \cos a(x, y) + y \sin a(x, y) + h(a(x, y)),$$

where $a(x, y)$ solves

$$-x \sin a + y \cos a + h'(a) = 0.$$

If we impose $h = 0$, then we obtain

$$\tan a(x, y) = \frac{y}{x}$$

and thus, a is the polar angle. Then, the solution is $\pm \sqrt{x^2 + y^2}$.

(3) Consider the Hamilton–Jacobi equation for a free particle in $\mathbb{R}^{n-1}$:

$$H(p) = \frac{p^2}{2m} : \qquad u_t + \frac{|\nabla u|^2}{2m} = 0.$$

A complete solution is

$$a \cdot x - \frac{a^2}{2m} t + b, \quad a \in \mathbb{R}^{n-1}, \quad b \in \mathbb{R}.$$

The general solution is obtained from

$$a(x, t) \cdot x - \frac{a(x, t)^2}{2m} t + h(a(x, t)),$$

where $a(x, t)$ is a solution of

$$x_i - \frac{1}{m} a_i t + h_{a_i}(a) = 0, \quad i \in \{1, \ldots, n-1\}.$$

If we set $h = 0$, then we get

$$a = m\frac{x}{t}$$

and hence, a solution of the Hamilton–Jacobi equation is

$$m\frac{|x|^2}{2t}.$$

§1.5.2. The Hodograph Method

Let's present a method that allows us to construct solutions for the following PDE

$$u_t = f(u)u_x,$$

for a given function $f : \mathbb{R} \to \mathbb{R}$. If $u = u(t,x)$ is a solution, then we can consider locally, whenever $u_x \neq 0$, the implicit function $x = x(t,u)$. This function satisfies $u = u(t, x(t,u))$, and therefore

$$1 = u_x x_u, \quad 0 = u_t + u_x x_t.$$

This implies

$$x_t = -\frac{u_t}{u_x} = -f(u)$$

whose solution is

$$\boxed{x = -f(u)t + g(u).}$$

This last equation is known as the **hodograph equation**. Suppose we have locally found a solution $u(x,t)$ of the hodograph equation, then

$$x + f(u(x,t))t = g(u(x,t))$$

and thus

$$1 + f'(u)tu_x = g'(u)u_x, \quad f'(u)tu_t + f(u) = g'(u)u_t.$$

Therefore,

$$f'(u)t - g'(u) = \frac{1}{u_x}, \quad f'(u)t - g'(u) = \frac{f(u)}{u_t}$$

and we have

$$u_t = f(u)u_x.$$

If $f = F'$, then the PDE is writen as

$$u_t = (F(u))_x.$$

These types of PDEs are known as conservations laws.

These equations encapsulate the fundamental principle that the alteration rate of conserved quantities within a particular region is precisely counterbalanced by the exchanges of those quantities across the boundaries of the same region, in addition to any internal sources or sinks of these quantities.

The realm of conservation laws extends across a diverse spectrum of applications, encompassing fields like fluid dynamics, electromagnetism, quantum mechanics, and beyond. Conservation laws serve as a cornerstone for explaining the behavior of physical systems, directing the course of numerical simulations, and furnishing profound insights into the foundational principles that underlie various natural phenomena.

One of the most illustrious instances of conservation laws, known as balance laws, is witnessed in the Euler equations within fluid dynamics. These equations intricately delineate the preservation of mass, momentum, and energy in the fluid flow context. Their capacity to accurately describe the intricate interplay of these fundamental attributes highlights the potency and universality of conservation laws in understanding and modeling a myriad of intricate physical processes.

Let us provide a multidimensional extension of this method. We will be following the approach outlined in our paper [48]. Our method commences with the following implicit equation for determining a scalar function, denoted as $u = u(x)$ which depends on n variables $x = (x_1, \ldots, x_n)$,

$$(1.21) \qquad X_0(u) + \sum_{i=1}^{n} x_i X_i(u) = 0.$$

Here, X_i, $i \in \{0, 1, \ldots, n\}$, are predefined functions of u. By representing $x = x_1, t_i = x_{i+1}(i = 1, \ldots, n-1)$, we can observe that Equation (1.21) serves as a hodograph transformation for a family of one-dimensional hydrodynamical systems; i.e., to:

$$u_{t_i} = C_i(u)u_x, \quad i \in \{1, \ldots, n-1\},$$

where

$$C_i(u) := \frac{X_{i+1}(u)}{X_1(u)}.$$

Our main observation is that (1.21) provides solutions for a family of nonlinear partial differential equations (PDEs) expressed as:

$$(1.22) \qquad \sum_{|\alpha|=m} c_\alpha D^\alpha \phi = D^\beta F(\phi), \quad |\beta| = m,$$

In this context, D^α and D^β represent partial differentiation operators of a given order m associated with n-component multi-indices α and β belonging to $\mathbb{N}_0^n$. Furthermore, $F = F(\phi)$ represents an arbitrary function, and c_α are arbitrary constants. Let's denote

$$(1.23) \qquad Q(u) := \frac{\sum_{|\alpha|=m} c_\alpha X^\alpha(u)}{X^\beta(u)}, \quad X^\sigma := X_{\sigma_1} \cdots X_{\sigma_n}.$$

We are now poised to demonstrate that there are solutions to Equation (1.22) that can be described by the function:

$$\phi(x) := G(Q(u)),$$

where $G := (F_u)^{-1}$ is the inverse function of the derivative F_u of F with respect to u. From (1.23) we can derive the following:

$$\phi_{x_i} = G'(Q(u))Q'(u)u_{x_i} = G'(Q(u))Q'(u)\frac{X_i(u)}{X_j(u)}u_{x_j}$$

$$= \frac{\partial}{\partial x_j}\int G'(Q(u))Q'(u)\frac{X_i(u)}{X_j(u)}\mathrm{d}u,$$

which leads us to the conclusion that:

$$D^\alpha\phi = D^\beta\int^u G'(Q(u))Q'(u)\frac{X^\alpha(u)}{X^\beta(u)}\mathrm{d}u.$$

From this relation, we conclude

$$\sum_{|\alpha|=m}c_\alpha D^\alpha\phi = D^\beta\int^u G'(Q(u))Q'(u)\frac{\sum_{|\alpha|=m}c_\alpha X^\alpha(u)}{X^\beta(u)}\mathrm{d}u$$

$$= D^\beta\int^u G'(Q(u))Q'(u)Q(u)\mathrm{d}u.$$

Now, if $H := F \circ G$, then

$$H'(Q) = \big(F' \circ G\big)(Q)G'(Q) = QG'(Q)$$

and hence

$$\sum_{|\alpha|=m}c_\alpha D^\alpha\phi = D^\beta\int^u H'(Q(u))Q'(u)\mathrm{d}u = D^\beta H(Q) = D^\beta F(\phi).$$

Further reading: The book *Introduction to Partial Differential Equations* by Olver on PDEs is a very good option for an introductory course on the subject [55]. Another highly recommended text is *Partial Differential Equations* by Evans [28], which is considered one of the best and most comprehensive resources on PDEs available today. We also recommend the inspiring short book by Arnold on the subject [4]. Finally, we recommend [29], an encyclopedic classic from the beginning of the last century, filled with valuable insights.

§1.6. Remarkable Lives and Achievements

Who was Laplace? Pierre-Simon Laplace, a French mathematician, astronomer, and physicist, was born on March 23, 1749, in Beaumont-en-Auge, Normandy, France. His remarkable contributions spanned various fields, earning him recognition as one of the most influential scientists of his time.

Laplace's groundbreaking work in celestial mechanics and the stability of the solar system revolutionized our understanding of planetary orbits. He formulated the Laplace–Lagrange secular theory, which explained the long-term variations in planetary orbits through gravitational interactions among celestial bodies.

Laplace also made significant advancements in probability theory and statistics, introducing the concept of probability as a mathematical discipline. He contributed to the theory of errors, including the method of least squares, which remains a crucial tool in data analysis.

Within potential theory, Laplace's work on scalar and vector fields generated by distributions of sources is noteworthy. His study of harmonic functions laid the foundation for this area of mathematics, leading to the term "harmonic functions" that is still used today.

In addition, Laplace explored the properties of the Laplace equation, a fundamental partial differential equation appearing in various areas of physics and engineering, including electromagnetism and fluid dynamics.

He also developed the Laplace transform, an integral transform technique widely used.

He passed away on March 5, 1827, in Paris, leaving behind a legacy of groundbreaking discoveries and influential contributions to the scientific community.

Who was Poisson? Siméon Denis Poisson was a French mathematician, physicist and engineer, born on June 21, 1781, in Pithiviers, France and passed away on April 25, 1840, in Paris. He made significant contributions to various branches of mathematics and physics, leaving behind a legacy of important mathematical and scientific discoveries.

One of Poisson's most notable achievements is his work in probability theory, particularly his development of the Poisson distribution. This distribution is fundamental for modeling the number of events occurring in a fixed interval of time or space, given the average rate of occurrence.

In addition to probability theory, Poisson made important contributions to potential theory, which deals with scalar and vector fields generated by sources. He developed the theory of harmonic functions and the Poisson's equation, which has essential applications in various areas of physics and engineering. Furthermore, he derived the Poisson integral formula, a vital tool in solving BVPs in potential theory.

Poisson's work also extended to partial differential equations, where he made significant advancements, including studies on the Laplace equation and the wave equation, among others. He made contributions to various areas of mathematical physics, including elasticity theory and the theory of heat conduction.

Who was Schrödinger? Erwin Schrödinger was an Austrian physicist who made significant contributions to the field of quantum mechanics. He was born on August 12, 1887, in Vienna, Austria, and passed away on January 4, 1961, in Vienna.

In 1926, Schrödinger formulated the famous Schrödinger equation, a fundamental equation in quantum mechanics that describes how the quantum state of a physical system changes over time. This equation played a crucial role in the development of quantum mechanics and helped to unify various quantum theories. Schrödinger's work led to the development of wave mechanics, which described particles as waves and provided a mathematical framework for understanding their behavior. His wave mechanics is equivalent to other formulations of quantum mechanics, such as matrix mechanics developed by Werner Heisenberg.

Erwin Schrödinger's work has had a profound impact on the understanding of quantum mechanics and laid the foundation for modern physics. His ideas continue to be studied and applied in various fields of science and technology.

Who was Dirichlet? Johann Peter Gustav Lejeune Dirichlet, commonly known as Dirichlet, was a German mathematician who made relevant achivements to various branches of mathematics, especially in the areas of number theory, analysis, and partial differential equations. He was born on February 13, 1805, in Düren, Prussia (now Germany), and passed away on May 5, 1859, in Göttingen, Germany.

One of Dirichlet's most famous results is his theorem on arithmetic progressions, that is fundamental in the study of prime numbers and has important implications in number theory. Dirichlet introduced what are now known as Dirichlet L-series, which are complex-valued functions that play a crucial role in number theory. They are used to study the distribution of primes in arithmetic progressions and have connections to many other areas of mathematics, including algebraic number theory.

Dirichlet is also credited with introducing the concept of Dirichlet's principle in analysis, which deals with the properties of functions defined on closed and bounded intervals. This principle laid the groundwork for modern approaches to calculus and mathematical analysis.

Dirichlet's work in number theory and analysis continues to be studied and applied in various areas of mathematics. He is considered one of the most influential mathematicians of the 19th century, and his contributions have had a lasting impact on the development of modern mathematics. For more information see [33].

Who was Neumann? Carl Gottfried Neumann was a German mathematician and physicist who made substantial contributions to various fields of mathematics and engineering. He was born on October 21, 1832, in Königsberg, Prussia (now Kaliningrad, Russia), and passed away on January 12, 1925, in Leipzig, Germany.

Neumann is best known for his work in the field of differential geometry. He made important contributions to the theory of minimal surfaces and the geometry of curves and surfaces. Additionally, he made significant advancements in the theory of partial differential equations, particularly in the study of the Laplace equation and the heat equation. Neumann's contributions to the theory of elasticity were also noteworthy, as he helped advance the mathematical understanding of stresses and strains in solid materials.

In the realm of complex analysis, Carl Neumann made important contributions to the theory of complex variables, especially in the study of elliptic functions and modular forms. He also made valuable contributions to potential theory, the theory of functions of a real variable, and the calculus of variations.

Who was Cauchy? Augustin-Louis Cauchy was a remarkable mathematician renowned for his groundbreaking contributions to various fields of mathematics. He lived during an era of profound intellectual development and played a pivotal role in the transformation of mathematical analysis into a rigorous and systematic discipline.

Born on August 21, 1789, in Paris, France, Cauchy exhibited exceptional mathematical talent from a young age. He pursued his education at the École Polytechnique, a prestigious institution known for nurturing numerous prominent mathematicians and scientists. Cauchy's academic journey led him to various European locations such as Paris, Berlin, Prague, and Turin due to political and academic reasons. During his travels, he engaged with and left an indelible influence on numerous mathematicians.

One of Cauchy's most significant accomplishments was his establishment of a rigorous foundation for calculus and analysis. He introduced the concept of limits and meticulously formulated precise definitions for derivatives and integrals, thereby laying the essential groundwork for the development of modern real analysis.

Cauchy's contributions to complex analysis were truly revolutionary. "Cauchy' theorem" and "Cauchy's integral formula" are foundational results in the field. These theorems serve as cornerstones for the study of analytic functions and possess wide-ranging applications. He also derived the Cauchy–Riemann equations, which are instrumental in characterizing holomorphic functions through their real and imaginary components.

Significantly contributing to the advancement of residue theory, Cauchy developed a potent technique for evaluating complex integrals. Residue theory's enduring significance spans across various mathematical and physical disciplines.

One of Cauchy's early contributions to group theory is related to permutation groups. He formulated and proved a special case of what is now known as Lagrange's theorem, which states that the order of a subgroup divides the order of the larger group. Cauchy's work laid the groundwork for the study of permutation groups and their properties.

His unwavering emphasis on mathematical rigor and his endeavor to establish robust foundations had a profound impact on the evolution of mathematical thought. He ardently championed the use of formal definitions and logical reasoning in mathematical proofs.

While his contributions were immense, Cauchy's life was not devoid of controversies and challenges. He engaged in mathematical disputes and occasionally held viewpoints that stirred debates among his contemporaries. Nonetheless, his dedication to rigorous mathematical reasoning and his wide-ranging accomplishments firmly established him as one of the most important mathematicians of his era, a foundational figure in the development of modern mathematics.

Cauchy's passing on May 23, 1857 marked the end of an era, yet his enduring legacy continues to shape the approach and understanding of mathematical concepts for mathematicians to this day.

For more information see [11, 33, 60].

Who was Kovalevskaya? Sofia Kovalevskaya (often referred to as Sonya Kovalevskaya) was a remarkable Russian mathematician and physicist who acomplished substantial achievements to the fields of mathematics and academia in the late 19th century. Her accomplishments and determination are particularly notable given the societal limitations placed on women during that era.

Born Sofia Vasilyevna Korvin-Krukovskaya on January 15, 1850, in Moscow, Russia, she displayed an early interest in mathematics. At the age of 11, she covered her room's walls with notes on differential and integral calculus by the Russian mathematician Mikhail Ostrogradski. These notes, remnants of her father's university years, introduced her to calculus. Her fascination was kindled by her uncle Pyotr Krukovsky, who initially taught her the basics, eventually leading her to develop a profound attraction to mathematics.

Her education was pursued under the guidance of private tutors employed by her father, who aimed to circumvent the societal barriers preventing women from pursuing mathematics. These challenges led her to marry Vladimir Kovalevsky, a young paleontologist, at 18, as a means to access higher education. From then on, she was known as Sofia Kovalevskaya.

This strategic marriage, a choice also made by her older sister Anna, brought her significant emotional turmoil and stress over the course of 15 years, culminating in Vladimir's tragic suicide. Nevertheless, it proved the sole avenue through which she could attain independence and continue her university studies.

Sofia's journey led her to Heidelberg in 1869, where she embarked on her academic pursuits. Completing her studies in 1871, she moved to Berlin, impressing her professors with her exceptional talents. In Berlin, she began her doctoral thesis under Karl Weierstrass's mentorship, despite being denied formal classes. Weierstrass personally provided her with private instruction.

In 1874, she successfully defended her doctoral thesis at the University of Göttingen, yet professorship remained elusive. Returning to Russia, she was offered only a secondary education position, which she rejected with a blend of bitterness and irony. To make ends meet, she wrote theater critiques and scientific articles for a St. Petersburg newspaper, as Vladimir struggled to secure an academic post.

In 1878, Sofia gave birth to a daughter, and 2 years later, she rekindled her passion for mathematics. In the spring of 1883, her husband Vladimir passed away by suicide. Swedish mathematician Gösta Mittag-Leffler, an acquaintance from her studies with Weierstrass, offered her a position in Stockholm. In 1884, she became the first female professor of natural sciences in Northern Europe. Subsequently, the Russian Imperial Academy of Sciences recognized her as an academician, though she remained barred from professorship in Russia. On February 10, 1891, at the zenith of her international prestige, she succumbed to influenza.

Sofia remained deeply engaged in the issues of her time. Shortly after beginning her studies in Heidelberg, she traveled to London with her husband, where she encountered luminaries such as Charles Darwin, Thomas Huxley, George Eliot, and Herbert Spencer. At just 19 years old, she engaged in a debate with Spencer about women's capacity for abstraction.

Kovalevskaya's contributions spanned various mathematical domains, including partial differential equations and the motion of rigid bodies. Her groundbreaking work, such as her solution to the problem of the rotation of a solid body around a fixed point, showcased her mathematical prowess. Furthermore, her focus on partial differential equations led her to prove what is now known as the Cauchy–Kovalevskaya theorem.

Her legacy includes the autobiographical novel *The Nihilist Girl* (MLA Translations). Kovalevskaya, in 1874, became the first European woman to receive a doctoral degree in mathematics. She continued to engage in research and scholarly pursuits, presenting papers at scientific gatherings and publishing her findings in reputable journals.

2. Linear PDEs

Contents

INEAR partial differential equations (PDEs) and corresponding BVP is the subject on this book. Linear PDEs are mathematical equations that involve linear combinations of partial derivatives of unknown functions. They are fundamental in physics, engineering, and various fields, describing phenomena like heat conduction, wave propagation, and quantum mechanics. Their linearity simplifies their solution and allows for modeling complex real-world processes. We discuss linear problems that are formulated naturally in terms of linear differential operators. The ideas of eigenvalue and eigenfunction are introduced at this point. Finally, we study the uniqueness of solutions for problems in electrostatics, geophysics, and fluid mechanics.

§2.1.   Linear Differential Operators

INEAR differential operators and linear boundary conditions are concepts commonly encountered in the field of differential equations, particularly in mathematics and physics. They play a crucial role in describing and solving various types of differential equations. Linear differential operators are important because many physical and mathematical systems can be modeled using linear differential equations. This makes them amenable to various analytical and numerical solution techniques. Boundary conditions are additional information provided to determine a specific solution to a differential equation. Linear boundary conditions are boundary conditions that involve linear relationships between the values of the function and its derivatives at specified points or along specified surfaces or curves.

In this section linear boundary values and/or initial condition problems, which are the subjects of study in this book, are discussed. Appropriate notations to deal with such problems are provided. Linear differential operators and some of their properties are briefly discussed.

The notion of Green functions is also considered. Green functions are fundamental mathematical tools used to solve partial differential equations and are crucial in various fields of science and engineering, playing a key role in solving boundary value problems, understanding physical processes, and finding solutions to differential equations in integral form.

§2.1.1. Linear Differential Operators

> **Linear Differential Operators**
>
> Let $C^\infty(\overline{\Omega})$ be the space of functions of class C^∞ in $\overline{\Omega}$. A differential operator L on $C^\infty(\overline{\Omega})$ is an application
> $$L : C^\infty(\overline{\Omega}) \longrightarrow C^\infty(\overline{\Omega})$$
> $$u \longmapsto Lu,$$
> defined as follows:
> $$Lu := \sum_\alpha{}' a_\alpha(x) D^\alpha u,$$
> where $\sum_\alpha'$ means that the sum extends to a finite set of multi-indices α with $|\alpha| \geq 0$, and the coefficients $a_\alpha(x)$ are supposed to be given functions in $C^\infty(\overline{\Omega})$.
>
> **Definition**

Every differential operator is a linear application, i.e.,
$$L(\lambda u + \mu v) = \lambda Lu + \mu Lv, \quad u, v \in C^\infty(\overline{\Omega}), \quad \lambda, \mu \in \mathbb{C}.$$

A common notation we will use to refer to a differential operator is
$$L := \sum_\alpha{}' a_\alpha(x) D^\alpha.$$

Examples:

(1) Let us consider the following differential operator acting on $C^\infty(\mathbb{R})$:
$$L = x^2 D + x.$$

Its action on
$$u = \cos x$$
is
$$Lu = x^2 Du + xu = -x^2 \sin x + x \cos x.$$

(2) The Laplacian operator
$$Lu := u_{xx} + u_{yy} + u_{zz},$$
can be written as
$$L = \frac{\partial^2}{\partial x^2} + \frac{\partial^2}{\partial y^2} + \frac{\partial^2}{\partial z^2}.$$

It is a differential operator over $C^\infty(\overline{\Omega})$, where Ω is any domain in $\mathbb{R}^3$. Its action on the function

$$u := e^{x^2+y^2+z^2},$$

is

$$Lu = \frac{\partial^2}{\partial x^2}e^{x^2+y^2+z^2} + \frac{\partial^2}{\partial y^2}e^{x^2+y^2+z^2} + \frac{\partial^2}{\partial z^2}e^{x^2+y^2+z^2}$$

$$= (4(x^2 + y^2 + z^2) + 6)e^{x^2+y^2+z^2}.$$

Differential operators are the appropriate mathematical objects to handle linear PDEs. Thus, a linear PDE of the form

$$\sum_\alpha' a_\alpha(x)D^\alpha u = f(x),$$

is written in condensed form as

$$Lu = f,$$

where L is the differential operator

$$Lu := \sum_\alpha' a_\alpha(x)D^\alpha u.$$

Therefore, the problem with linear PDE consists of finding functions

$$u \in C^\infty(\overline{\Omega})$$

whose image, under the linear application L, matches the function f.

Relevant cases of these PDEs are the following:

(1) **Laplace equation** in three dimensions:

$$Lu = 0, \quad Lu := \Delta u.$$

(2) **Poisson equation** in three dimensions:

$$Lu = f, \quad Lu := \Delta u.$$

(3) **Wave equation** in 1+3 dimensions:

$$Lu = 0, \quad Lu := u_{tt} - c^2\Delta u.$$

(4) **Schrödinger equation** in 1+3 dimensions:

$$Lu = 0, \quad Lu := i\hbar u_t + \frac{\hbar^2}{2m}\Delta u - Vu.$$

(5) **Heat equation** in 1+3 dimensions:

$$Lu = 0, \quad Lu := u_t - a^2\Delta u.$$

Properties of Differential Operators

Differential operators have very important algebraic properties. The following natural operations are relevant:

(1) Addition of operators:
$$(L + M)u := Lu + Mu.$$

(2) Product of complex numbers and operators:
$$(\lambda L)u := \lambda(Lu).$$

(3) Product of operators:
$$(LM)u := L(Mu).$$

(4) Commutator of operators:
$$[L, M]u := L(Mu) - M(Lu).$$

Two operators commute if and only if their commutator is the zero operator, denoted as $[L, M] = 0$.

With respect to the addition and the product by complex numbers, the differential operators form a linear space. The zero operator, defined as the differential operator with all coefficients equal to zero, is denoted $L = 0$ (obviously in such a case $Lu = 0$ for all u). Regarding the operation of operators' product, which coincides with the composition of operators as applications, the most outstanding peculiarity is that it is not a commutative operation.

Examples:

(1) If we take
$$L = D, \quad M = D + x,$$

then we have that
$$L(Mu) = D^2u + xDu + u, \quad M(Lu) = D^2u + xDu.$$

Hence $LM \neq ML$ and $[L, M] = 1$.

(2) Let us now calculate an example in three dimensions with
$$L_x = y\partial_z - z\partial_y, \quad L_z = x\partial_y - y\partial_x,$$

the commutator is
$$[L_x, L_z] = [y\partial_z - z\partial_y, x\partial_y - y\partial_x]$$
$$= [y\partial_z, x\partial_y] + [y\partial_z, -y\partial_x] + [-z\partial_y, x\partial_y] + [-z\partial_y, -y\partial_x]$$
$$= [y\partial_z, x\partial_y] + [z\partial_y, y\partial_x] = z\partial_x - x\partial_z =: L_y.$$

This set of three operators (L_x, L_y, L_z) is essentially the ***orbital angular momentum operator*** that appears in **quantum physics** (a $-i\hbar$ factor is missing).

In quantum mechanics, angular momentum is a fundamental property that arises due to the rotational symmetry of magnitudes of physical systems. It is one of the key quantities that characterize the behavior of particles on a quantum level, and it plays a crucial role in explaining various phenomena, particularly in the context of atomic, molecular, and subatomic systems.

§2.1.2. Initial Condition and Boundary Operators

Boundary Operators

Let $C^\infty(\overline{\Omega})$ be the space of smooth functions over $\overline{\Omega}$. A boundary operator l on $C^\infty(\overline{\Omega})$ is a map

$$l : C^\infty(\overline{\Omega}) \longrightarrow C(S)$$
$$u \longmapsto l(u),$$

where $C(S)$ is the set of continuous functions on a hypersurface $S \subset \partial\Omega$ that belongs to the boundary of Ω, defined as follows:

$$l(u) := \sum_\alpha{}' b_\alpha(x) D^\alpha u|_S.$$

Here $\sum_\alpha'$ means that the sum extends to a finite set of multi-indices α with $|\alpha| \geq 0$, and the coefficients $b_\alpha(x)$ are given functions in $C(S)$.

Definition

Initial Condition Operators

When one of the independent variables is a time type coordinate,

$$x = (t, x), \quad x := (x_1, \ldots, x_{n-1}),$$

and S denotes the subset intersection of a hyperplane $t = t_0$ with $\overline{\Omega}$, we define an initial condition operator as an application of the form

$$l : C^\infty(\overline{\Omega}) \longrightarrow C(S)$$
$$u \longmapsto l(u),$$

given by

$$l(u) := \frac{\partial^k u}{\partial t^k}\bigg|_{t=t_0}, \quad k \in \mathbb{N}_0.$$

Definition

Both boundary operators and initial condition operators are linear maps. That is, they satisfy

$$l(\lambda u + \mu v) = \lambda l(u) + \mu l(v), \qquad u, v \in C^\infty(\overline{\Omega}), \qquad \lambda, \mu \in \mathbb{C}.$$

It is clear that these initial condition operators are also linear.

Boundary and initial condition operators are the appropriate mathematical objects to handle linear boundary conditions and initial conditions. Thus, a linear boundary condition can be written in a condensed manner as follows: $l(u) = g$, where l is the boundary operator defined as $l(u) := \sum_\alpha' b_\alpha(x) D^\alpha u \mid_S$. Relevant cases of boundary operators associated with boundary conditions $l(u) = g$ are:

(1) **Dirichlet condition:** $l(u) := u \mid_S$.

(2) **Neumann condition:** $l(u) := \dfrac{\partial u}{\partial \boldsymbol{n}} \Big|_S$.

(3) **Robin condition:** For $a \neq 0$, $l(u) := \left(au + \dfrac{\partial u}{\partial \boldsymbol{n}} \right) \Big|_S$.

BVPs

The paradigmatic problem that we will consider in this book is to characterize functions $u \in C^\infty(\overline{\Omega})$ that are solutions of a system of equations of the form

$$(2.1) \quad \begin{cases} Lu = f, \\ l_i(u) = g_i, \quad i \in \{1, \ldots, m\}, \end{cases}$$

where L is a differential operator and l_i are a series of boundary and/or initial condition operators.

Definition

Associated Subspace to BVPs

Given a set of boundary operators l_i, $i \in \{1, \ldots, m\}$ acting on the space $C^\infty(\overline{\Omega})$, we define the **associated subspace** $\mathfrak{D}$ as the set of smooth functions satisfying

$$l_i(u) = 0, \quad i \in \{1, \ldots, m\}.$$

Definition

Domain of a Linear Differential Operator

Let L be a differential operator over $C^\infty(\overline{\Omega})$, and let $\mathfrak{D}$ be the associated subspace to the boundary operators $\{l_i\}_{i=1}^m$ on $C^\infty(\overline{\Omega})$. When we restrict ourselves to considering the action of L only on the elements of $\mathfrak{D}$, we can view L as a linear map that takes us from $\mathfrak{D}$ to $C^\infty(\overline{\Omega})$, denoted as

$$L : \mathfrak{D} \longrightarrow C^\infty(\overline{\Omega})$$
$$u \longmapsto Lu.$$

Definition

In this context, we can say that we are considering the operator L defined on the **domain** $\mathfrak{D}$ of $C^\infty(\overline{\Omega})$.

§2.1.3. Green Functions

The linearity exhibited by both the differential operator L and the operators associated with boundary and/or initial conditions, denoted as l_i, signifies a fundamental property in shaping the solutions of problem (2.1):

> **Linear Superposition Principle**
>
> The solutions to (2.1), denoted as u, exhibit a linear dependence on the inhomogeneous data (f, g_i) of the problem.

Hence, if u and $\tilde{u}$ represent solutions of (2.1) corresponding to the inhomogeneous data (f, g_i) and $(\tilde{f}, \tilde{g}_i)$, respectively:

$$\begin{cases} Lu = f, \\ l_i(u) = g_i, \end{cases}$$

$$\begin{cases} L\tilde{u} = \tilde{f}, \\ l_i(\tilde{u}) = \tilde{g}_i, \end{cases}$$

then, for all constants c and $\tilde{c}$, the linear superposition $cu + \tilde{c}\tilde{u}$ serves as a solution to problem (2.1) with a linear superposition of inhomogeneous data $(cf + \tilde{c}\tilde{f}, cg_i + \tilde{c}\tilde{g}_i)$:

$$\begin{cases} L(cu + \tilde{c}\tilde{u}) = cf + \tilde{c}\tilde{f}, \\ l_i(cu + \tilde{c}\tilde{u}) = cg_i + \tilde{c}\tilde{g}_i. \end{cases}$$

Specifically, this property implies that:

> **Green Function**
>
> The contribution of the inhomogeneous term f to the solution $u = u(x)$ of the differential equation
>
> $$Lu = f,$$
>
> assumes the form of an integral:
>
> $$\int_\Omega G(x, y) f(y) d^n y,$$
>
> where G is a specific function referred to as the Green function of the problem. This function depends on the domain Ω and the operators associated with boundary or initial conditions, denoted as l_i.

Examples:

(1) Let's consider the boundary value problem defined as:

$$\begin{cases} u_{xx} = f, & 0 < x < 1; \\ \begin{cases} u(0) = c_1, \\ u(1) = c_2. \end{cases} \end{cases}$$

Its solution is expressed as:

$$u(x) = \int_0^1 G(x,y)f(y)\mathrm{d}y + c_1(x-1) + c_2 x,$$

where the Green function of the problem, denoted as $G(x,y)$, takes the following form:

$$G(x,y) = \begin{cases} x(y-1), & x < y, \\ y(x-1), & x > y. \end{cases}$$

(2) The boundary problem for the Poisson equation with a Dirichlet boundary condition, as presented in [43], is

$$\begin{cases} \Delta u = f, & x \in \Omega, \\ u|_{\partial\Omega} = g. \end{cases}$$

The solution to this problem is expressed as:

$$u(x) = \int_\Omega G(x,y)f(y)\mathrm{d}^3 y + \int_{\partial\Omega} \frac{\partial G}{\partial \boldsymbol{n}} g(y)\mathrm{d}S.$$

Here, the function $G(x,y)$ represents the Green function, which is dependent on the domain Ω.

It can be proved [74] that

Inverse of a Differential Operator

Given a differential operator L with domain $\mathfrak{D}$ specified by a set

$$\{l_i\}_{i=1}^m$$

of boundary operators such that $u \equiv 0$ is the unique element in $\mathfrak{D}$ satisfying

$$Lu = 0,$$

then the inverse operator L^{-1} of L exists and it is given by an integral operator

$$L^{-1}v(x) = \int_\Omega G(x,y)v(y)\mathrm{d}^n y,$$

with an appropriate Green function $G(x,y)$.

Theorem

The solution of the boundary problem with homogeneous boundary conditions

$$\begin{cases} Lu = f, \\ l_i(u) = 0, \quad i \in \{1,\dots,m\}, \end{cases}$$

is given by

$$u(x) = \int_\Omega G(x,y)f(y)\mathrm{d}^n y.$$

Note that we get

$$Lu = \int_\Omega LG(x,y)f(y)\mathrm{d}^n y = f(x).$$

Hence, the linear functional represented by the integration in Ω with weight $LG(x,y)$ must be the Dirac distribution $\delta(x-y)$, so that the Green function satisfies

(2.2) $$LG(x-y) = \delta(x-y).$$

Here L is a differential operator acting on the x variable.

In Chapter 4, we will delve into this aspect in detail, particularly for one-dimensional second-order problems.

Observations:

(1) A solution of Equation (2.2) is known as a fundamental solution [38, 72].
(2) In some applications, Equation (2.2) for the Green function sometimes has a minus sign.

§2.1.4. Eigenvalues and Eigenfunctions

One of the basic ingredients of our methods for solving linear problems such as (2.1) will be to use eigenvalues and eigenfunctions of differential operators.

Eigenvalues of differential operators are values associated with a differential equation associated with the corresponding eigenfunctions. These are special functions that satisfy the equation in a way similar to how eigenvectors satisfy linear algebraic equations for matrices. In the context of differential operators, eigenvalues provide information about the nature and behavior of solutions to the corresponding differential equation.

Eigenvalues and their associated eigenfunctions are commonly found in various areas of science and engineering, particularly in the study of partial differential equations, quantum mechanics, and other fields that involve mathematical modeling of physical phenomena. They are essential for understanding the spectrum of possible solutions and characterizing the stability and behavior of systems described by differential equations. The eigenvalues help identify specific patterns, oscillations, and modes of response in these systems.

Eigenvalues of a Linear Differential Operator

Let L be a differential operator defined on a **domain** $\mathfrak{D}$ of $C^\infty(\overline{\Omega})$. We'll say that a complex number λ is an **eigenvalue** of L over $\mathfrak{D}$ if there exists some function $u \neq 0$ in $\mathfrak{D}$ such that $Lu = \lambda u$.

In that case, we will say that such a function u is an **eigenfunction** of L corresponding to the eigenvalue λ. The set $\sigma(L)$ of all eigenvalues of L is called the **spectrum** of L.

For each eigenvalue $\lambda \in \sigma(L)$, we define the corresponding **eigensubspace** as the following linear subspace:

$$\mathfrak{D}_\lambda := \{u \in \mathfrak{D} : Lu = \lambda u\}.$$

When $\dim \mathfrak{D}_\lambda = 1$, we say that the eigenvalue is **simple**, and if $\dim \mathfrak{D}_\lambda \geq 2$, then we say that it is **degenerate**.

Definition

In general, solving an eigenvalue and eigenfunction problem $Lu = \lambda u$ for $u \in \mathfrak{D}$ in $\mathbb{R}^n$ requires finding the solutions of the homogeneous linear PDE $Lu - \lambda u = 0$ in n independent variables. Until the next chapter, we will not have a method to treat this type of PDE with $n \geq 2$. So, for now, we'll limit ourselves to problems with a single variable.

Examples:

(1) Let $Lu = u_x$ in the domain

$$\mathfrak{D} = \{u \in C^\infty([0,1]) \mid u(0) = 0\}.$$

The eigenvalue problem is

$$u_x = \lambda u,$$
$$u(0) = 0.$$

The solution to the differential equation is

$$u(x) = ce^{\lambda x},$$

and the boundary condition implies that $c = 0$. Therefore, $u = 0$, and there is no nontrivial eigenvalue. Hence,

$$\sigma(L) = \varnothing.$$

(2) Let us consider the previous operator but now in a different domain associated with periodic boundary conditions. That is,

$$Lu = u_x$$

with

$$\mathfrak{D} = \{u \in C^\infty([0,1]) : u(0) = u(1)\}.$$

The solution to the differential equation is

$$u(x) = ce^{\lambda x},$$

and the boundary condition implies

$$e^{\lambda} = 1.$$

Therefore, $\lambda = \log 1 = 2n\pi i$,

$$\sigma(L) = \{2n\pi i\}_{n \in \mathbb{Z}},$$

with eigensubspaces

$$\mathfrak{D}_{2n\pi i} = \mathbb{C}e^{2n\pi i x}$$

and simple eigenvalues, i.e.,

$$\dim \mathfrak{D}_{2n\pi i} = 1.$$

The attached figure shows the distribution of the spectrum in the complex plane.

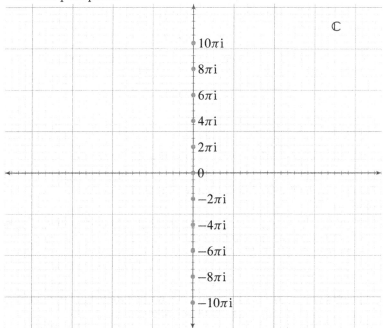

(3) Let us consider the operator

$$Lu = -u_{xx}$$

in the domain

$$\mathfrak{D} = \{u \in C^{\infty}([0, 1]) : u(0) = 0, u(1) = 0\}.$$

The eigenvalue problem is

$$\begin{cases} -u_{xx} = \lambda u, \\ u(0) = 0, \\ u(1) = 0. \end{cases}$$

For $\lambda \neq 0$, the solution to the differential equation is

$$u(x) = c_1 \cos \sqrt{\lambda} x + c_2 \sin \sqrt{\lambda} x.$$

The boundary conditions are now reduced to

$$c_1 = 0,$$

$$c_1 \cos \sqrt{\lambda} + c_2 \sin \sqrt{\lambda} = 0.$$

For there to be a nontrivial solution $(c_1, c_2) \neq (0, 0)$ of this system, it is necessary and sufficient that the determinant of the matrix of coefficients of the system is zero

$$\begin{vmatrix} 1 & 0 \\ \cos \sqrt{\lambda} & \sin \sqrt{\lambda} \end{vmatrix} = \sin \sqrt{\lambda} = 0.$$

Therefore, the eigenvalues are given by

$$\lambda_n = n^2 \pi^2,$$

for $n \in \mathbb{N}$. For each of these eigenvalues, the solutions of the system are $c_1 = 0$ and c_2 arbitrary. Thus, the eigensubspaces are

$$\mathfrak{D}_{n^2\pi^2} = \mathbb{C} \sin n\pi x.$$

Eigenvalues and Domains

The eigenvalues and eigenfunctions of a differential operator depend on its domain $\mathfrak{D}$. In general, different choices of $\mathfrak{D}$ can lead to different sets of eigenvalues and corresponding eigenfunctions for the same operator L.

Therefore, the domain $\mathfrak{D}$ plays a crucial role in determining the spectrum of L and the associated eigensubspaces.

---

§2.2.   One-Dimensional Eigenvalue Problems

STRATEGIES for solving eigenvalue problems for differential operators in one variable in suitable domains, i.e., smooth functions that fulfill as many boundary conditions as the order of the differential operator, are discussed now. We will provide some examples for ordinary differential operators with constant coefficients.

Let L be an N-th order one-dimensional linear differential operator in a domain $\mathfrak{D}$ determined by a series of linear boundary conditions on the closed interval $[a, b]$, $l_j(u) = 0$ for $j \in \{1, \ldots, N\}$. To solve the spectral problem $Lu = \lambda u$ with $u \in \mathfrak{D}$, we may proceed as follows.

First, we seek a fundamental set of solutions:

Eigenvalues of Linear Differential Operator I

To solve the linear ODE, i.e., $Lu - \lambda u = 0$, considering λ as a parameter, we can seek a solution in the form of a linear combination of functions $u_i(\lambda, x)$, where $\{u_i\}_{i=1}^{N}$ is a maximal set of linearly independent solutions, i.e., a fundamental set. Therefore, the general solution will be of the form:

$$(2.3) \qquad u(\lambda, x) = c_1 u_1(\lambda, x) + \cdots + c_N u_N(\lambda, x).$$

Associated with this fundamental set of solutions and the BVP, we have the following matrix $\mathscr{L}$ and vector $\boldsymbol{c}$ given as follows

$$\mathscr{L}(\lambda) := \begin{bmatrix} l_1(u_1) \cdots\cdots l_1(u_N) \\ \vdots \qquad\qquad \vdots \\ l_N(u_1) \cdots\cdots l_N(u_N) \end{bmatrix}, \qquad \boldsymbol{c} := \begin{bmatrix} c_1 \\ \vdots \\ c_N \end{bmatrix}.$$

Eigenvalues of a Linear Differential Operator II

After substituting the general solution $u(\lambda, x)$ into each boundary condition $l_j(u) = 0$, for $j \in \{1, \ldots, N\}$, and exploiting the linearity of the boundary operators, we obtain a homogeneous linear system:

$$(2.4) \qquad\qquad \mathscr{L}\boldsymbol{c} = \boldsymbol{0}.$$

The elements λ of the spectrum of L are characterized by the property of admitting nontrivial solutions $u \in \mathfrak{D}$, i.e., solutions of

$$Lu = \lambda u.$$

This occurs if and only if the homogeneous linear system (2.4) for $\boldsymbol{c}$ has nontrivial solutions $\boldsymbol{c} \neq \boldsymbol{0}$. This, in turn, happens if and only if the determinant of the coefficient matrix in the system vanishes:

$$(2.5) \qquad\qquad \det \mathscr{L}(\lambda) = 0.$$

This condition constitutes an equation of the type

$$f(\lambda) = 0,$$

and its solutions form the spectrum $\sigma(L)$.

Observations: On one hand, the matrix $\mathscr{L}(\lambda)$ compiles the effects of the various boundary operators $\{l_1, \ldots, l_N\}$ on the distinct elements within the fundamental set

$$\{u_1, \ldots, u_N\}$$

under consideration. On the other hand, the vector c consists of coefficients that correspond to the solution when it is represented using the fundamental set of solutions.

Thus, we have derived the spectral Equation (2.5), which is determined solely by the boundary operators. Importantly, a change of basis—a different selection of the fundamental set—will not alter this equation. It remains changeless, as it is represented by a determinant that remains invariant regardless of changes in the basis.

Finally, we find the corresponding eigenfunctions:

> ## Eigenvalues of a Linear Differential Operator III
>
> For each $\lambda \in \sigma(L)$, we solve the linear system (2.4) for $(c_1, \ldots, c_N)$ and determine the general solution $u = u(\lambda, x)$ in (2.3). This general solution will constitute the associated eigensubspace $\mathfrak{D}_\lambda$, which is the set of all functions $u \in \mathfrak{D}$ that satisfy $Lu = \lambda u$. In other words, $\mathfrak{D}_\lambda$ is the subspace of $\mathfrak{D}$ spanned by the eigenfunctions corresponding to the eigenvalue λ.

§2.2.1. Constant Coefficients Operators

A typical case we will encounter is a problem with a N-th order differential operator L with constant coefficients. We will describe how to proceed in the two simplest cases:

(1) **First-Order Operators**

$$Lu = \alpha u_x + \beta u,$$

$$\mathfrak{D} = \{u \in C^\infty([a,b]) : \gamma u(a) - \nu u(b) = 0\},$$

where, to avoid trivial cases, we assume that the constants $\alpha, \beta, \gamma, \nu \in \mathbb{C}$ satisfy $\alpha \neq 0$ and $\gamma \neq 0$. The eigenvalue equation, $Lu = \lambda u$, becomes

$$\alpha u_x + (\beta - \lambda) u = 0,$$

whose general solution is

$$u(x) = c \, \exp\left(\frac{\lambda - \beta}{\alpha} x\right),$$

where $c \in \mathbb{C}$ is arbitrary. By imposing the boundary condition, we obtain the equation that characterizes the eigenvalues:

$$c \left(\gamma \exp\left(\frac{\lambda - \beta}{\alpha} a\right) - \nu \exp\left(\frac{\lambda - \beta}{\alpha} b\right)\right) = 0.$$

Simplifying the constant factor c, we have

$$\gamma \exp\left(\frac{\lambda - \beta}{\alpha}(a - b)\right) = v.$$

Since $\gamma \neq 0$, we find

$$\exp\left(\frac{\lambda - \beta}{\alpha}(a - b)\right) = \frac{v}{\gamma} \implies \lambda = \beta + \frac{\alpha}{a - b}\log\left|\frac{v}{\gamma}\right|,$$

thus the eigenvalues are

$$\lambda_n = \beta + \frac{\alpha}{a - b}\left(\log\left|\frac{v}{\gamma}\right| + i\left(\arg\left(\frac{v}{\gamma}\right) + 2n\pi\right)\right),$$

where $n \in \mathbb{Z}$, and the corresponding eigenfunctions are given by

$$u_n(x) = c \exp\left(\frac{\lambda_n - \beta}{\alpha}x\right),$$

with $c \in \mathbb{C}$ arbitrary.

(2) **Second-Order Operators**

$$Lu = \alpha u_{xx} + \beta u_x + \gamma u,$$

$$\mathfrak{D} = \{u \in C^\infty([a, b]) : l_i(u) = 0, \, i \in \{1, 2\}\},$$

with boundary operators of the form

$$l_i(u) = v_i\,u(a) + \mu_i\,u_x(a) + v_i'\,u(b) + \mu_i'\,u_x(b), \quad i \in \{1, 2\}.$$

The general solution of the differential equation

$$\alpha\,u_{xx} + \beta\,u_x + \gamma\,u = \lambda\,u$$

is of the form

$$u = c_1 e^{r_1(\lambda)x} + c_2 e^{r_2(\lambda)x},$$

where $r_1(\lambda)$ and $r_2(\lambda)$ are two different solutions of the characteristic equation

$$\alpha r^2 + \beta r + \gamma = \lambda.$$

The boundary conditions require

$$\begin{cases} c_1 l_1(e^{r_1(\lambda)x}) + c_2 l_1(e^{r_2(\lambda)x}) = 0, \\ c_1 l_2(e^{r_1(\lambda)x}) + c_2 l_2(e^{r_2(\lambda)x}) = 0. \end{cases}$$

For there to be a nontrivial solution

$$(c_1, c_2) \neq (0, 0)$$

of this system, it is necessary and sufficient that the determinant of the matrix of coefficients of the system is zero:

$$\begin{vmatrix} l_1(e^{r_1(\lambda)x}) & l_1(e^{r_2(\lambda)x}) \\ l_2(e^{r_1(\lambda)x}) & l_2(e^{r_2(\lambda)x}) \end{vmatrix} = 0.$$

This condition takes the form of $f(\lambda) = 0$, whose solutions are the operator's eigenvalues. Once these eigenvalues are determined, the system is solved for c_1 and c_2, and the respective eigenfunctions are determined.

§2.3. Linear BVPs in Physics

⟨∘⟩

BOTH in geophysics and electrostatics, the Poisson equation provides a key tool for predicting and understanding the behavior of scalar fields in response to source distributions, and it is an integral part of modeling and analyzing physical systems in these domains. In some cases, the Laplace equation can be used to model the behavior of potential fields associated with certain physical systems, including those involving fluids. For instance, when dealing with irrotational flow (fluid flow without vorticity), the velocity potential of a fluid can satisfy the Laplace equation.

Electrostatics/Geophysics Equations

Given the domain $\Omega \subset \mathbb{R}^3$, if ρ represents the distribution of electric charge (or mass density in the case of geophysics), and a_i are boundary operators, then the electrostatic potential (or gravitational potential) u satisfies the boundary value problem:

$$\begin{cases} \Delta u = \rho, \\ a_i(u) = g_i, \quad i \in \{1, \ldots, m\}. \end{cases}$$

This problem is encountered in various fields such as electrostatics and geophysics, and other areas where the behavior of the electric or gravitational potential is of interest. The equation $\Delta u = \rho$ represents the Laplace or Poisson equation, depending on the nature of the problem. The boundary conditions $a_i(u) = g_i$ provide constraints on the potential u on the boundaries of the domain Ω. The overall goal is to find a solution for u that satisfies both the partial differential equation and the specified boundary conditions, providing a complete description of the potential within the domain.

Perfect Fluid Equations

In the context of perfect fluid mechanics, the potential for velocities is determined by a similar type of problem, but with $\rho \equiv 0$. The problem can be formulated as follows.
Given the domain $\Omega \subset \mathbb{R}^3$, we seek to find the velocity potential u for the fluid flow. The potential u satisfies the boundary value problem:

$$\begin{cases} \Delta u = 0, \\ a_i(u) = g_i, \quad i \in \{1, \ldots, m\}. \end{cases}$$

In this context, the Laplace equation $\Delta u = 0$ finds its application as the governing equation describing the potential for the velocity field of an incompressible fluid devoid of vorticity. To prescribe the behavior of such a fluid, boundary conditions $a_i(u) = g_i$ are imposed. These conditions act as constraints on the velocity potential at the boundaries of the domain Ω, offering insight into the fluid's interaction with those boundaries.

The successful solution of this boundary value problem not only reveals the velocity potential u, but also it affords a comprehensive understanding of fluid flow within the defined domain. It serves as a fundamental instrument in the study of ideal fluid motion and offers insight into a wide array of phenomena within the realm of fluid mechanics.

§2.3.1. Uniqueness

Within the scope of this book, we delve into methodologies tailored to solving BVPs of this nature. Notably, the question of uniqueness emerges from a physical perspective: Are these solutions distinctive, or could there exist alternative solutions? Unraveling this uniqueness aspect forms a pivotal thread in our exploration.

Let us explore how we can address this question in a straightforward manner, focusing on the Dirichlet BVP

$$\begin{cases} \Delta u = \rho, \\ u|_{\partial\Omega} = g, \end{cases}$$

and the Neumann BVP

$$\begin{cases} \Delta u = \rho, \\ \dfrac{\partial u}{\partial \boldsymbol{n}}\bigg|_{\partial\Omega} = g, \end{cases}$$

for the Poisson equation. We are looking for $u \in C^\infty(\overline{\Omega})$. However, there may be two cases:

- **Inner Problem**: Ω is a bounded set of $\mathbb{R}^3$.
- **Exterior Problem**: Ω is an unbounded set, and its complement is bounded.

We must remember the **theorem of divergence**, which assures that for a smooth vector field $\boldsymbol{A}$, we have

$$\boxed{\int_\Omega \nabla \cdot \boldsymbol{A}\,\mathrm{d}^3x = \int_{\partial\Omega} \boldsymbol{A} \cdot \mathrm{d}\boldsymbol{S}.}$$

Let's consider $\boldsymbol{A} = u\nabla u$, where u is a smooth function, so that:

$$\nabla \cdot (u\nabla u) = \nabla u \cdot \nabla u + u\Delta u.$$

Then, the divergence theorem leads to an instance of the first Green's identity:

$$\int_{\Omega}(|\nabla u|^2 + u\Delta u)\mathrm{d}^3x = \int_{\partial\Omega} u\frac{\partial u}{\partial \boldsymbol{n}}\mathrm{d}S,$$

where $\partial\Omega$ represents the boundary of the domain Ω, and

$$\frac{\partial u}{\partial \boldsymbol{n}}$$

is the outward normal derivative of u on the surface.

Uniqueness for the inner problem: Let's consider two solutions u_1 and u_2 of either the Dirichlet or Neumann BVPs. The difference function $u = u_1 - u_2$ satisfies the following conditions:

$$\begin{cases} \Delta u = 0, \\ u|_{\partial\Omega} = 0, \end{cases}$$

or

$$\begin{cases} \Delta u = 0, \\ \dfrac{\partial u}{\partial \boldsymbol{n}}\bigg|_{\partial\Omega} = 0, \end{cases}$$

respectively, where $\partial\Omega$ is the boundary of the domain Ω. Applying Green's identity to this function, we find

$$\int_{\Omega} |\nabla u|^2 \mathrm{d}^3x = 0,$$

which implies that $\nabla u = \boldsymbol{0}$ in the connected set Ω. Consequently, u is constant in Ω. For the Dirichlet BVP, as u is zero at the boundary, this constant is zero. Therefore, we conclude:

Uniqueness for the Inner Problem

- The inner Dirichlet problem admits at most one solution.
- In the inner Neumann problem, the difference between two distinct solutions is a constant.

Theorem

Uniqueness of the exterior problem: Now, Ω is unbounded, and we cannot apply Green's identity directly to the difference function $u = u_1 - u_2$ of two solutions.

For this reason, we change to a different domain denoted as Ω_R as described below.

The complementary set

$$\Omega^{\mathrm{c}} := \mathbb{R}^3\backslash\Omega$$

of Ω is bounded.

Thus, there exists an $R_0 > 0$ such that if $R > R_0$, then

$$\Omega^c \subset B(0, R),$$

where

$$B(0, R) := \{x \in \mathbb{R}^3 : |x| < R\}$$

is the ball of radius R centered at the origin.

We denote by

$$\Omega_R := B(0, R) \backslash \Omega^c,$$

with the boundary given by

$$\partial \Omega_R = S_1 \cup S_2 \cup \cdots \cup S_m \cup S(0, R),$$

where

$$S(0, R) := \{x \in \mathbb{R}^3 : \|x\| = R\} = \partial B(0, R)$$

is the sphere of radius R centered at the origin, and

$$\partial \Omega = S_1 \cup S_2 \cup \cdots \cup S_m.$$

In the following illustration, the geometry of this set is depicted:

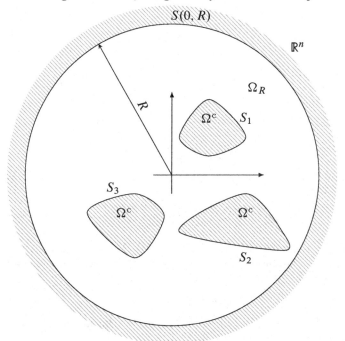

Applying first Green's identity to u in Ω_R, we get

$$\int_{\Omega_R} |\nabla u|^2 \mathrm{d}^3 x = \int_{S(0,R)} u \frac{\partial u}{\partial n} \mathrm{d}S.$$

Remembering that the normal derivative on the sphere is the radial derivative u_r, we are led to:

$$\int_{\Omega_R} |\nabla u|^2 \mathrm{d}^3 x = \int_{S(0,R)} u u_r \mathrm{d}S.$$

The surface integral can be bounded as follows:

$$\left| \int_{S(0,R)} u u_r \mathrm{d}S \right| \leq 4\pi R^2 \max_{x \in S(0,R)} |u u_r|.$$

Recall that given two functions f and g, we say that

$$f = \mathrm{O}(|g|), \quad r \to \infty,$$

if one can find M and $r_0 > 0$ such that

$$|f| \leq M |g|, \quad \forall r \geq r_0.$$

This O notation is called asymptotic or Landau notation. For a good introduction of asymptotic analysis, see [54]. Assuming that

(2.6)
$$\begin{cases} u = \mathrm{O}\left(\dfrac{1}{r}\right), \\ u_r = \mathrm{O}\left(\dfrac{1}{r^2}\right), \end{cases}$$

for $r \to \infty$, we get:

$$\int_{\Omega} |\nabla u|^2 \, \mathrm{d}^3 x = \lim_{R \to \infty} \int_{\Omega_R} |\nabla u|^2 \, \mathrm{d}^3 x \leq C \lim_{R \to \infty} \frac{1}{R} = 0.$$

A harmonic function u is said to be regular at infinity if (2.6) is fulfilled. Hence, a harmonic function u regular at infinity satisfies

$$\lim_{R \to \infty} \int_{S(0,R)} u u_r \mathrm{d}S = 0.$$

A harmonic function that approaches zero uniformly at infinity is regular at infinity. See, for example, [71].

Uniqueness for the Exterior Problem

To ensure uniqueness in the Dirichlet BVP and uniqueness up to a constant in the Neumann BVP, it is sufficient to demand that the solutions at infinity satisfy

$$u = f + u_0, \quad r \to \infty,$$

where f is a given fixed function and u_0 is a harmonic function that tends to zero uniformly at infinity.

Theorem

Further reading: To gain a deeper understanding of the concepts discussed in the final section of this chapter, we suggest consulting the comprehensive works by Olver [55], Evans [28] and the resource by Tijonov and Samarski [71]. These texts offer valuable insights and additional material to enhance your knowledge.

Of course, the authoritative text on linear differential operators is Hörmander [38]. Additionally, Naimark [52] and Trèves [72] contains information that is challenging to obtain from other sources.

For a comprehensive exploration of Green functions and their applications, we direct your attention to the authoritative guide by Stakgold and Holst [69] and that by Duffy [23]. These resources provide a thorough examination of Green functions, including their mathematical foundations and practical utility in various contexts. Reading these sources will help you gain a deeper insight into this important mathematical concept.

§2.4. Exercises

§2.4.1. Exercises with Solutions

(1) Show that the commutator of two differential operators is a derivation, i.e., satisfies the Leibniz property

$$[A, BC] = [A, B]C + B[A, C].$$

Solution: On the one hand, we have $[A, BC] = ABC - BCA$,
and on the other hand, we have $[A, B]C = ABC - BAC$ and $B[A, C] =$
$BAC - BCA$, so the result follows.

(2) Determine the commutator $[L, M]$ of the operators
   (a)

$$L := \frac{\partial^2}{\partial x^2} + \frac{\partial^2}{\partial y^2} + \frac{\partial^2}{\partial z^2}, \quad M := x\frac{\partial}{\partial y} - y\frac{\partial}{\partial x} + z.$$

   (b)

$$L := \frac{\partial^2}{\partial x \partial y}, \quad M := x^2 + y^2 + z^2.$$

Solution:

(a) To evaluate the commutator, we use the Leibniz property and
also that $\left[\frac{\partial}{\partial x_i}, x_j\right] = \delta_{i,j}$. Hence, we have

$$[L, M] = \left[\frac{\partial^2}{\partial x^2}, x\frac{\partial}{\partial y}\right] - \left[\frac{\partial^2}{\partial y^2}, y\frac{\partial}{\partial y}\right] + \left[\frac{\partial^2}{\partial z^2}, z\right]$$

$$= 2\frac{\partial}{\partial x}\left[\frac{\partial}{\partial x}, x\right]\frac{\partial}{\partial y} - 2\frac{\partial}{\partial y}\left[\frac{\partial}{\partial y}, y\right]\frac{\partial}{\partial x} + 2\frac{\partial}{\partial z}\left[\frac{\partial}{\partial z}, z\right]$$

$$= 2\frac{\partial^2}{\partial x \partial y} - 2\frac{\partial^2}{\partial y \partial x} + 2\frac{\partial}{\partial z}.$$

Consequently,

$$\boxed{[L, M] = 2\frac{\partial}{\partial z}.}$$

(b) Now we have

$$[L, M] = \left[\frac{\partial^2}{\partial x \partial y}, x^2 + y^2 + z^2\right] = \left[\frac{\partial^2}{\partial x \partial y}, x^2\right] + \left[\frac{\partial^2}{\partial x \partial y}, y^2\right]$$

$$= \left[\frac{\partial}{\partial x}, x^2\right]\frac{\partial}{\partial y} + \left[\frac{\partial}{\partial y}, y^2\right]\frac{\partial}{\partial x} = 2x\frac{\partial}{\partial y} + 2y\frac{\partial}{\partial x}.$$

Exercises

(1) Prove that the associated subspaces $\mathfrak{D}$ to sets $\{l_i\}_{i=1}^{m}$ of linear boundary operators are linear subspaces of the linear space $C^\infty(\overline{\Omega})$. That is,

$$u, v \in \mathfrak{D}, \quad \lambda, \mu \in \mathbb{C} \quad \Longrightarrow \quad \lambda u + \mu v \in \mathfrak{D}.$$

(2) Determine the order of the following PDEs. Indicate which are linear and which are homogeneous linear:

$$u_x - x\,u_y = 0,$$
$$u + u_x\,u_y = 0,$$
$$\sqrt{1 + x^2}\,\cos y\,u_x + u_{xy} - e^{x/y}u = x^2,$$
$$u_t + u_{xxx} + u\,u_x = 0,$$
$$iu_t + u_{xx} + |u|^2 u = 0,$$
$$u_x + e^y u_y = x^2,$$
$$u_{xy} = e^u,$$
$$u_{tt} - \Delta u = xyz.$$

(3) Consider the heat equation in one spatial dimension:

$$u_t = a^2\,u_{xx}.$$

(a) Show that the function

$$u = \frac{M}{2\sqrt{\pi a^2 t}}\,\exp\left(-\frac{x^2}{4a^2 t}\right),$$

where M is an arbitrary constant, is a solution to the equation.
(b) For each fixed value of $t \neq 0$, represent u as a function of x. Determine its maximum and inflection points as functions of t.
(c) Determine the value of the integral

$$\int_{-\infty}^{\infty} u(t, x)\,dx.$$

Hint: $\displaystyle\int_{-\infty}^{\infty} e^{-x^2}\,dx = \sqrt{\pi}.$

(d) Describe how the shape of the u graph changes when it grows t.
(e) If $u(t, x)$ represents the concentration of a substance per unit length, interpret what this solution describes.

(4) Considering the Schrödinger equation in one spatial dimension:

$$i u_t = -C\,u_{xx}, \quad C > 0.$$

(a) Prove that the function

$$u = \frac{M}{2\sqrt{\pi C t}}\,\exp\left(i\frac{x^2}{4C t}\right),$$

where M is an arbitrary constant, is a solution to the equation.

(b) For each fixed value of $t \neq 0$, represent the real and imaginary parts of u as functions of x.

(c) For each fixed value of $t \neq 0$, represent $|u|$ as a function of x.

(5) Determine a general solution of the equations:

$$u_{xy} = 0,$$
$$u_{xxyy} = 0,$$
$$3u_y + u_{xy} = 0,$$
$$u_{xy} = x,$$
$$u_x - u_y = 0,$$
$$u_x + u_y = 0,$$
$$u_{tt} - c^2 u_{xx} = 0,$$
$$u_{tt} - c^2 u_{xx} = t.$$

(6) Determine unit normal outgoing vectors $\boldsymbol{n}$ for the following curves and surfaces:

the circle: $(x - a)^2 + (y - b)^2 = r^2$,

the sphere: $(x - a)^2 + (y - b)^2 + (y - c)^2 = r^2$,

the cylinder : $(x - a)^2 + (y - b)^2 = r^2$, $0 \leq z \leq h$,

the inner rectangle: $a \leq x \leq b$, $c \leq y \leq d$,

the interior parallelepiped: $a \leq x \leq b$, $c \leq y \leq d$, $e \leq z \leq f$.

(7) Let the following differential operators be:

$$L = \frac{\partial^2}{\partial x \partial y}, \quad M = \sin(xy)\frac{\partial}{\partial y}.$$

Calculate the result $LM\, u$ of applying the product operator LM to the function $u = xe^y$.

(8) Be $\Delta = \partial_{xx} + \partial_{yy} + \partial_{zz}$ the Laplacian operator in three dimensions. Compute the result Δu of applying the operator Δ to the function

$$u = \cos r, \quad r = \sqrt{x^2 + y^2 + z^2}.$$

(9) Consider the differential operator $Lu = -iu_x$ over the domain $\mathfrak{D}$ of the functions $u \in C^\infty([-1,1])$ that satisfy the boundary condition $u(1) = iu(-1)$. Determine the eigenvalues and eigenfunctions of L.

(10) Find operator eigenvalues and eigenfunctions, $Lu = -u_{xx}$, over domains $\mathfrak{D}$ of functions in $C^\infty([-1,1])$ that meet boundary conditions:

- $u(0) = u(1), u_x(0) = u_x(1)$.   • $u_x(0) = 0, u_x(1) = 0$.

3. Separation of Variables Method

Contents

EPARATION of variables is one of the oldest methods used to generate solutions for linear PDEs. Essentially, it enables us to simplify the process of finding solutions for specific types of PDEs by breaking them down into solving ordinary differential equations (ODEs). Moreover, this method serves as a fundamental building block for the eigenfunction expansion technique, which helps to determine solutions for problems involving boundary conditions and/or initial conditions.

§3.1. Separable Homogeneous Linear PDEs

E will now discuss the concept of a "separable differential operator" and explore how to construct solutions using the method of separation of variables for homogeneous linear PDEs. This approach will guide us toward obtaining a "general" solution for the equation.

§3.1.1. Separable Differential Operators

Let L be a linear differential operator that acts on $C^\infty(\overline{\Omega})$, where Ω is a domain in $\mathbb{R}^n$:

$$Lu = \sum_\alpha{}' a_\alpha(x) D^\alpha u.$$

For the purpose of our discussion, it is convenient to express L in the following form:

$$L = L\left(x_0, x_1, \ldots, x_{n-1}; \frac{\partial}{\partial x_0}, \frac{\partial}{\partial x_1}, \ldots, \frac{\partial}{\partial x_{n-1}}\right),$$

where it is implied that L operates with derivatives with respect to the variables

$$(x_0, x_1, \ldots, x_{n-1}),$$

and its coefficients

$$a_\alpha = a_\alpha(x)$$

depend on these variables.

Separable Operators

An operator L is considered **separable** with respect to the variable x_0 if it can be decomposed into the sum of two operators as follows:

$$L = A + B,$$

where A and B have the forms:

$$A = A\left(x_0; \frac{\partial}{\partial x_0}\right), \quad B = B\left(x_1, \ldots, x_{n-1}; \frac{\partial}{\partial x_1}, \ldots, \frac{\partial}{\partial x_{n-1}}\right).$$

In other words, A exclusively involves derivatives with respect to x_0, and its coefficients solely depend on x_0, while B involves derivatives with respect to the variables $(x_1, \ldots, x_{n-1})$, and its coefficients solely depend on $(x_1, \ldots, x_{n-1})$.

Definition

Examples:

(1) The operator

$$Lu = xu_x + yu_{yy} + yzu_z + x^2 u,$$

is separable with respect to x since $L = A + B$ being

$$A\left(x; \frac{\partial}{\partial x}\right) = x\frac{\partial}{\partial x} + x^2, \quad B\left(y, z; \frac{\partial}{\partial y}, \frac{\partial}{\partial z}\right) = y\frac{\partial^2}{\partial y^2} + yz\frac{\partial}{\partial z}.$$

(2) The Laplacian operator

$$Lu := \Delta u = u_{xx} + u_{yy} + u_{zz},$$

is separable from any of its three variables (x, y, z).

(3) The d'Alembertian operator that appears in the 1+3 dimensional wave equation:

$$Lu := u_{tt} - c^2 \Delta u,$$

is separable with respect to any of its four variables (t, x, y, z).

(4) The operator who appears in the Schrödinger equation in 1+3 dimensions,

$$Lu := i\hbar u_t + \frac{\hbar^2}{2m}\Delta u - V(x, y, z)u,$$

is separable in the variable t. It will be separable with respect to the variable x only when the function $V = V(x, y, z)$ is

$$V = u(x) + v(y, z).$$

Analogously for variables y and z.

§3.1.2. Separable Homogeneous Linear PDEs

The Separation of Variables Method (SVM) applies to homogeneous linear PDEs where the operator L is separable. Separation of variables is most applicable to linear PDEs, where the PDE can be written as a linear combination of partial derivatives and the function itself. Linear PDEs are amenable to separation because the assumption of separable solutions is consistent with the linearity of the equation.

Separation of Variables Method

Given a homogeneous linear PDE of the form

$$A\left(x_0; \frac{\partial}{\partial x_0}\right)u + B\left(x_1, \ldots, x_{n-1}; \frac{\partial}{\partial x_1}, \ldots, \frac{\partial}{\partial x_{n-1}}\right)u = 0,$$

any pair of functions

$$v = v(x_0), \quad w = w(x_1, \ldots, x_{n-1}),$$

which satisfy the system of equations

$$Av = \lambda v,$$
$$Bw = -\lambda w,$$

with arbitrary $\lambda \in \mathbb{C}$, determine a solution given by

$$u = v(x_0)w(x_1, \ldots, x_{n-1}).$$

Theorem

Given two nonzero eigenfunctions

$$v = v(x_0), \quad w = w(x_1, \ldots, x_{n-1}),$$

the form of the operators A and B makes it clear that $u = vw$ satisfies the following equations:

$$Au = A(vw) = wAv = \lambda vw = \lambda u, \quad Bu = B(vw) = vBw = -\lambda vw = -\lambda u.$$

Thus, we have:

$$Au + Bu = \lambda u - \lambda u = 0.$$

Observations:

(1) The utility of the SVM lies in the fact that it reduces the problem of searching for solutions of a PDE of the form:

$$Au + Bu = 0,$$

with n independent variables $(x_0, x_1, \ldots, x_{n-1})$, to a system which consists of an ODE, $Av = \lambda v$, and a PDE, $Bw + \lambda w = 0$, with $(n-1)$ independent variables $x_1, \ldots, x_n$ which is, in turn, another homogeneous linear PDE. If we can apply the SVM to this PDE, meaning it is separable with respect to some of its independent variables, then we will reduce it to an ODE and a homogeneous linear PDE with $n-2$ independent variables. By iterating this process, we can conclude that if we can apply the SVM $n-1$ times, we will have managed to reduce the PDE to a system of n ordinary differential equations. When this happens, we say that the PDE is **solvable through the SVM**.

(2) The functions v and w that appear in the SVM are eigenfunctions of the operators A and B with eigenvalues λ and $-\lambda$, respectively.

(3) When we apply the SVM, we assume that λ is an arbitrary complex parameter. Then, the obtained solutions u will depend on that parameter λ. In a PDE with n independent variables, solvable through SVM, we must apply SVM $n-1$ times, and thus, the solutions obtained will depend on the $n-1$ parameters $\boldsymbol{\lambda} = (\lambda_1, \ldots, \lambda_{n-1})$ introduced by SVM. In other words, the SVM provides a family of $(n-1)$-parametric solutions: $u = u(\boldsymbol{\lambda}, x)$.

(4) As the PDE, $Au + Bu = 0$, is linear and homogeneous, given any family of solutions:

$$\{u = u(\boldsymbol{\lambda}, x), \boldsymbol{\lambda} \in \Lambda\},$$

any linear combination of elements in the family,

$$\sum_{k=1}^{N} c_k u(\boldsymbol{\lambda}_k, x),$$

is also a solution of the PDE. Under appropriate conditions, we will also be able to build solutions through generalized linear combinations of the infinite series type:

$$\sum_{k=1}^{\infty} c_k u(\boldsymbol{\lambda}_k, x),$$

or an integral expression:

$$\int_{\Lambda_0} c(\boldsymbol{\lambda}) u(\boldsymbol{\lambda}, x) \mathrm{d}^{n-1}\boldsymbol{\lambda}, \quad \Lambda_0 \subset \Lambda.$$

It is also possible to consider a superposition of both forms:

$$\sum_{k=1}^{\infty} c_k u(\boldsymbol{\lambda}_k, x) + \int_{\Lambda_0} c(\boldsymbol{\lambda}) u(\boldsymbol{\lambda}, x) \mathrm{d}^{n-1}\boldsymbol{\lambda}.$$

Although we will not use the concept of a general solution, it can be shown in concrete examples that this scheme to generate solutions leads, under appropriate conditions, to general solutions of the problem.

§3.2. Separable Homogeneous BVP

 EPARATION of variables can also be applied in situations where a linear equation in partial derivatives must be satisfied in a domain and, in addition, m boundary conditions must be met at the boundary

$$\begin{cases} Lu = 0, \\ l_i(u) = 0, \quad i \in \{1, \ldots, m\}. \end{cases}$$

Boundary Operators in One Variable

A boundary operator is said to act only on the variable x_0 if, for all pairs of functions $v = v(x_0)$ and $w = w(x_1, \ldots, x_{n-1})$, the following condition is satisfied:

$$l(vw) = w(x_1, \ldots, x_{n-1})l(v).$$

Similarly, it is said that l *acts only on variables* $(x_1, \ldots, x_{n-1})$ when, for all pairs of functions $v = v(x_0)$ and $w = w(x_1, \ldots, x_{n-1})$, the following condition is satisfied:

$$l(vw) = v(x_0)l(w).$$

Definition

Examples:

(1) Let $\Omega \subset \mathbb{R}^3$. The boundary operators

$$l(u) = u|_{x=1}, \quad l(u) = u_x|_{x=1},$$

act only on the variable x. However,

$$l(u) = (u + u_y)|_{z=1}, \quad l(u) = u|_{y+z=3},$$

act on variables (y, z).

(2) Let $\Omega \subset \mathbb{R}^2$ and Γ be the unit circle. The boundary operator

$$l(u) = u|_\Gamma,$$

is not separable with respect to either x or y. However, if we take polar coordinates (r, θ), then the boundary operator becomes separable with respect to the radial variable r since

$$l(u) = u|_{r=1}.$$

The separation of variables method can be applied to homogeneous BVPs when the linear partial differential equation is separable, and the boundary conditions are appropriate. This means that the boundary conditions are expressed as linear combinations of the function and its derivatives, equated to zero. This allows for the separation of variables to be applied effectively.

At the end, separation of variables works by assuming that the solution can be expressed as a product of functions, each depending on a single variable.

We can summarize the favorable situation in the following statement.

Homogeneous Boundary Value Problems

Let us consider a BVP of the form

$$\begin{cases} A\left(x_0; \dfrac{\partial}{\partial x_0}\right)u + B\left(x_1, \ldots, x_{n-1}; \dfrac{\partial}{\partial x_1}, \ldots, \dfrac{\partial}{\partial x_{n-1}}\right)u = 0, \\ a_i(u) = 0, \quad i \in \{1, \ldots, r\}, \\ b_j(u) = 0, \quad j \in \{1, \ldots, s\}, \end{cases}$$

such that

(1) Boundary operator a_i acts only on the variable x_0.

(2) Boundary operator b_j acts only on the variables $(x_1, \ldots, x_{n-1})$.

Then, every pair of functions

$$v = v(x_0), \quad w = w(x_1, \ldots, x_{n-1}),$$

that satisfy the systems of equations

$$\begin{cases} Av = \lambda v, \\ a_i(v) = 0, \quad i \in \{1, \ldots, r\}, \end{cases}$$

$$\begin{cases} Bw = -\lambda w, \\ b_j(w) = 0, \quad j \in \{1, \ldots, s\}, \end{cases}$$

with $\lambda \in \mathbb{C}$, determine a solution given by

$$u = v(x_0)w(x_1, \ldots, x_{n-1}).$$

Theorem

The only thing left to prove is that if $v = v(x_0)$ satisfies the first r boundary conditions, and $w = w(x_1, \ldots, x_{n-1})$ satisfies the s second boundary conditions, then $u = vw$ is a solution to our problem. However, this property is an immediate consequence of the properties of the boundary operators a_i and b_j, since we have $a_i(u) = a_i(vw) = wa_i(v) = 0$ and $b_j(u) = b_j(vw) = vb_j(w) = 0$.

Examples:

(1) Let $v(x, z)$ be the velocity field of a stationary fluid moving in the plane xz above an impervious background at depth $z = -h$.

The fluid velocity potential $u = u(x, z)$ is defined by the relationship $v = \nabla u$, and it satisfies the boundary value problem for the planar Laplace equation

$$\begin{cases} u_{xx} + u_{zz} = 0, \\ u_z|_{z=-h} = 0. \end{cases}$$

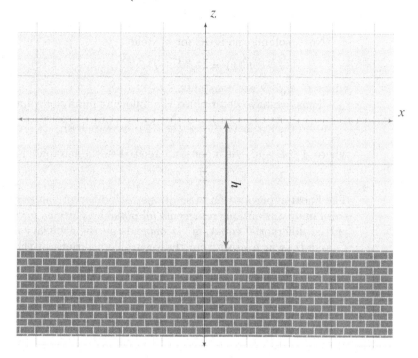

Horizontal plane

We can apply the SVM to find solutions of the form $u = v(z)w(x)$, where

$$\begin{cases} v_{zz} = \lambda v, \\ v_z|_{z=-h} = 0, \end{cases}$$

and

$$w_{xx} = -\lambda w.$$

By introducing the change of variable $\lambda = k^2$, the ODE for v becomes

$$v''(z) = k^2 v(z),$$

and its solution is

$$v(z) = Ae^{kz} + Be^{-kz},$$

where A and B are constants. Imposing the boundary value condition to v, we obtain

$$v'(-h) = Ae^{-kh} - Be^{kh} = 0,$$

$$B = Ae^{-2kh}.$$

Substituting this back into the expression for $v(z)$, we get

$$v(z) = A(e^{kz} + e^{-2kh}e^{-kz})$$
$$= 2Ae^{-kh}\cosh k(z+h).$$

Next, solving the ODE for w yields

$$w(x) = Ce^{ikx} + De^{-ikx},$$

where C and D are constants.

Thus, we have determined the following family of solutions:

$$u(x,z) = 2\tilde{A}(Ce^{ikx} + De^{-ikx})\cosh k(z+h),$$

where $\tilde{A} \in \mathbb{C}$ and k is a constant related to the parameter λ by

$$\lambda = k^2.$$

The linear superposition of solutions with different values of k from this family of solutions represents all possible solutions for the given partial differential equation. It depends on the arbitrary constants C and D, which arise from the separable solutions of the original problem.

To clarify, the general solution $u(x,z)$ can be expressed as a superposition of solutions with different values of k:

$$u(x,z) = \int (\tilde{C}(k)e^{ikx} + \tilde{D}(k)e^{-ikx})\cosh k(z+h)dk,$$

where

$$C = \tilde{A}(k)C(k),$$
$$\tilde{D} = \tilde{A}(k)D(k),$$

represents the arbitrary constant associated with each value of k.

Each term in the integral corresponds to a specific separable solution of the original problem, where $\tilde{C}(k)$ and $\tilde{D}(k)$ are determined by the boundary conditions and initial conditions. The linear combination of these separable solutions gives the complete solution $u(x,z)$ that satisfies the partial differential equation and the given boundary and/or initial conditions.

(2) It is very frequent to encounter BVPs in which the SVM is not initially applicable, but it becomes applicable after making an appropriate change of independent variables. For example, let's consider the problem of the velocity field of a stationary perfect fluid moving in the xz plane over an impervious bottom. However, let's suppose now that this bottom is not horizontal but is described by a line with equation

$$z = -mx, \qquad\qquad m := \tan \alpha.$$

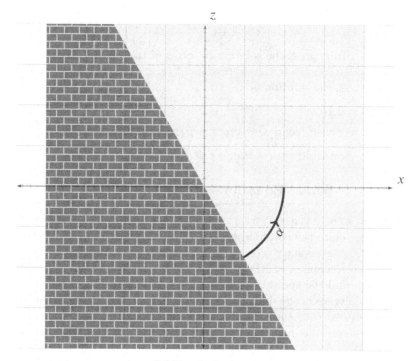

Inclined plane

The velocity potential of the fluid $u = u(x, z)$ then satisfies the BVP:

$$\begin{cases} u_{xx} + u_{zz} = 0, \\ \dfrac{\partial u}{\partial n}\bigg|_{z=-mx} = 0. \end{cases}$$

Considering that a unit normal at the bottom is given by:

$$n = \sin\alpha\, i + \cos\alpha\, k,$$

the BVP can be rewritten as:

$$\begin{cases} u_{xx} + u_{zz} = 0, \\ \left(\sin\alpha\, u_x + \cos\alpha\, u_z\right)\big|_{z=-mx} = 0. \end{cases}$$

It is clear that the boundary value condition does not allow for a direct application of SVM. However, if we introduce the change of coordinates:

$$x' = \cos\alpha\, x - \sin\alpha\, z, \quad z' = \sin\alpha\, x + \cos\alpha\, z,$$

then the problem is reduced to:

$$\begin{cases} u_{x'x'} + u_{z'z'} = 0, \\ u_{z'}\big|_{z'=0} = 0. \end{cases}$$

Now, we can apply SVM, and as we saw in the previous example:

$$u(x', z') = 2A(Ce^{ikx'} + De^{-ikx'}) \cosh kz'.$$

To express the solution in terms of the variables (x, z), it is enough to substitute the expressions of (x', z') in terms of (x, z) into u. That is, the solution is

$$u(x, z) = 2A(Ce^{ik(\cos\alpha\, x - \sin\alpha\, z)} + De^{-ik\cos\alpha\, x - \sin\alpha\, z}) \cosh k(\sin\alpha\, x + \cos\alpha\, z)$$
$$= c\cos(k(\cos\alpha\, x - \sin\alpha\, z + \beta))\cosh k(\sin\alpha\, x + \cos\alpha\, z),$$

for arbitrary constants c, β.

§3.3. Equations of Mathematical Physics

MATHEMATICAL PHYSICS problems solved by the SVM frequently require the consideration of a linear equation in partial derivatives of evolutionary type, in which a time type variable is clearly separated from the remaining spatial type variables. The variable *time* belongs to an interval, and the spatial variables belong to a domain in a multidimensional space. Therefore, the domain for the problem is the Cartesian product of the time interval with the spatial region. Additionally, initial conditions for the temporal variable and boundary conditions at the spatial boundary will be satisfied

(3.1)
$$\begin{cases} \alpha\dfrac{\partial^N u}{\partial t^N} - Lu = f, \quad t \in (a, b), \quad \pmb{x} \in \Lambda \subseteq \mathbb{R}^{n-1}, \\ \left.\dfrac{\partial^i u}{\partial t^i}\right|_{t=t_0} = f_i(\pmb{x}), \quad \pmb{x} \in \Lambda, \quad i \in \{0, \dots N-1\}, \\ l_j(u) = g_j(t, \pmb{x}), \quad t \in (a, b), \quad \pmb{x} \in \partial\Lambda, \quad j \in \{1, \dots, s\}, \end{cases}$$

where $\alpha \in \mathbb{C}$ and L is an operator in the spatial variables $\pmb{x} = (x_1, \dots, x_{n-1})$ represented as

$$L = L\left(x_1, \dots, x_{n-1}; \frac{\partial}{\partial x_1}, \dots, \frac{\partial}{\partial x_{n-1}}\right),$$

and the l_j are boundary operators in the spatial variables, such problems arise naturally in electromagnetism, quantum mechanics; and mechanics of continuous media. It should be noted that in general, the problem is not homogeneous, even though the first member of the PDE

$$\alpha\frac{\partial^N u}{\partial t^N} - Lu$$

is a separable operator with respect to the variable t.

As we will see later, the usual method to solve these problems is the **eigenfunction expansion method**, which is based on constructing the solution through families of solutions of the **associated spectral problem**

(3.2)
$$\begin{cases} Lw = \lambda w, \\ l_j(w) = 0, \quad j \in \{1, \ldots, s\}. \end{cases}$$

The latter is a homogeneous problem that, when the appropriate separability conditions are met, can be solved by the SVM. The solutions $w(x)$ of (3.2) determine the so-called **standing waves** of the physical problem described by (3.1). These standing waves are functions of the form

$$u = e^{-i\omega t} w(x), \quad \omega \in \mathbb{R},$$

where $w(x)$ is a solution of (3.2), and the time factor $e^{-i\omega t}$ is introduced to ensure that u satisfies the PDE in (3.1). In quantum mechanics, *standing waves* are known as *stationary states* or *energy stationary states*. These are states in which the probability of finding the particle at a specific position does not vary with time. Stationary states are a fundamental part of quantum theory and play an important role in describing the properties of subatomic particles. This condition determines the ***frequency*** ω of the wave, since by replacing the function u in the PDE, we obtain

$$\alpha(-i\omega)^N e^{-i\omega t} w = \lambda e^{-i\omega t} w,$$

which, upon simplification, yields the *dispersion relation*

$$\boxed{\alpha(-i\omega)^N = \lambda,}$$

which links ω to λ. Only if there are real solutions ω to this relationship can we obtain standing waves in the corresponding physical phenomenon.

In many interesting physical problems, $L = \Delta$, so the corresponding eigenvalue equation is

$$\boxed{-\Delta w = \lambda w,}$$

denoting $\lambda = k^2$, this equation becomes the **Helmholtz equation**

$$\boxed{\Delta w + k^2 w = 0,}$$

which will be fundamental in what follows.

Examples:

(1) Let's consider a typical problem for the wave equation in $1 + 1$ dimensions:

$$\begin{cases} c^{-2} u_{tt} = u_{xx}, \quad t \in (a, b), \quad x \in (0, l), \\ \begin{cases} u|_{t=t_0} = f_1(x), \\ u_t|_{t=t_0} = f_2(x); \end{cases} \\ \begin{cases} u|_{x=0} = g_1(t), \\ u|_{x=l} = g_2(t). \end{cases} \end{cases}$$

The associated spectral problem is:

$$\begin{cases} w_{xx} = \lambda w, \\ \begin{cases} w(0) = 0, \\ w(l) = 0. \end{cases} \end{cases}$$

For $n \in \mathbb{N}$, the solutions to this spectral problem are given by:

$$w_n(x) = \sin k_n x, \quad k_n = \frac{\pi}{l} n,$$

with eigenvalues $\lambda_n = -k_n^2$. The dispersion relation is:

$$c^{-2}(-i\omega)^2 = -k^2,$$

which allows us to find the frequencies of the standing waves:

$$\omega_n = \pm c k_n.$$

Therefore, the standing wave solutions of the original problem are:

$$u_n(t, x) = e^{-i\omega_n t} \sin k_n x, \quad \omega_n = \pm c k_n.$$

(2) Let us now consider a problem with the free Schrödinger equation in $1 + 1$ dimensions:

$$\begin{cases} i\hbar u_t = -\frac{\hbar^2}{2m} u_{xx}, \quad t \in (a,b), \quad x \in (0,l), \\ \begin{cases} u|_{t=t_0} = f_1(x), \\ \begin{cases} u|_{x=0} = g_1(t), \\ u|_{x=l} = g_2(t). \end{cases} \end{cases} \end{cases}$$

The associated spectral problem is:

$$\begin{cases} -\frac{\hbar^2}{2m} w_{xx} = \lambda w, \\ \begin{cases} w(0) = 0, \\ w(l) = 0. \end{cases} \end{cases}$$

For $n \in \mathbb{N}$, the solutions to this spectral problem are

$$w_n(x) = \sin k_n x, \quad k_n = \frac{\pi}{l} n,$$

where the corresponding eigenvalues are:

$$\boxed{\lambda_n = \frac{\hbar^2 k_n^2}{2m}.}$$

The dispersion relation is:

$$\boxed{\omega = \frac{\hbar k^2}{2m}.}$$

For $n \in \mathbb{N}$, the standing waves of the problem are the following *stationary states*:

$$u_n(t, x) = \mathrm{e}^{-\mathrm{i}\omega_n t} \sin \frac{\pi}{l} n x,$$

with frequencies

$$\omega_n = \frac{\hbar \pi^2}{2m l^2} n^2.$$

§3.4. Helmholtz Equation

HELMHOLTZ equation in Cartesian coordinates, also known as rectangular coordinates, can be considered as a spectral problem in three dimensions and, moreover, being a homogeneous linear PDE separable with respect to the three Cartesian coordinates to which the SVM can be applied to find solutions. For $k \in \mathbb{C}$, it reads

$$\frac{\partial^2 u}{\partial x^2} + \frac{\partial^2 u}{\partial y^2} + \frac{\partial^2 u}{\partial z^2} + k^2 u = 0.$$

In this way, we begin by looking for a solution of the form:

$$u(x, y, z) = X(x)w(y, z),$$

which leads to the following equations:

$$\begin{cases} \dfrac{\mathrm{d}^2 X}{\mathrm{d}x^2} = \lambda X, \\ w_{yy} + w_{zz} + k^2 w = -\lambda w, \end{cases}$$

where we denote

$$\lambda = -k_1^2 \in \mathbb{C}.$$

By separating variables again, we seek a solution to the second equation of the form:

$$w(y, z) = Y(y)Z(z),$$

resulting in the following equations:

$$\begin{cases} \dfrac{\mathrm{d}^2 Y}{\mathrm{d}y^2} = \lambda' Y, \\ \dfrac{\mathrm{d}^2 Z}{\mathrm{d}z^2} + (k^2 - k_1^2)Z = -\lambda' Z, \end{cases}$$

where

$$\lambda' = -k_2^2 \in \mathbb{C}.$$

Introducing

$$k_3^2 = k^2 - k_1^2 - k_2^2,$$

we observe that the sought solution has the form:

$$
\begin{aligned}
&u_{k_1,k_2,k_3}(x,y,z) = X_{k_1}(x)Y_{k_2}(y)Z_{k_3}(z),\\[6pt]
&\left\{
\begin{aligned}
X_{k_1}(x) &= \begin{cases} a_1 x + b_1, & k_1 = 0,\\ a_1 e^{ik_1 x} + b_1 e^{-ik_1 x}, & k_1 \neq 0, \end{cases}\\[4pt]
Y_{k_2}(y) &= \begin{cases} a_2 y + b_2, & k_2 = 0,\\ a_2 e^{ik_2 y} + b_2 e^{-ik_2 y}, & k_2 \neq 0, \end{cases}\\[4pt]
Z_{k_3}(z) &= \begin{cases} a_3 x + b_3, & k_3 = 0,\\ a_3 e^{ik_3 z} + b_3 e^{-ik_3 z}, & k_3 \neq 0. \end{cases}
\end{aligned}
\right.
\end{aligned}
$$

(3.3)

with

$$
k^2 = k_1^2 + k_2^2 + k_3^2.
$$

These solutions are suitable for boundary conditions where the boundary is formed by planes parallel to the coordinate planes. Let's analyze three examples of this type.

Quantum particle in an impenetrable box. The stationary states of energy E of a quantum particle in an impenetrable box with sides L_1, L_2, and L_3 are described by the following Dirichlet BVP:

$$
\begin{cases}
-\dfrac{\hbar^2}{2m}\left(\dfrac{\partial^2 u}{\partial x^2} + \dfrac{\partial^2 u}{\partial y^2} + \dfrac{\partial^2 u}{\partial z^2}\right) = Eu,\\[8pt]
u\big|_{\text{walls}} = 0.
\end{cases}
$$

In this problem, u represents the wave function of the quantum particle, Δ is the Laplacian operator, $\hbar$ is the reduced Planck's constant, and m is the mass of the particle. The equation

$$
-\frac{\hbar^2}{2m}\left(\frac{\partial^2 u}{\partial x^2} + \frac{\partial^2 u}{\partial y^2} + \frac{\partial^2 u}{\partial z^2}\right) = Eu
$$

is the time-independent Schrödinger equation, and the Dirichlet boundary condition

$$
u\big|_{\text{walls}} = 0
$$

enforces that the wave function vanishes on the walls of the impenetrable box.

Solving this Dirichlet problem allows us to find the allowed energy levels and corresponding wave functions for the quantum particle confined to the box. The solutions to this problem are highly important in quantum mechanics and provide insights into the behavior of particles in confined systems.

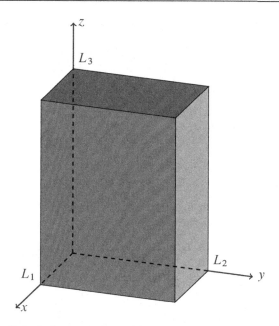

The PDE is a Helmholtz equation with

$$k^2 = \frac{2mE}{\hbar^2},$$

and the boundary value conditions break down into the following conditions:

$$u(0, y, z) = 0,$$
$$u(L_1, y, z) = 0;$$
$$u(x, 0, z) = 0,$$
$$u(x, L_2, z) = 0;$$
$$u(x, y, 0) = 0,$$
$$u(x, y, L_3) = 0.$$

Let us impose these boundary conditions on the solution

$$u_{k_1, k_2, k_3}(x, y, z) = X_{k_1}(x) Y_{k_2}(y) Z_{k_3}(z)$$

from (3.3). One can check that when any of k_1, k_2, or k_3 vanishes, the solution becomes trivial. Hence, we require

$$k_1, k_2, k_3 \in \mathbb{N}$$

to be positive integers. For X_{k_1}, we have:

$$a_1 + b_1 = 0,$$
$$a_1 e^{ik_1 L_1} + b_1 e^{-ik_1 L_1} = 0.$$

This implies:

$$a_1 = -b_1,$$
$$k_1 = \frac{\pi}{L_1} n_1$$

and $n_1 \in \mathbb{N}$, and consequently:

$$X_{k_1}(x) = A_1 \sin \frac{\pi n_1}{L_1} x.$$

For $n_1, n_2, n_2 \in \mathbb{N}$, following an analogous procedure for the variables y and z, we can conclude that the solution is:

$$\boxed{u_{n_1,n_2,n_3}(x, y, z) = a \sin \frac{\pi n_1}{L_1} x \, \sin \frac{\pi n_2}{L_2} y \, \sin \frac{\pi n_3}{L_3} z,}$$

For $n_1, n_2, n_3 \in \mathbb{N}$,

$$k^2 = k_1^2 + k_2^2 + k_3^2$$

is, therefore, of the form:

$$k^2 = \pi^2 \left(\frac{n_1^2}{L_1^2} + \frac{n_2^2}{L_2^2} + \frac{n_3^2}{L_3^2} \right),$$

and the following energy "quantization" is obtained:

$$\boxed{E_{n_1,n_2,n_3} = \frac{\hbar^2 \pi^2}{2m} \left(\frac{n_1^2}{L_1^2} + \frac{n_2^2}{L_2^2} + \frac{n_3^2}{L_3^2} \right).}$$

Now, let's show the graph of

$$|u(x, y)|^2,$$

which represents the probability density of the presence of the quantum particle at the point (x, y) of the rectangle

$$[0, 2] \times [0, 4].$$

We also present the corresponding probability density plot in the rectangle.

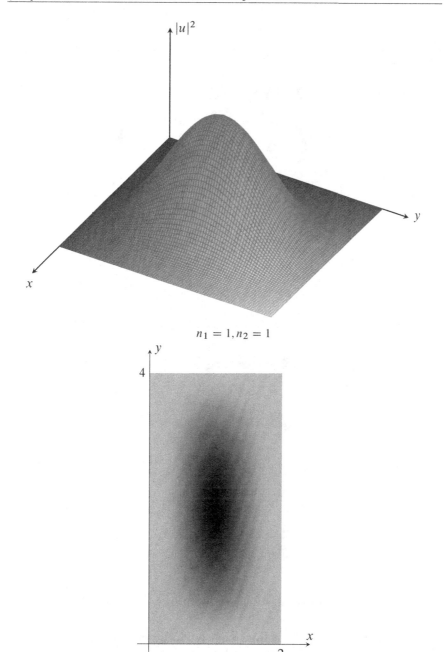

$n_1 = 1, n_2 = 1$

Probability density $n_1 = 1$, $n_2 = 1$

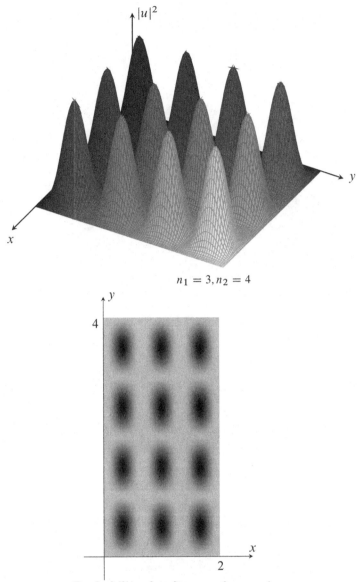

$n_1 = 3, n_2 = 4$

Probability density $n_1 = 3$, $n_2 = 4$

Quantum particle in a box with periodic conditions. We now impose periodic boundary conditions, which are given by:

$$
\begin{aligned}
u(0, y, z) &= u(L_1, y, z), & u_x(0, y, z) &= u_x(L_1, y, z), \\
u(x, 0, z) &= u(x, L_2, z), & u_y(x, 0, z) &= u_y(x, L_2, z), \\
u(x, y, 0) &= u(x, y, L_3), & u_z(x, y, 0) &= u_z(x, y, L_3).
\end{aligned}
$$

Applying these conditions to the eigenfunctions u_{k_1,k_2,k_3} from (3.3), we find that for $k_i = 0$, $a_i = 0$, and when $k_i \neq 0$, we have:

$$a_i + b_i = a_i e^{ik_i L_i} + b_i e^{-ik_i L_i},$$

$$a_i - b_i = a_i e^{ik_i L_i} - b_i e^{-ik_i L_i},$$

with $i = 1, 2$, and 3. For these linear systems to have nontrivial solutions $(a_i, b_i) \neq (0,0)$, we must demand that:

$$\begin{vmatrix} 1 - e^{ik_i L_i} & 1 - e^{-ik_i L_i} \\ 1 - e^{ik_i L_i} & -(1 - e^{-ik_i L_i}) \end{vmatrix} = 0.$$

It simplifies to:

$$(e^{ik_i L_i/2} - e^{-ik_i L_i/2})^2 = 0.$$

This condition is satisfied when:

$$\boxed{k_i = \frac{2\pi}{L_i} n_i, \quad n_i \in \mathbb{Z}, \quad i \in \{1,2,3\}.}$$

Therefore, the possible energies are:

$$\boxed{E_{n_1,n_2,n_3} = \frac{\hbar^2}{2m} 4\pi^2 \left(\frac{n_1^2}{L_1^2} + \frac{n_2^2}{L_2^2} + \frac{n_3^2}{L_3^2} \right).}$$

The solutions u are those of (3.3) with these discretized values of $k_i = \frac{2\pi}{L_i} n_i$.

Fluid in a parallelepipedic pipe. The fluid flow in a pipe is commonly modeled as the set $[0, L] \times \mathbb{R} \times [0, L'] \subset \mathbb{R}^3$, as illustrated in the following illustration.

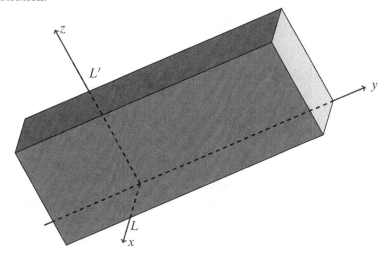

The fluid in a pipe has a velocity potential u that is determined by the following Neumann BVP for the Laplace equation:

$$\begin{cases} \Delta u = 0, \\ \begin{cases} u_x(0, y, z) = 0, \\ u_x(L, y, z) = 0, \\ u_z(x, y, 0) = 0, \\ u_z(x, y, L') = 0. \end{cases} \end{cases}$$

Let us impose these conditions on the eigenfunctions

$$u_{k_1, k_2, k_3} = X_{k_1}(x) Y_{k_2}(y) Z_{k_3}(z).$$

For $k_1 = 0$, we have $X_0(x) = a_1 x + b_1$, where a_1 and b_1 are constants to be determined by the boundary conditions. From $X_0'(0) = X_0'(L) = 0$, we deduce that $a_1 = 0$, but b_1 is free.

For $k_1 \neq 0$, the function $X_{k_1} = a_1 e^{ik_1 x} + b_1 e^{-ik_1 x}$ satisfies

$$X'(0) = X'(L) = 0,$$

which leads to the system

$$a_1 - b_1 = 0,$$

$$a_1 e^{ik_1 L} - b_1 e^{-ik_1 L} = 0.$$

The existence of nontrivial solutions $(a_1, b_1) \neq (0, 0)$ requires

$$\begin{vmatrix} 1 & -1 \\ e^{ik_1 L} & -e^{-ik_1 L} \end{vmatrix} = 0,$$

which implies $\sin 2k_1 L = 0$, and consequently,

$$k_1 = \frac{\pi}{L} n,$$

for $n \in \mathbb{N}_0$. As $a_1 = b_1$, the eigenfunctions are

$$X_n(x) = \cos \frac{\pi}{L} n x.$$

A similar analysis for the z variable leads to

$$k_3 = \frac{\pi}{L'} n', \quad n' \in \mathbb{N}_0,$$

with associated eigenfunctions

$$Z_{n'}(z) = \cos \frac{\pi}{L'} n' z.$$

For the variable y, we do not have any boundary condition to impose. However, since we are now solving Laplace's equation, it becomes the Helmholtz equation with

$$k^2 = 0.$$

So, as $k_1^2 + k_2^2 + k_3^2 = k^2 = 0$, we deduce that

$$k_2 = \pm i\pi \sqrt{\frac{n^2}{L^2} + \frac{n'^2}{L'^2}},$$

and for $k_2 \neq 0$, we obtain the following solutions:

$$u_{n,n'}(x, y, z) = \cos \frac{n\pi x}{L} \cos \frac{n'\pi z}{L'} \cosh \left(\pi \sqrt{\frac{n^2}{L^2} + \frac{n'^2}{L'^2}} y - \beta \right).$$

For $n = n' = 0$, the solution is given by

$$u_{n,n'}(x, y, z) = Ay + B,$$

where A and B are arbitrary constants.

Further reading: To gain a deeper understanding of the technique of  separation of variables, we recommend exploring the insights and techniques presented in Olver's work [55]. Additionally, Tijonov and Samarski's comprehensive resource [71] offers valuable perspectives and examples that can further enhance your comprehension of this fundamental method.

§3.5. Remarkable Lives and Achievements

Who was Helmholtz? Hermann Ludwig Ferdinand von Helmholtz, commonly known as Hermann von Helmholtz, was a prominent German physicist and physiologist. He was born on August 31, 1821, in Potsdam, Prussia (now part of Germany), and he passed away on September 8, 1894, in Charlottenburg, Germany.

Helmholtz made significant contributions to various scientific fields, including physics, physiology, and mathematics. He is often regarded as one of the most versatile scientists of the 19th century. Throughout his lifetime, Helmholtz demonstrated a deep understanding of the natural world, and his work laid the foundation for many scientific disciplines.

In the field of physics, Helmholtz is best known for his studies in electromagnetism and optics. He formulated the law of the conservation of energy, also known as the first law of thermodynamics, independently of other researchers. Helmholtz's work in optics led to his development of the ophthalmoscope, a device used to examine the inside of the human eye, and his research on color vision and the physiology of the eye greatly advanced the understanding of human vision.

As a physiologist, Helmholtz made significant contributions to the understanding of nerve impulses and the mechanics of human muscle movement. He conducted pioneering research on the speed of nerve conduction and is credited with formulating the theory of the "specific energies of nerves," which suggested that different nerve fibers convey distinct sensations to the brain.

Helmholtz also made substantial contributions to mathematics, particularly in the area of differential equations and mathematical physics. His work in this field influenced the development of mathematical techniques used in various scientific disciplines.

Beyond his scientific contributions, Helmholtz was known for his humility and dedication to teaching and mentoring younger scientists. He held various academic positions, including professorships at the University of Königsberg, the University of Bonn, and the University of Heidelberg. He also served as the director of the Physiological Institute in Berlin and as the president of the Physikalisch-Technische Reichsanstalt (Imperial Physical-Technical Institute).

Helmholtz's work and ideas have had a lasting impact on the fields of physics, physiology, and mathematics, and he is considered one of the most influential scientists in history. His legacy continues to inspire researchers and scientists to this day.

§3.6. Exercises

§3.6.1. Exercises with Solutions

(1) Apply the MSV to the following linear and first-order PDEs

$$u_x + u_y = 0,$$
$$x u_x + u_y = 0,$$
$$x u_x - y u_y = 0.$$

Provide a general solution candidate by linear superposition.

Solution: For the first equation, we have the operators:

$$A = \frac{\partial}{\partial x}, \quad B = \frac{\partial}{\partial y},$$

and we are looking for factorized solutions in the form:

$$u(x, y) = v(x) w(y),$$

where $v(x)$ and $w(y)$ are eigenfunctions such that:

$$A v = \lambda v, \quad B w = -\lambda w.$$

That is,

$$v' = \lambda v, \quad w' = -\lambda w.$$

Therefore, we can write:

$$v(x) = c_1 \exp(\lambda x), \quad w(y) = c_2 \exp(-\lambda y),$$

where c_1 and c_2 are arbitrary constants, and the eigenvalues λ can take any complex value. This leads to the factored solution:

$$\boxed{u(x, y) = c \exp(\lambda(x - y)),}$$

where we have combined the constants c_1 and c_2 into the single constant $c = c_1 c_2$. Therefore, the method of separation of variables gives us the following solution:

$$\boxed{u(x, y) = \int c(\lambda) \exp(\lambda(x - y)) d\lambda,}$$

which can be considered a general solution.

We observe that the general solution to this equation is in the form of

$$\boxed{u(x, y) = f(x - y),}$$

for any smooth function f.

In the second PDE, we have the operators:

$$A = x \frac{\partial}{\partial x}, \quad B = \frac{\partial}{\partial y},$$

and the factors of the equation:
$$u(x, y) = v(x)w(y),$$
will be eigenfunctions satisfying:
$$Av = \lambda v, \quad Bw = -\lambda w.$$
Hence, we obtain the following conditions:
$$xv' = \lambda v, \quad w' = -\lambda w.$$
Solving these equations, we find:
$$v(x) = c_1 x^\lambda, \quad w(y) = c_2 \exp(-\lambda y),$$
which leads to the factored solution:
$$\boxed{u(x, y) = c \exp\left(\lambda(\log x - y)\right),}$$
where c is a constant that absorbs the factors c_1 and c_2. Furthermore, for a given function $c(\lambda)$, the general solution is given by:

$$\boxed{u(x, y) = \int c(\lambda) \exp\left(\lambda(\log x - y)\right) d\lambda.}$$

In particular, if $f(x)$ is a smooth function and $x > 0$, the general solution of the equation is
$$\boxed{u(x, y) = f\left(\log x - y\right).}$$

Finally, for the third equation, we consider the operators:
$$A = x\frac{\partial}{\partial x}, \quad B = -y\frac{\partial}{\partial y},$$
and the factorized solutions:
$$u(x, y) = v(x)w(y),$$
which require their factors to satisfy the equations:
$$xv' = \lambda v, \quad yw' = \lambda w.$$
As a result, we can express the factors as:
$$v(x) = c_1 x^\lambda, \quad w(y) = c_2 y^\lambda.$$
Therefore, the factored solution is given by:
$$\boxed{u(x, y) = cx^\lambda y^\lambda.}$$
The general solution is obtained by considering the integral:

$$\boxed{\int c(\lambda)(xy)^\lambda d\lambda.}$$

Furthermore, for any smooth function $f(x)$, the function
$$\boxed{u(x, y) = f(xy)}$$
serves as a general solution.

(2) Apply the SVM to the Laplace equation in two dimensions

$$u_{xx} + u_{yy} = 0.$$

Solution: We seek solutions in factorized form

$$u(x, y) = v(x)w(y),$$

which will satisfy the following eigenvalue problems:

$$v'' = k^2 v, \quad w'' = -k^2 w.$$

Consequently, they can be expressed as follows:

$$v(x) = ae^{kx} + be^{-kx}, \quad w(y) = ce^{iky} + de^{-iky},$$

where a, b, c, d are arbitrary complex constants in $\mathbb{C}$. By using the complex variable $z = x + iy$, we can write:

$$u(x, y) = Ae^{kz} + Be^{-kz} + Ce^{k\bar{z}} + De^{-k\bar{z}},$$

where A, B, C, D are arbitrary constants in $\mathbb{C}$. This leads us to the following general solution:

$$\boxed{u(x, y) = \int A(k)e^{kz}dk + \int C(k)e^{k\bar{z}}dk.}$$

In fact, the most general harmonic function is the sum of a holomorphic function $f(z)$ (i.e., one that does not depend on $\bar{z}$) and an antiholomorphic function $g(\bar{z})$ (i.e., one that does not depend on z). In other words,

$$\boxed{u(x, y) = f(z) + g(\bar{z})}$$

is a general solution.

(3) Apply the MSV to the two-dimensional Laplace equation

$$u_{xx} + u_{yy} = 0,$$

subject Dirichlet boundary conditions

$$\begin{cases} u(0, y) = 0, \\ u(1, y) = 0. \end{cases}$$

Solution: We will use the solutions discussed in the previous problem and impose the corresponding boundary conditions:

$$\begin{cases} v(0) = 0, \\ v(1) = 0. \end{cases}$$

This leads to the system of equations:

$$\begin{bmatrix} 1 & 1 \\ e^k & e^{-k} \end{bmatrix} \begin{bmatrix} a \\ b \end{bmatrix} = \begin{bmatrix} 0 \\ 0 \end{bmatrix},$$

which possesses nontrivial solutions when:

$$0 = \begin{vmatrix} 1 & 1 \\ e^k & e^{-k} \end{vmatrix} = -e^k + e^{-k},$$

and hence $e^{2k} = 1$, whose solutions are:

$$\boxed{k = n\pi i, \quad n \in \mathbb{N}.}$$

Additionally, we must have $a + b = 0$, so:

$$\boxed{v(x) = A \sin n\pi x,}$$

for $A \in \mathbb{C}$ an arbitrary complex constant and $n \in \mathbb{N}$.

The solution to the Laplace equation will be:

$$u(x, y) = (ce^{-n\pi y} + de^{n\pi y}) \sin n\pi x, \quad c, d \in \mathbb{C}.$$

(4) Apply the MSV to the two-dimensional Laplace equation

$$u_{xx} + u_{yy} = 0,$$

with boundary conditions

$$\begin{cases} u(0, y) = 0, \\ u(1, y) = 0, \\ u(x, 0) = 0. \end{cases}$$

Solution: We use the solutions discussed in the previous problem and impose the corresponding additional boundary value condition:

$$w(0) = 0.$$

So, we get $c + d = 0$, and therefore

$$\boxed{w(y) = B \sinh n\pi y.}$$

We conclude that

$$\boxed{u(x, y) = A \sin n\pi x \sinh n\pi y}$$

is the solution to our problem. The most general solution that the SVM provides will take the following form:

$$\boxed{u(x, y) = \sum_{n=1}^{\infty} c_n \sin n\pi x \sinh n\pi y}$$

for some adequate sequence $\{c_n\}_{n=1}^{\infty}$.

(5) Apply the method of separation of variables to the one-dimensional heat equation, with $D > 0$,

$$u_t = Du_{xx}.$$

Solution: The separation of variables is done with the operators:

$$A = \frac{d}{dt}, \quad B = -D\frac{d^2}{dx^2},$$

looking for a solution in the form $u(t,x) = v(t)w(x)$, each factor being an eigenfunction of the corresponding differential operator, i.e.,

$$Av = -k^2 v, \quad Bw = k^2 w,$$

such that we've written the eigenvalue as $\lambda = k^2$ with $k \in \mathbb{C}$.

Therefore, the following equations must be fulfilled:

$$v' = -k^2 v, \quad Dw'' = -k^2 w,$$

whose solutions are:

$$\boxed{v(t) = ce^{-k^2 t}, \quad w(x) = a\exp\left(\frac{ikx}{\sqrt{D}}\right) + b\exp\left(-\frac{ikx}{\sqrt{D}}\right).}$$

Thus, the solution of the heat equation is:

$$\boxed{u(t,x) = a\exp\left(-k^2 t + \frac{ikx}{\sqrt{D}}\right) + b\exp\left(-k^2 t - \frac{ikx}{\sqrt{D}}\right);}$$

and the principle of linear superposition leads to the following general solution:

$$\boxed{u(t,x) = \int \left(a(k)\exp\left(-k^2 t + \frac{ikx}{\sqrt{D}}\right) + b(k)\exp\left(-k^2 t - \frac{ikx}{\sqrt{D}}\right)\right)dk.}$$

(6) Apply the SVM to the one-dimensional free Schrödinger equation

$$i\hbar u_t = -\frac{\hbar^2}{2m}u_{xx}.$$

Solution: To apply the SVM, we consider the operators:

$$A = i\hbar\frac{d}{dt}, \quad B = \frac{\hbar^2}{2m}\frac{d^2}{dx^2},$$

and look for solutions in the form of $u(t,x) = v(t)w(x)$, where

$$Av = k^2 v, \quad Bw = -k^2 w,$$

with $\lambda = k^2$ and $k \in \mathbb{C}$.

Hence, we obtain the following ODEs:

$$i\hbar v' = k^2 v, \quad \frac{\hbar^2}{2m}w'' = -k^2 w,$$

whose solutions are:

$$v(t) = c \exp\left(-i\frac{k^2 t}{\hbar}\right),$$

$$w(x) = a \exp\left(\frac{i\sqrt{2m}kx}{\hbar}\right) + b \exp\left(-\frac{i\sqrt{2m}kx}{\hbar}\right).$$

The corresponding solution of the Schrödinger equation is:

$$u(t,x) = a \exp\left(-i\frac{k^2 t}{\hbar} + \frac{i\sqrt{2m}kx}{\hbar}\right) + b \exp\left(-i\frac{k^2 t}{\hbar} - \frac{i\sqrt{2m}kx}{\hbar}\right),$$

which leads to the following general solution:

$$u(t,x) = \int \left(a(k) \exp\left(-i\frac{k^2 t}{\hbar} + \frac{i\sqrt{2m}kx}{\hbar}\right) + b(k) \exp\left(-i\frac{k^2 t}{\hbar} - \frac{i\sqrt{2m}kx}{\hbar}\right) \right) dk.$$

(7) Apply the SVM to the one-dimensional wave equation

$$u_{tt} - c^2 u_{xx} = 0.$$

Solution: SVM can be effectively utilized with the differential operators:

$$A = \frac{d^2}{dt^2}, \quad B = -c^2 \frac{d^2}{dx^2}$$

alongside the proposed ansatz $u(t,x) = v(t)w(x)$. This leads to the following eigenvalue problems (with eigenvalues expressed as $\lambda = -k^2$):

$$Av = -k^2 v, \quad Bw = k^2 w$$

which correspond to the following ordinary differential equations:

$$v'' = -k^2 v, \quad -c^2 w'' = k^2 w$$

The solutions to these equations take the forms:

$$v(t) = P e^{ikt} + Q e^{-ikt},$$

$$w(x) = R e^{ik\frac{x}{c}} + S e^{-ik\frac{x}{c}},$$

here

$$P, Q, R, S \in \mathbb{C}.$$

Consequently, the SVM solution becomes:

$$u(t,x) = \alpha e^{ik\left(t-\frac{x}{c}\right)} + \beta e^{-ik\left(t-\frac{x}{c}\right)} + \gamma e^{ik\left(t+\frac{x}{c}\right)} + \delta e^{-ik\left(t+\frac{x}{c}\right)},$$

where
$$\alpha, \beta, \gamma, \delta \in \mathbb{C}.$$

Finally, the solution is:

$$u(t, x) = \int \left(\alpha(k) e^{ik\left(t - \frac{x}{c}\right)} + \gamma(k) e^{ik\left(t + \frac{x}{c}\right)} \right) dk.$$

In summary, we recover the general solution of the wave equation:

$$u(t, x) = f(x - ct) + g(x - ct).$$

(8) A one-dimensional quantum particle enclosed in a box of length L is described by a Dirichlet BVP:

$$\begin{cases} iu_t = -u_{xx}, & t > 0, \quad 0 < x < L, \\ \begin{cases} u|_{x=0} = 0, \\ u|_{x=L} = 0. \end{cases} \end{cases}$$

Determine the solutions provided by the SVM.

Solution: To apply the SVM, we write:

$$A = i\frac{d}{dt}, \quad B = \frac{d^2}{dx^2},$$

and look for solutions in the form $u(t, x) = v(t)w(x)$, with each factor being an eigenfunction:

$$Av = k^2 v, \quad Bw = -k^2 w,$$

where we've written the eigenvalue in the form of $\lambda = k^2$. Therefore, the following ODEs must be satisfied:

$$iv' = k^2 v, \quad w'' = -k^2 w,$$

which have the solutions:

$$v(t) = c e^{-ik^2 t}, \quad w(x) = a e^{ikx} + b e^{-ikx},$$

where a, b, and c are arbitrary complex constants.

The boundary conditions act only on the x variable, so we must impose that:

$$\begin{cases} w(0) = 0, \\ w(1) = 0. \end{cases}$$

This leads to the system:

$$\begin{bmatrix} 1 & 1 \\ e^{ikL} & e^{-ikL} \end{bmatrix} \begin{bmatrix} a \\ b \end{bmatrix} = \begin{bmatrix} 0 \\ 0 \end{bmatrix}.$$

The existence of nontrivial solutions requires:

$$\begin{vmatrix} 1 & 1 \\ e^{ikL} & e^{-ikL} \end{vmatrix} = 0,$$

i.e., $k = \frac{n\pi}{L}$, with $n \in \mathbb{N}$, and the eigenfunctions are:

$$w_n(x) = \sin \frac{n\pi x}{L}.$$

Hence, the solutions that the SVM provides for this problem are of the following form:

$$u(t,x) = \sum_{n=1}^{\infty} c_n \exp\left(-i\frac{n^2\pi^2 t}{L^2}\right) \sin \frac{n\pi x}{L},$$

for appropriate sequences of numbers $\{c_n\}_{n=1}^{\infty}$.

(9) Determine the solutions provided by the SVM for the Dirichlet BVP:

$$\begin{cases} tu_t = u_{xx} + 2u, & t > 0, \quad 0 < x < 1, \\ u|_{x=0} = 0, \\ u|_{x=\pi} = 0. \end{cases}$$

Solution: The separation of variables is considered in this case with the following differential operators:

$$A = t\frac{\mathrm{d}}{\mathrm{d}t} - 2, \quad B = -\frac{\mathrm{d}^2}{\mathrm{d}x^2},$$

which, for factorized solutions in the form $u(t,x) = v(t)w(x)$, lead to the following eigenvalue problems with $\lambda = -k^2$:

$$Av = -k^2 v, \quad Bw = k^2 w.$$

Thus, we must solve the following ordinary differential equations:

$$tv' = (2 - k^2)v, \quad w'' = -k^2 w,$$

whose solutions are, for $a, b, c \in \mathbb{C}$, given by

$$v(t) = ct^{2-k^2}, \quad w(x) = a\sin kx + b\cos kx,$$

The boundary conditions act only on the variable x, so they will have to be applied to $w(x)$, i.e.,

$$\begin{cases} w(0) = 0, \\ w(1) = 0, \end{cases}$$

which leads to the system:

$$\begin{bmatrix} 0 & 1 \\ \sin k & \cos k \end{bmatrix} \begin{bmatrix} a \\ b \end{bmatrix} = \begin{bmatrix} 0 \\ 0 \end{bmatrix}.$$

For nontrivial solutions to exist, we need the system determinant to vanish, which gives $\sin k = 0$, and consequently:

$$k = n\pi, \quad n \in \mathbb{N}.$$

Therefore, the eigenvalues are:

$$\lambda_n = -n^2\pi^2, \quad n \in \mathbb{N},$$

and the corresponding eigenfunctions are given by:

$$w_n(x) = \sin n\pi x, \quad n \in \mathbb{N}.$$

Hence, we arrive at the following solution:

$$u(t,x) = ct^{2-n^2\pi^2}\sin n\pi x, \quad c \in \mathbb{C}, \quad n \in \mathbb{N},$$

and the corresponding series yields the solution:

$$u(t,x) = \sum_{n=1}^{\infty} c_n t^{2-n^2\pi^2}\sin n\pi x,$$

where $\{c_n\}_{n=1}^{\infty} \subset \mathbb{C}$ represents suitable sequence of complex numbers.

(10) Determine the solutions provided by the SVM for the BVP:

$$\begin{cases} u_t = u_{xx}, & t > 0, \quad 0 < x < 1, \\ u|_{x=0} = 0, \\ (u_x + \alpha u)|_{x=1} = 0, \end{cases}$$

Solution: Before solving the problem, let's provide its physical interpretation. We have a problem of heat diffusion in a thread of unit length, with a prescribed zero temperature at one edge, and at the other edge, there is a mechanism that cools or warms in proportion to the temperature at which it is found.

To apply the SVM, we consider eigenvalue problems for the operators:

$$A = \frac{d}{dt}, \quad B = -\frac{d^2}{dx^2},$$

which yield the following equations:

$$Av = -k^2 v, \quad Bw = k^2 w,$$

allowing us to find solutions in the factorized form $u(t,x) = v(t)w(x)$. Therefore, the following ordinary differential equations need to be solved:

$$v' = -k^2 v, \quad w'' = -k^2 w,$$

whose solutions are:

$$v(t) = ce^{-k^2 t}, \quad w(x) = a\sin kx + b\cos kx, \quad a, b, c \in \mathbb{C}.$$

The boundary conditions act only on the x variable, so we impose them on $w(x)$, i.e.,

$$\begin{cases} w(0) = 0, \\ w'(1) + \alpha w(1) = 0, \end{cases}$$

which, considering that

$$w'(x) = ka \cos kx - kb \sin kx,$$

and that due to this

$$w'(x) + \alpha w(x) = a(\alpha \sin kx + k \cos kx) + b(\alpha \cos kx - k \sin kx),$$

can be written as the following linear system:

$$\begin{bmatrix} 0 & 1 \\ \alpha \sin k + k \cos k & \alpha \cos k - k \sin k \end{bmatrix} \begin{bmatrix} a \\ b \end{bmatrix} = \begin{bmatrix} 0 \\ 0 \end{bmatrix}.$$

For nontrivial solutions to exist, the determinant of this system must be zero, leading to the eigenvalues in the form $\lambda_n = -k_n^2$, where k_n is one of the infinite solutions of the following transcendental equation:

$$\boxed{\tan k = -\frac{k}{\alpha}.}$$

The eigenfunctions will be:

$$\boxed{w_n(x) = c \sin k_n x,}$$

and the solutions of the BVP for the heat equation are:

$$\boxed{u(t, x) = c \exp(-k_n^2 t) \sin k_n x, \quad n \in \mathbb{N}.}$$

Finally, the most general solution provided by SVM will be in the form:

$$\boxed{u(t, x) = \sum_{n=1}^{\infty} c_n \exp\left(-k_n^2 t\right) \sin k_n x,}$$

where $\{c_n\}_{n=1}^{\infty} \subset \mathbb{C}$ represents a conveniently chosen sequence of numbers.

The table of the initial k_n is:

Initial zeros of $\tan k + \frac{k}{3}$						
2.4556	5.2329	8.2045	11.2560	14.3434	17.4490	20.5652

For $\alpha = 3$, let's represent the transcendental equation $\tan k = -\frac{k}{3}$ and its initial solutions in the graph below:

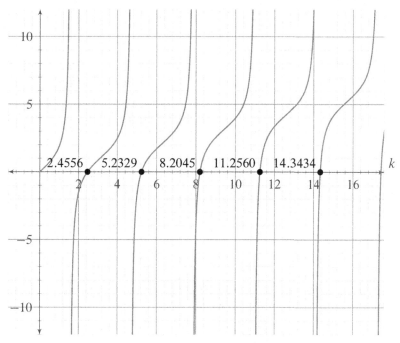

Graph of $f(k) = \tan(k) + \frac{k}{3}$ and the initial five zeros

From the graph, we can observe the values of k at which the function crosses the x-axis, and those values correspond to the zeros of the transcendental equation. These zeros determine the values k_n for which the solutions of the BVP are given by $w_n(x) = c \sin k_n x$, where c is a constant and $n \in \mathbb{N}$ represents the different eigenfunctions.

§3.6.2. Exercises

(1) An infinite string held by one of its ends is described by the BVP:

$$\begin{cases} u_{tt} = c^2 u_{xx}, & -\infty < x < 0,\ t > 0, \\ u|_{x=0} = 0. \end{cases}$$

Determine the standing waves of the problem and interpret them dynamically.

(2) Under certain conditions, some components of the electromagnetic field inside a rectangular section waveguide are described by the BVP:

$$\begin{cases} u_{tt} = c^2 \Delta u, & 0 < x < L_1,\ 0 < y < L_2,\ -\infty < z < +\infty, \\ u|_{\text{walls}} = 0. \end{cases}$$

Determine the standing waves of the model.

(3) Under certain conditions, the components of an electromagnetic field confined inside a parallelepipedic box are described by the BVP:

$$\begin{cases} u_{tt} = c^2 \Delta u, & 0 < x < L_1,\ 0 < y < L_2,\ 0 < z < L_3, \\ u|_{\text{walls}} = 0. \end{cases}$$

Determine the standing waves of the model and interpret them dynamically.

(4) The height above the equilibrium position of the membrane of a rectangular drum obeys the BVP:

$$\begin{cases} u_{tt} = c^2 \Delta u, & 0 < x < L_1,\ 0 < y < L_2, \\ u|_{\text{walls}} = 0. \end{cases}$$

Determine the standing waves of the model and interpret them dynamically.

4. Symmetric Differential Operators

Contents

HILBERT spaces and orthogonal sets, which allow for eigenfunction series expansions of solutions to boundary value problems, are introduced in this chapter. Orthogonal sets and Hilbert spaces are foundational tools in comprehending and resolving boundary value problems related to linear differential equations. They furnish a robust mathematical framework for the precise representation and analysis of solutions to intricate phenomena.

Within these Hilbert spaces of functions, we delve into the realm of symmetric differential operators. Furthermore, we introduce a highly significant category of second-order differential operators referred to as Sturm–Liouville operators. These operators assume a pivotal role in the resolution of separable linear partial differential equations, especially when dealing with boundary value problems. They are closely linked to second-order linear differential equations that surface in various mathematical and physical contexts.

§4.1. Hilbert Spaces

HILBERT space is a fundamental concept in functional analysis, a branch of mathematics that deals with vector spaces of functions equipped with additional structures, which for a Hilbert space is a scalar (or inner) product. Hilbert spaces provide a rigorous framework for studying the properties of functions and vectors in a way that generalizes the notions of length and orthogonality from finite-dimensional spaces to infinite-dimensional ones.

Notable examples of Hilbert spaces include L^2 spaces of square-integrable functions, Sobolev spaces used in the study of partial differential equations, and sequence spaces such as ℓ^2, which consist of square-summable sequences.

In essence, Hilbert spaces provide a rich mathematical structure for understanding and analyzing functions and solutions to BVPs.

§4.1.1. Scalar Product of Functions

Scalar Product in Vector Spaces

Suppose we have a complex vector space V. A scalar product (or inner product) in V is a mapping $(\cdot,\cdot): V \times V \to \mathbb{C}$ that satisfies the following properties:

(1) $(u,v) = \overline{(v,u)}$ for $u,v \in V$.

(2) $(u, \alpha v_1 + \beta v_2) = \alpha(u,v_1) + \beta(u,v_2)$ for all $u, v_1, v_2 \in V$ and for $\alpha, \beta \in \mathbb{C}$.

(3) $(u,u) \geq 0$ for all $u \in V$; $(u,u) = 0$ if and only if $u = 0$.

Definition

From properties 1 and 2, we deduce that $(\alpha v_1 + \beta v_2, u) = \bar{\alpha}(v_1, u) + \bar{\beta}(v_2, u)$. We observe that $(\cdot, \cdot)$ is almost a bilinear form, sometimes called a sesquilinear form. Property 3 tells us that the scalar product of a nonzero vector by itself is always a positive number. The scalar product allows us to assign a length to vectors, which we will call the norm:

$$\|u\| := +\sqrt{(u,u)}.$$

A fundamental inequality, known as the Cauchy–Schwarz inequality, relates the norm and the scalar product as follows:

$$|(u,v)| \leq \|u\|\|v\|.$$

Three important properties are fulfilled by the norm:

(1) $\|u\| = 0 \Leftrightarrow u = 0$.

(2) $\|\alpha u\| = |\alpha|\|u\|$, for $\alpha \in \mathbb{C}$ and for $u \in V$.

(3) $\|u + v\| \leq \|u\| + \|v\|$ (triangle inequality), for $u, v \in V$.

These properties allow us to define a distance in V as follows:

$$d(u,v) := \|u - v\| = \sqrt{(u - v, u - v)}.$$

The notion of a scalar product as outlined here can be applied to functional spaces, as we shall now see.

Scalar Product in Functional Spaces

Let's consider an open set $\Omega \subset \mathbb{R}^n$ and a function $\rho : \Omega \to \mathbb{R}$ that is smooth and positive, $\rho(x) > 0, x \in \Omega$. Given two complex-valued functions u and v defined on Ω, their scalar product corresponding to the weight function ρ is defined by:

$$(u, v) = \int_\Omega \bar{u}(x)v(x)\rho(x)\mathrm{d}^n x.$$

Definition

We must remember that the integral of functions on $\overline{\Omega}$ with complex values is understood as:

$$\int_{\overline{\Omega}} u(x)\rho(x)\mathrm{d}^n x := \int_{\overline{\Omega}} \mathrm{Re}(u(x))\rho(x)\mathrm{d}^n x + \mathrm{i} \int_{\overline{\Omega}} \mathrm{Im}(u(x))\rho(x)\mathrm{d}^n x.$$

The Riemann integral is not entirely adequate in this context. In fact, it is necessary to use a more sophisticated integration theory due to Lebesgue.

Obviously, for this scalar product to make sense, it is necessary that the corresponding integral exists. This condition is guaranteed if we work in the space of functions

$$\mathcal{L}^2(\overline{\Omega}, \rho\,\mathrm{d}^n x) := \left\{ u : \overline{\Omega} \to \mathbb{C} : \|u\|^2 = \int_{\overline{\Omega}} |u(x)|^2\, \rho(x)\mathrm{d}^n x < \infty \right\}.$$

The inequality of Cauchy–Schwarz implies that all the scalar products between functions of this space exist.

In fact, the $(\cdot, \cdot)$ operation we just defined is not strictly a scalar product. What fails is that there are functions $u(x) \neq 0$ in Ω that have a zero norm $\|u\| = 0$. For example, the function of Dirichlet that is 1 on rationals and 0 on irrationals is Lebesgue integrable and has norm 0. To solve this problem, we must consider a function to be the zero function if its norm is zero. Likewise, two functions will be considered to be the same if and only if their difference has a zero norm. What we are doing here is nothing more than considering equivalence classes of functions: $u \sim v$ provided that $\|u - v\| = 0$. The quotient space $\mathcal{L}^2(\overline{\Omega}, \rho\,\mathrm{d}^n x)/ \sim$ is what we denote by $L^2(\overline{\Omega}, \rho\,\mathrm{d}^n x)$. The scalar product in $L^2(\overline{\Omega}, \rho\,\mathrm{d}^n x)$ endows this infinite dimensional space of functions with a mathematical structure that is known by the name of **Hilbert space**.

The elements of $L^2(\overline{\Omega}, \rho\,\mathrm{d}^n x)$ are known as **square integrable functions** with weight ρ over $\overline{\Omega}$. If $\overline{\Omega}$ is a compact set (which is equivalent to Ω being a bounded set), then it is clear that any smooth function in $\overline{\Omega}$ is square integrable in $\overline{\Omega}$. That is to say, if $\overline{\Omega}$ is compact, then $C^\infty(\overline{\Omega}) \subset L^2(\overline{\Omega}, \rho\,\mathrm{d}^n x)$. If $\overline{\Omega}$ is not compact, then such an inclusion is not true, and only those smooth functions that are decaying adequately at infinity will be of integrable square.

Examples:

(1) Let $\Omega = (0, +\infty) \subset \mathbb{R}$; $u(x) = x$ and $v(x) = 1$. If the weight function is $\rho(x) = 1$, then the scalar product of u and v does not exist due to

$$(u, v) = \int_0^\infty x \, dx = \infty.$$

If we now take the weight $\rho(x) = e^{-x^2}$, then the scalar product of u and v exists, and it is given by

$$(u, v) = \int_0^\infty x e^{-x^2} dx = \frac{1}{2}.$$

(2) If we take the domain $\Omega = (0, 1)$ and weight $\rho(x) = 1$, the scalar product of the functions $u(x) = \frac{1}{x}$ and $v(x) = 1$ is not defined due to the singularity of the integrand at $x = 0$, that is $(u, v) = \int_0^1 \frac{1}{x} dx = \infty$. However, for the weight $\rho(x) = x$, we get the following scalar product:

$$\int_0^1 x \frac{1}{x} dx = 1.$$

(3) Now, we take $\Omega = (0, 1)$, $u(x) = x + ix^2$, and $v(x) = 1 - ix$ with $\rho(x) = 1$. Then, the scalar product is given by:

$$(u, v) = \int_0^1 (x - ix^2)(1 - ix) dx$$

$$= \int_0^1 (x - x^3 - 2ix^2) dx = \frac{1}{4} - \frac{2i}{3},$$

and the norm of u is:

$$\|u\|^2 = (u, u) = \int_0^1 |x + ix^2|^2 \, dx = \int_0^1 (x^2 + x^4) dx = \frac{8}{15}.$$

(4) If $\Omega = (0, 2\pi)$, $u(x) = e^{ix}$ and $v(x) = e^{2ix}$ with $\rho(x) = 1$, then we obtain:

$$(u, v) = \int_0^{2\pi} e^{-ix} e^{2ix} dx = 0,$$

and the square of the length of u will be:

$$\|u\|^2 = \int_0^{2\pi} |e^{ix}|^2 \, dx = \int_0^{2\pi} 1 dx = 2\pi.$$

§4.1.2. Change of Variables

We have defined the scalar product of functions on domains in $\mathbb{R}^n$ using Cartesian coordinates. However, due to the theorem of the change of variables in integral calculus, using the Jacobian of the transformation, these scalar products can be expressed in other types of coordinates. Let us consider two important cases:

- If the domain of definition is the plane $\mathbb{R}^2$, we can consider **polar coordinates** and obtain the following expression for the scalar product:

$$
(u, v) = \int_{\mathbb{R}^2} \bar{u}(x, y)v(x, y)\rho(x, y)dxdy
$$
$$
= \int_0^\infty \int_0^{2\pi} \bar{u}(r, \theta)v(r, \theta)\rho(r, \theta)rdrd\theta.
$$

In polar coordinates, the weight function transforms as $\rho(x, y) \to r\rho(r, \theta)$. Note that, for the sake of simplicity, we use $\rho(r, \theta)$ to represent $\tilde{\rho}(r, \theta) = \rho(x(r, \theta), y(r, \theta))$. This convention is commonly employed in applications.

- In $\mathbb{R}^3$, we use **spherical coordinates**. In this coordinate system, the scalar product is given by:

$$
(u, v) = \int_{\mathbb{R}^3} \bar{u}(x, y, z)v(x, y, z)\rho(x, y, z)dxdydz
$$
$$
= \int_0^\infty \int_0^\pi \int_0^{2\pi} \bar{u}(r, \theta, \phi)v(r, \theta, \phi)\rho(r, \theta, \phi)r^2 \sin\theta drd\theta d\phi.
$$

and $\rho(x, y, z) \to r^2 \sin\theta\rho(r, \theta, \phi)$.

Example: Let the weight be $\rho = 1$, and consider the functions

$$
u(r, \theta, \phi) = \frac{e^{\frac{i\phi}{2}}}{r^2 + 1}, \quad v(r, \theta, \phi) = \frac{\sin\theta}{r^2}.
$$

Then, their scalar product is given by:

$$
(u, v) = \int_0^\infty \int_0^\pi \int_0^{2\pi} \frac{e^{-\frac{i\phi}{2}}}{(r^2 + 1)} \frac{\sin\theta}{r^2} r^2 \sin\theta drd\theta d\phi
$$
$$
= \left(\int_0^\infty \frac{dr}{1 + r^2}\right)\left(\int_0^\pi \sin^2\theta d\theta\right)\left(\int_0^{2\pi} e^{-\frac{i\phi}{2}} d\phi\right)
$$
$$
= [\arctan r]_0^\infty \left[\frac{\theta}{2} - \frac{\sin 2\theta}{4}\right]_0^\pi \left[2ie^{-\frac{i\phi}{2}}\right]_0^{2\pi}
$$
$$
= \left(\frac{\pi}{2}\right)\left(\frac{\pi}{2}\right)\left(2i(-2)\right)
$$
$$
= -i\pi^2.
$$

§4.2. Orthogonal Sets of Functions

ORTHOGONALITY, complete sets of functions, and orthogonal bases of a functional space are discussed in this section. As we will see, these concepts are fundamental elements for this book and are instrumental in the resolution of some important PDEs in mathematical physics.

Orthogonal Set

Given a set of functions $\{u_n(x)\}_{n\in J}$ in $L^2(\overline{\Omega}, \rho\, d^n x)$, where $J \subset \mathbb{Z}$ is a subset of finite or infinite indices, we say that it is **orthogonal** if $(u_n, u_m) = 0$ for all $n \neq m$, where n and m are elements of the set J. In other words, the scalar product of any two distinct functions from the set is equal to zero.

Definition

This orthogonality property is an essential concept in functional analysis and has wide-ranging applications in mathematics and physics. It allows us to represent functions as linear combinations of orthogonal components, simplifying the analysis and solving differential equations in a structured and systematic manner.

Examples:

(1) The set
$$\left\{ \sin \frac{n\pi x}{\ell} \right\}_{n=1}^{\infty}$$
is orthogonal on the interval $[0, \ell]$ with weight function $\rho(x) = 1$. To see this, we can observe that
$$\int_0^\ell \sin \frac{n\pi x}{\ell} \sin \frac{m\pi x}{\ell} dx = \int_0^\ell \frac{1}{2}\left(\cos \frac{(n-m)\pi x}{\ell} - \cos \frac{(n+m)\pi x}{\ell} \right) dx$$
$$= 0,$$
for $n \neq m$.

(2) The set
$$\left\{ \cos \frac{n\pi x}{\ell} \right\}_{n=0}^{\infty}$$
is orthogonal on the interval $[0, \ell]$ with weight $\rho(x) = 1$. Similarly, for $n \neq m$, we have
$$\int_0^\ell \cos \frac{n\pi x}{\ell} \cos \frac{m\pi x}{\ell} dx = \int_0^\ell \frac{1}{2}\left(\cos \frac{(n-m)\pi x}{\ell} + \cos \frac{(n+m)\pi x}{\ell} \right) dx$$
$$= 0.$$

These orthogonality properties are important in the study of Fourier series and other related applications, where we can express functions as generalized linear combinations of orthogonal basis functions to simplify their representation and analysis.

In the integral calculation, we have used the sine and cosine addition formulas:

$$\cos(A \pm B) = \cos A \cos B \mp \sin A \sin B,$$

$$\sin(A \pm B) = \sin A \cos B \pm \cos A \sin B,$$

which allow us to rewrite the product of sines as a linear combination of cosines, leading to the following addition formulas

$$\sin A \sin B = \frac{1}{2}(\cos(A - B) - \cos(A + B)),$$

$$\sin A \cos B = \frac{1}{2}(\sin(A - B) + \sin(A + B)),$$

$$\cos A \cos B = \frac{1}{2}(\cos(A - B) + \cos(A + B)).$$

§4.2.1. Orthogonal Function Series Expansions

Given an orthogonal set

$$\{u_n(x)\}_{n \in J},$$

an important question is to determine the type of functions that can be expanded in the form

$$u(x) = \sum_{n \in J} c_n u_n(x).$$

When the set of indices J that labels the orthogonal set is infinite, we must be cautious about the meaning of the sum extended to the indices in J. If $J = \mathbb{N}$, the expression

$$u(x) = \sum_{n=1}^{\infty} c_n u_n(x)$$

is understood as the limit

$$\lim_{N \to \infty} \sum_{n=1}^{N} c_n u_n(x).$$

Similarly, when $J = \mathbb{Z}$,

$$u(x) = \sum_{n=-\infty}^{\infty} c_n u_n(x) = \lim_{N \to \infty} \sum_{n=0}^{N} c_n u_n(x) + \lim_{M \to \infty} \sum_{n=1}^{M} c_{-n} u_{-n}(x).$$

However, when we deal with functions, there are several different notions of limit. From this moment on, unless stated otherwise, we will only refer to series of functions that converge strongly, i.e., in L^2 mean. The following different types of convergence will play a fundamental role.

Types of Functional Limits

We will utilize three notions of limit:

(1) Pointwise limit: For all $x_0 \in \overline{\Omega}$, we have

$$u(x_0) = \lim_{N \to \infty} \sum_{n=1}^{N} c_n u_n(x_0).$$

(2) Uniform limit:

$$\lim_{N \to \infty} \sup_{x \in \Omega} \left| u(x) - \sum_{n=1}^{N} c_n u_n(x) \right| = 0.$$

(3) Strong limit (L^2 mean limit):

$$\lim_{N \to \infty} \left\| u - \sum_{n=1}^{N} c_n u_n \right\| = 0.$$

Definition

Uniform convergence implies pointwise convergence; however, in general, the converse is not true. On the other hand, when the closure of Ω is compact, uniform convergence implies strong convergence, and again, the reciprocal is not satisfied in general.

The following properties are fulfilled:

Coefficients

If u admits an expansion $u = \sum_{n \in J} c_n u_n$, then this expansion is unique and its coefficients c_n are determined by

$$c_n = \frac{(u_n, u)}{\|u_n\|^2}.$$

Theorem

In what follows, the sums within the scalar product are taken out. This manipulation is obviously licit as long as the sums are finite. However, it also holds true in the case of strongly convergent series. Since

$$u = \sum_{m \in J} c_m u_m$$

and $\{u_n\}_{n \in J}$ is an orthogonal set, we find

$$(u_n, u) = \left(u_n, \sum_{m \in J} c_m u_m \right) = \sum_{m \in J} c_m (u_n, u_m) = c_n (u_n, u_n).$$

Scalar product

If $u = \sum_{n \in J} c_n u_n$ and $v = \sum_{n \in J} c'_n u_n$, then their scalar product is

$$(u, v) = \sum_{n \in J} \bar{c}_n c'_n \, \|u_n\|^2 \, .$$

Theorem

The corresponding scalar product is

$$(u, v) = \sum_{n, m \in J} \bar{c}_n c'_m (u_n, u_m) = \sum_n \bar{c}_n c'_n \, \|u_n\|^2 \, .$$

Parseval identity

If

$$u = \sum_{n \in J} c_n u_n,$$

its norm fulfills the following

$$\|u\|^2 = \sum_{n \in J} \frac{|(u_n, u)|^2}{\|u_n\|^2} \, .$$

Theorem

This is a consequence of the two previous properties by taking $v = u$.

A relevant property that allows finding orthogonal sets of functions in several variables is the following:

Product Spaces

If $\{u_n(x_1)\}_{n \in J_1}$ is orthogonal in the domain Ω_1 with respect to the weight $\rho_1(x)$, and $\{v_m(x_2)\}_{m \in J_2}$ is orthogonal in the domain Ω_2 with respect to the weight $\rho_2(x)$, then

$$\{u_n(x_1) v_m(x_2)\}_{(n,m) \in J_1 \times J_2}$$

is orthogonal in the domain $\Omega_1 \times \Omega_2$ with respect to the weight $\rho_1(x_1) \rho_2(x_2)$.

Theorem

According to Fubini's theorem, for $n \neq n'$ or $m \neq m'$, we have:

$$\int_{\Omega_1 \times \Omega_2} \bar{u}_n(x_1) \bar{v}_m(x_2) u_{n'}(x_1) v_{m'}(x_2) \rho_1(x_1) \rho_2(x_2) \mathrm{d}x_1 \mathrm{d}x_2$$

$$= \left(\int_{\Omega_1} \bar{u}_n(x_1)u_{n'}(x_1)\rho_1(x_1)\mathrm{d}x_1 \right)\left(\int_{\Omega_2} \bar{v}_m(x_2)v_{m'}(x_2)\rho_2(x_2)\mathrm{d}x_2 \right) = 0.$$

Complete Orthogonal Sets

Given an orthogonal set
$$\{u_n\}_{n\in J} \subset L^2(\overline{\Omega}, \rho\, \mathrm{d}^n x),$$
we say that it is **complete** when all functions
$$u \in L^2(\overline{\Omega}, \rho\, \mathrm{d}^n x)$$
can be expanded in a series
$$u = \sum_{n\in J} c_n u_n.$$
In that case, it is said that $\{u_n\}_{n\in J}$ forms an **orthogonal basis** of the space $L^2(\overline{\Omega}, \rho\, \mathrm{d}^n x)$.

Definition

Complete orthogonal sets are fundamental tools in various branches of mathematics and physics. They provide a way to approximate and work with complex functions using simpler orthogonal functions, making computations, analysis, and solving problems more manageable. These complete orthogonal sets are known as Hilbert bases. In fact, its existence in a general Hilbert space implies that is a separable Hilbert space [58]. In particular, in quantum mechanics, the space of pure states is described by a separable Hilbert space. The elements of a complete orthogonal set of the Hamiltonian operator of the system describes the stationary states.

§4.3. Green Functions

REEN functions for second-order ordinary differential operators and its role in the construction of the inverse of second-order differential operators will be studied in this section. Green's functions are essential for solving second-order linear differential equations with boundary conditions.

They provide a powerful method for finding solutions to such equations and are especially useful when dealing with inhomogeneous terms or boundary conditions.

Green's function effectively acts as an "inverse" of the differential operator L, in the sense that it allows you to find $u(x)$ when you know $f(x)$ by performing the convolution of $f(x)$ with $G(x, y)$. In other words, it "inverts" the effect of the operator L on the solution.

Boundary conditions are essential when dealing with second-order differential equations. In many cases, Green's function depends on these boundary conditions. It ensures that the solutions satisfy the prescribed conditions at the boundaries.

Green's functions are versatile tools with applications that span numerous scientific and engineering disciplines. In electromagnetism, they aid in analyzing and solving problems related to electromagnetic fields, including scenarios involving antennas, waveguides, and electromagnetic scattering. In quantum mechanics, Green's functions play a fundamental role in comprehending particle interactions, quantum scattering, and the behavior of quantum systems when subjected to external potentials. In solid-state physics, they assist in understanding electron behavior in solid-state materials, facilitating calculations pertaining to electronic structure and electron transport.

In fluid dynamics, Green's functions are indispensable for solving problems related to fluid motion, covering applications such as aerodynamics, hydrodynamics, and wave propagation. For structural analysis, these functions are essential in examining how structures respond to applied loads, enabling the determination of stresses and deformations in materials. In seismology, they contribute to modeling the propagation of seismic waves, advancing our comprehension of earthquake mechanisms and predictions of ground motion.

Green's functions find significant utility in heat transfer, resolving heat conduction problems in thermal physics and engineering, offering insights into the mechanisms of heat transfer within materials. They also assist in control theory, facilitating the analysis and design of feedback control systems in control systems engineering, providing insights into how systems respond to various input signals.

In the realm of environmental science, Green's functions play a crucial role in modeling the transport of substances within the environment, aiding assessments of the impact of human activities on natural systems. Additionally, they are employed in engineering dynamics, where they aid in the study of how mechanical and structural systems respond to dynamic loads, contributing to the analysis of vibrations and oscillations. In the field of chemical kinetics, Green's functions assist in understanding the dynamics of chemical reactions, shedding light on the behavior of reactants, products, and intermediates.

In quantum field theory, Green's functions, often referred to as propagators, are central to understanding quantum fields and their interactions. They describe the probability amplitudes for particles to propagate from one space-time point to another, facilitating calculations related to particle interactions, creation, and annihilation. These functions are used to compute scattering amplitudes, providing essential information about particle interactions and momentum changes during collisions. They also reveal quantum fluctuations in the vacuum state, giving rise to phenomena like the Casimir effect and the concept of zero-point energy.

In quantum field theory, calculations frequently encounter divergences. Green's functions are instrumental in the process of renormalization, which eliminates these divergences, ensuring the theory's well-defined nature. They are represented using Feynman diagrams, graphical tools that aid in visualizing and computing complex particle interactions. Green's functions are also crucial for calculating quantum corrections to classical field equations, accounting for the influence of quantum fluctuations on field behavior.

In summary, Green's functions are an essential framework for precise calculations and predictions in a wide array of scientific and engineering domains, including quantum field theory. They have played a significant role in advancing our understanding of fundamental forces, particles, and complex phenomena across the universe.

They are particularly valuable when solving partial differential equations with complex geometries and boundary conditions. While analytical expressions for Green's functions exist for simple cases, numerical methods are often used to compute Green's functions for more complex problems. These numerical approaches are essential when dealing with practical applications.

§4.3.1. Inverting Differential Operators

Let L be a one-dimensional second-order ordinary differential operator defined as:

$$L = p_0(x)\frac{\mathrm{d}^2}{\mathrm{d}x^2} + p_1(x)\frac{\mathrm{d}}{\mathrm{d}x} + p_2(x)$$

with continuous coefficients, and such that

$$p_0(x) \neq 0$$

for

$$x \in [a,b].$$

This operator operates on a domain $\mathfrak{D}$ defined as:

$$\mathfrak{D} = \left\{ u \in C^\infty([a,b]) : l_1(u) = 0, l_2(u) = 0 \right\}$$

where the boundary operators l_1 and l_2 take the form:

$$l_1(u) = \alpha_1 u(a) + \tilde{\alpha}_1 u(b) + \beta_1 u'(a) + \tilde{\beta}_1 u'(b),$$
$$l_2(u) = \alpha_2 u(a) + \tilde{\alpha}_2 u(b) + \beta_2 u'(a) + \tilde{\beta}_2 u'(b).$$

Suppose that the only solution

$$u \in \mathfrak{D}$$

of

$$Lu = 0$$

is the zero function

$$u \equiv 0.$$

In this case, denoting by $\mathfrak{R}$ the image set,

$$L\mathfrak{D},$$

we can assert the existence of the inverse operator

$$L^{-1}$$

defined as:

$$L^{-1} : \mathcal{R} \longrightarrow \mathcal{D}.$$

Inverse of a Differential Operator

The operator L^{-1} can be described as an integral operator:

$$(4.1) \qquad L^{-1}u(x) = \int_a^b G(x, y)u(y)\mathrm{d}y,$$

with a continuous kernel $G(x, y)$ for x and y in the interval $[a, b]$. The properties of $G(x, y)$ are as follows:

(1) For a fixed y in the open interval (a, b), the functions G_x and G_{xx} are continuous as functions of x in the intervals $[a, y)$ and $(y, b]$. Furthermore, G_x experiences a jump at $x = y$ given by:

$$(4.2) \qquad G_x(y + 0, y) - G_x(y - 0, y) = \frac{1}{p_0(y)}.$$

(2) As functions of x, it holds that for any fixed y within the open interval (a, b):

$$\begin{cases} LG(x, y) = 0, & x \in [a, y) \cup (y, b], \\ l_1(G(\cdot, y)) = 0, \\ l_2(G(\cdot, y)) = 0. \end{cases}$$

These properties define the nature of the kernel $G(x, y)$ and its behavior within the given interval.

Theorem

Let's prove the above statements. Consider taking a basis of independent solutions, denoted as $\{u_1, u_2\}$, for the differential equation

$$Lu = 0.$$

In this case, the determinant of their Wronskian matrix, given by:

$$\begin{vmatrix} u_1(x) & u_2(x) \\ u_1'(x) & u_2'(x) \end{vmatrix} \neq 0, \qquad a \leq x \leq b,$$

does not vanish within the interval $[a, b]$.

We seek the function G in the following form:

$$(4.3) \qquad G(x, y) = \begin{cases} a_1(y)u_1(x) + a_2(y)u_2(x), & a \leq x < y, \\ b_1(y)u_1(x) + b_2(y)u_2(x), & y < x \leq b. \end{cases}$$

To ensure the continuity of G at $x = y$, it is necessary that

$$a_1(y)u_1(y) + a_2(y)u_2(y) = b_1(y)u_1(y) + b_2(y)u_2(y),$$

and Equation (4.2) implies

$$a_1(y)u_1'(y) + a_2(y)u_2'(y) - b_1(y)u_1'(y) - b_2(y)u_2'(y) = \frac{1}{p(y)}.$$

These two equations together form a linear system for the variables c_i, defined as $c_i := b_i - a_i$ for $i \in \{1, 2\}$. The determinant of this system is the Wronskian of the basis $\{u_1, u_2\}$ of solutions for $Lu = 0$, and, therefore, it does not equal zero.

Now, by imposing the conditions

$$(4.4) \qquad \begin{cases} l_1(G(\cdot, y)) = 0, \\ l_2(G(\cdot, y)) = 0, \end{cases}$$

we obtain:

$$\sum_{j=1}^{2}(b_j(y) - c_j(y))l_i(a; u_j) + \sum_{j=1}^{2} b_j(y)l_i(b, u_j) = 0, \qquad i \in \{1, 2\},$$

where we define:

$$l_i(a; u) := \alpha_i u(a) + \beta_i u'(a),$$
$$l_i(b; u) := \tilde{\alpha}_i u(b) + \tilde{\beta}_i u'(b).$$

Consequently, we obtain the following system of equations

$$(4.5) \qquad \sum_{j=1}^{2} b_j(y)l_i(b; u_j) - \sum_{j=1}^{2} c_j(y)l_i(a, u_j) = 0, \qquad i \in \{1, 2\}.$$

This system is linear with respect to $\{b_1(y), b_2(y)\}$, and its determinant does not vanish. This is due to our assumption that the only solution $u \in \mathfrak{D}$ of $Lu = 0$ is the zero function ($u \equiv 0$). Therefore, Equation (4.5) has a unique solution, namely, $\{b_1(y), b_2(y)\}$.

In this manner, we have established the existence of a unique function $G(x, y)$ that is continuous for x and y in the interval $[a, b]$ and satisfies $l_i(G(\cdot, y)) = 0$ for $i \in \{1, 2\}$. Furthermore, from Equation (4.3), it becomes evident that for a fixed y in the open interval (a, b), the functions G_x and G_{xx} are continuous as functions of x in the intervals $[a, y)$ and $(y, b]$.

We shall now prove Equation (4.1). Let's define:

$$w(x) = \int_a^b G(x, y)v(y)dy$$
$$= \int_a^x G(x, y)v(y)dy + \int_x^b G(x, y)v(y)dy, \qquad a < x < b.$$

This leads to:

$$w_x(x) = \int_a^x G_x(x, y)v(y)dy + G(x, x + 0)v(x) + \int_x^b G_x(x, y)v(y)dy$$
$$- G(x, x - 0)v(x)$$

$$= \int_a^x G_x(x, y)v(y)dy + \int_x^b G_x(x, y)v(y)dy.$$

Taking into account Equation (4.2) and the fact that $LG(x, y) = 0$ for $x \in [a, y) \cup (y, b]$, we further have:

$$w_{xx}(x) = \int_a^x G_{xx}(x, y)v(y)dy + G_x(x, x + 0)v(x)$$

$$+ \int_x^b G_{x,x}(x, y)v(y)dy - G_x(x, x - 0)v(x)$$

$$= \int_a^b G_{xx}(x, y)v(y)dy + \frac{v(x)}{p_0(x)}.$$

Therefore, due to the property $LG(x, y) = 0$ for $x \in [a, y) \cup (y, b]$, it follows that:

$$Lw(x) = v(x).$$

Moreover, w belongs to $\mathfrak{D}$ since, as a consequence of Equation (4.4), we have:

$$l_i(w) = \int_a^b l_i(G(\cdot, y)v(y)dy = 0,$$

for $i \in \{1, 2\}$.

Example: An important case of Equation (4.1) arises when the domain $\mathfrak{D}$ is defined by two boundary conditions $l_i(u) = 0$, $i \in \{1, 2\}$, in the following form (separate boundary conditions):

$$l_1(u) = \alpha_1 u(a) + \beta_1 u'(a) = 0,$$
$$l_2(u) = \alpha_2 u(b) + \beta_2 u'(b) = 0.$$

We shall now demonstrate that the Green function is given by:

$$G(x, y) = \frac{1}{p_0(y)W(u_1, u_2)(y)} \begin{cases} u_1(x)u_2(y), & x < y, \\ u_1(y)u_2(x), & x > y, \end{cases}$$

where u_i represent nonzero solutions of $Lu = 0$ satisfying the conditions

$$\begin{cases} l_1(u_1) = 0, \\ l_2(u_2) = 0, \end{cases}$$

and

$$W(u_1, u_2) = u_1 u_2' - u_1' u_2$$

denotes the Wronskian of these two solutions. It is readily apparent that $G(x, y)$ satisfies the hypotheses of the aforementioned theorem. In particular, we can observe:

$$G_x(y + 0, y) - G_x(y - 0, y) = \frac{1}{p_0(y)W(u_1, u_2)(y)}(u_1(y)u_2'(y) - u_1'(y)u_2(y))$$

$$= \frac{1}{p_0(y)}.$$

The solution to the nonhomogeneous boundary value problem expressed in

$$\begin{cases} Lu = f, \\ l_1(u) = c_1, \\ l_2(u) = c_2, \end{cases}$$

is given by:

$$u(x) = \int_a^b G(x,y)f(y)dy + \frac{c_2}{l_2(u_1)}u_1(x) + \frac{c_1}{l_1(u_2)}u_2(x).$$

§4.4. Symmetric Differential Operators

SYMMETRIC differential operators, a quite important notion in this context, are introduced in this section. A symmetric differential operator is a type of linear differential operator that possesses a specific symmetry property related to the associated inner product structure.

Symmetric differential operators are particularly important in functional analysis, differential equations, and various areas of mathematical and physical research.

It should be stressed here that the concept of a symmetric differential operator is closely linked to the fundamental concept of self-adjoint operator in quantum mechanics, representing the physical observables of the quantum system.

Consider the linear differential operator L defined as follows:

$$L = \sum_{\alpha}{}' a_\alpha(x)D^\alpha,$$

where the sum is taken over certain finite set of multi-indices α. This operator is defined over a linear subspace of functions

$$\mathfrak{D} \subset L^2(\overline{\Omega}, \rho\, d^n x),$$

which we'll refer to as the operator's domain.

Symmetric Differential Operators

We will say that a differential operator L is **symmetric** over a domain $\mathfrak{D}$ if and only if:

$$(v, Lw) = (Lv, w),$$

for $v, w \in \mathfrak{D}$.

Definition

The natural way to obtain domains of functions in which an operator is symmetric is to characterize the domain by imposing appropriate boundary conditions on its elements.

Examples:

(1) Let us consider the operator

$$L = -\frac{d^2}{dx^2},$$

acting on functions defined in a bounded set $\Omega = (a, b)$ of the real line. To determine the domains

$$\mathfrak{D} \subset L^2([a, b], dx)$$

in which this operator is symmetric, let's observe that for all pairs of smooth functions v and w in $[a, b]$, the following Lagrange's identity is fulfilled:

(4.6)
$$(v, Lw) = (Lv, w) + I,$$
$$I = \left[-\bar{v}(x)w'(x) + \bar{v}'(x)w(x) \right]_a^b.$$

This identity can be easily obtained by integrating by parts twice:

$$(v, Lw) = \int_a^b \bar{v}(x)(-w''(x))dx$$

$$= \left[-\bar{v}(x)w'(x) \right]_a^b + \int_a^b \bar{v}'(x)w'(x)dx$$

$$= \left[-\bar{v}(x)w'(x) + \bar{v}'(x)w(x) \right]_a^b + \int_a^b (-\bar{v}''(x)w(x))dx$$

$$= \left[-\bar{v}(x)w'(x) + \bar{v}'(x)w(x) \right]_a^b + (Lv, w).$$

Therefore, a domain for a symmetric operator L must satisfy the condition:

(4.7)
$$\boxed{\left[-\bar{v}(x)w'(x) + \bar{v}'(x)w(x) \right]_a^b = 0, \quad v, w \in \mathfrak{D}.}$$

Using condition (4.7), it is immediate to see that for this operator there are at least three types of boundary conditions that determine domains where the operator is symmetric.

(a) Homogeneous Dirichlet

$$\boxed{\mathfrak{D} := \left\{ u \in C^\infty([a, b]) : u(a) = u(b) = 0 \right\}.}$$

(b) Homogeneous Neumann

$$\boxed{\mathfrak{D} := \left\{ u \in C^\infty([a, b]) : u'(a) = u'(b) = 0 \right\}.}$$

(c) Periodic

$$\boxed{\mathfrak{D} := \left\{ u \in C^\infty([a, b]) : u(a) = u(b), \ u'(a) = u'(b) \right\}.}$$

(2) Let us now look at the following example in three dimensions: $Lu = -\Delta u = -\mathrm{div}(\nabla u)$. To determine domains where this operator is symmetric, consider the space $L^2(\overline{\Omega}, \mathrm{d}^3 x)$, where Ω is a bounded domain in $\mathbb{R}^3$. Using the scalar product in $L^2(\overline{\Omega})$ and applying the divergence theorem, we obtain

$$(Lv, w) - (v, Lw) = \int_{\overline{\Omega}} \left(\left(-\mathrm{div}\big(\nabla \bar{v}(x)\big)\right)w(x) + \bar{v}(x)\big(\mathrm{div}\big(\nabla w(x)\big)\big)\right)\mathrm{d}^3 x$$

$$= \int_{\overline{\Omega}} \mathrm{div}(\bar{v}(\nabla w) - (\nabla \bar{v})w)\mathrm{d}^3 x$$

$$= \int_{\partial \Omega} (\bar{v}(\nabla w) - (\nabla \bar{v})w) \cdot \mathrm{d}S$$

$$= \int_{\partial \Omega} \left(\bar{v}\frac{\partial w}{\partial \boldsymbol{n}} - \frac{\partial \bar{v}}{\partial \boldsymbol{n}}w \right) \mathrm{d}S.$$

It is then immediate to deduce from this identity that L is symmetric in the following domains:

(a) Homogeneous Dirichlet

$$\boxed{\mathfrak{D} := \big\{ u \in C^\infty(\overline{\Omega}) : u|_{\partial \Omega} = 0 \big\}.}$$

(b) Homogeneous Neumann

$$\boxed{\mathfrak{D} := \left\{ u \in C^\infty(\overline{\Omega}) : \frac{\partial u}{\partial \boldsymbol{n}}\bigg|_{\partial \Omega} = 0 \right\}.}$$

§4.4.1. Eigenvalues and Eigenfunctions

Symmetric operators are essential in the resolution methods of PDEs. This is because of the properties of their eigenfunctions.

Spectrum of Symmetric Operators

If L is a symmetric differential operator on a domain $\mathfrak{D}$ in an L^2 space, then the following properties are satisfied:
  (1) Eigenvalues are real numbers.
  (2) If $v, w \in \mathfrak{D}$ are eigenfunctions corresponding to different eigenvalues, then they are orthogonal.

Theorem

(1) If $\lambda \in \sigma(L)$, then there's a nonzero function $u \in \mathfrak{D}$ such that $Lu = \lambda u$. For this reason,

$$(Lu, u) = (\lambda u, u) = \bar{\lambda} \, \|u\|^2$$

and

$$(u, Lu) = (u, \lambda u) = \lambda \, \|u\|^2 .$$

As

$$(Lu, u) = (u, Lu)$$

and $\|u\| \neq 0$, we must have $\bar{\lambda} = \lambda$. Then λ is a real number.

(2) Given two different elements λ, μ of the spectrum of L (which we already know belongs to $\mathbb{R}$), there are eigenfunctions v and w such that $Lv = \lambda v$ and $Lw = \mu w$; hence,

$$(Lv, w) = \lambda(v, w), \quad (v, Lw) = \mu(v, w).$$

Since L is symmetric, $(Lv, w) = (v, Lw)$, and thus

$$(\lambda - \mu)(v, w) = 0,$$

which implies, as $\lambda \neq \mu$, that $(v, w) = 0$.

### §4.4.2.	Spectral Representation of Green Functions

Let L be a symmetric operator defined on a domain $\mathfrak{D}$ within a Hilbert space $L^2(\bar{\Omega}, \rho \, d^n x)$, characterized by a set of homogeneous boundary conditions:

$$l_j(u) = 0, \qquad\qquad j \in \{1, \ldots, m\}.$$

Assuming the existence of a complete set $\{u_n\}_{n=1}^{\infty}$ of eigenfunctions for L, satisfying:

$$Lu_n = \lambda_n u_n, \qquad u_n \in \mathfrak{D}, \qquad (u_i, u_j) = \delta_{ij},$$

with the condition that $\lambda = 0$ is not an eigenvalue of L, we can establish the existence of the inverse operator L^{-1}. This inverse operator is represented as an integral operator:

$$L^{-1}v(x) = \int_{\Omega} G(x, y)v(y)\rho(y)d^n y,$$

where the Green function $G(x, y)$ is given by:

$$\boxed{G(x, y) = \sum_{n=1}^{\infty} \frac{\bar{u}_n(y)u_n(x)}{\lambda_n}.}$$

Indeed, we have:

$$LL^{-1}v(x) = \int_{\Omega} \sum_{n=1}^{\infty} \frac{\bar{u}_n(y)Lu_n(x)}{\lambda_n} v(y)\rho(y)d^n y$$

$$= \sum_{n=1}^{\infty} \left(\int_{\Omega} \bar{u}_n(y)v(y)\rho(y)d^n y \right) u_n(x)$$

$$= v(x).$$

Specifically, as a consequence of the spectral representation of $G(x, y)$, we can determine that the solution to the boundary value problem given in equation

$$\begin{cases} Lu = f, \\ l_i(u) = 0, \end{cases}$$

which includes an inhomogeneous term:

$$f(x) = \sum_{n=1}^{\infty} f_n u_n(x),$$

is provided by:

$$u(x) = \int_{\Omega} G(x, y) f(y) \rho(y) \mathrm{d}^n y = \sum_{n=1}^{\infty} \frac{f_n}{\lambda_n} u_n(x).$$

Example: For $c > 0$, the linear differential operator

$$Lu = u_{xx} + c^2 u, \qquad\qquad x \in (0, 1),$$

with its domain defined by the boundary conditions

$$\begin{cases} u(0) = 0, \\ u(1) = 0, \end{cases}$$

possesses a complete set of normalized eigenfunctions, which are the trigonometric Fourier basis:

$$u_n(x) = \sqrt{2} \sin n\pi x,$$

for $n \in \mathbb{N}$, accompanied by corresponding eigenvalues

$$\lambda_n = c^2 + \pi^2 n^2.$$

Hence, the Green function for the inverse operator L^{-1} of L can be represented spectrally as:

$$G(x, y) = \sum_{n=1}^{\infty} \frac{2 \sin n\pi x \, \sin n\pi y}{c^2 + \pi^2 n^2}.$$

§4.5.   Sturm–Liouville Differential Operators

STURM–LIOUVILLE differential operators are a specific class of second-order linear differential operators that play a significant role in the study of ordinary differential equations, particularly in the context of eigenvalue problems and boundary value problems. They are named after the French mathematicians Jacques Charles François Sturm and Joseph Liouville, who contributed to their development.

These operators are a fundamental tool in the study of ordinary differential equations with boundary conditions. They provide insights into the behavior of eigenvalues and eigenfunctions, enabling the analysis of complex problems and the development of techniques for solving a wide range of mathematical and physical phenomena.

§4.5.1. One-Dimensional Sturm–Liouville Operators

One-Dimensional Sturm–Liouville Operators

A second-order differential operator defined on a domain $\mathfrak{D} \subset L^2([a,b], \rho\, dx)$ is of the Sturm–Liouville type if it has the form:

$$Lu = \frac{1}{\rho}\left(-\frac{d}{dx}\left(p\frac{du}{dx}\right) + qu \right),$$

where ρ, p, and q are real functions in $C^\infty((a,b))$ such that ρ and p are positive in the open interval (a,b).

Definition

Note that, in general, the functions ρ, p, and q can be singular at $x = a$ and $x = b$, and the functions ρ and p can be zero at $x = a$ and $x = b$.

Regular Sturm–Liouville Operators

A Sturm–Liouville differential operator on a domain

$$\mathfrak{D} := \big\{ u \in C^\infty([a,b]) : l_1(u) = 0, l_2(u) = 0 \big\},$$

with boundary operators of the form

$$\begin{cases} l_1(u) = \alpha_1 u(a) + \tilde{\alpha}_1 u(b) + \beta_1 \dfrac{du}{dx}(a) + \tilde{\beta}_1 \dfrac{du}{dx}(b) = 0, \\[2mm] l_2(u) = \alpha_2 u(a) + \tilde{\alpha}_2 u(b) + \beta_2 \dfrac{du}{dx}(a) + \tilde{\beta}_2 \dfrac{du}{dx}(b) = 0, \end{cases}$$

where $\tilde{\alpha}_1, \tilde{\beta}_2, \tilde{\beta}_1, \tilde{\alpha}_2, \alpha_1, \beta_2, \beta_1, \alpha_2 \in \mathbb{R}$ and (linearly independent boundary conditions)

$$\operatorname{rank} \begin{bmatrix} \alpha_1 & \tilde{\alpha}_1 & \beta_1 & \tilde{\beta}_1 \\ \alpha_2 & \tilde{\alpha}_2 & \beta_2 & \tilde{\beta}_2 \end{bmatrix} = 2,$$

is regular if and only if

  (i) The interval (a,b) is bounded ($a \neq -\infty$ and $b \neq \infty$).
  (ii) The functions ρ, p, q belong to the space $C^\infty([a,b])$.
  (iii) The functions ρ and p are strictly positive in the closed interval $[a,b]$.

Definition

Domains for Symmetric Operators

A regular Sturm–Liouville differential operator L on a domain

$$\mathfrak{D} \subset L^2([a,b], \rho \, dx)$$

is symmetric if and only if:

(4.8) $$p(a) \begin{vmatrix} \bar{v}(a) & w(a) \\ \bar{v}_x(a) & w_x(a) \end{vmatrix} = p(b) \begin{vmatrix} \bar{v}(b) & w(b) \\ \bar{v}_x(b) & w_x(b) \end{vmatrix},$$

for $v, w \in \mathfrak{D}$.

Theorem

To prove it, it is enough to show that the difference

$$(Lv, w) - (v, Lw)$$

vanishes for all $v, w \in \mathfrak{D}$ when the identity (4.8) is satisfied. This is true given that if $v, w \in \mathfrak{D}$,

$$(Lv, w) - (v, Lw) = \int_a^b dx \left(-\left(\frac{d}{dx}\left(p(x)\frac{d\bar{v}}{dx}(x) \right) \right) w(x) \right.$$

$$\left. + \bar{v}(x) \frac{d}{dx}\left(p(x)\frac{dw}{dx}(x) \right) \right)$$

$$= -\int_a^b dx \, \frac{d}{dx}\left(p(x)\left(\frac{d\bar{v}}{dx} w - \bar{v}\frac{dw}{dx} \right) \right)$$

$$= -[p(x)(\bar{v}_x w - \bar{v} w_x)]_a^b$$

$$= p(b) \begin{vmatrix} \bar{v}(b) & w(b) \\ \bar{v}_x(b) & w_x(b) \end{vmatrix} - p(a) \begin{vmatrix} \bar{v}(a) & w(a) \\ \bar{v}_x(a) & w_x(a) \end{vmatrix}$$

$$= 0.$$

Note that the contribution of q is canceled.

Therefore, the symmetric character of L:

$$(Lv, w) = (v, Lw),$$

for $v, w \in \mathfrak{D}$, is equivalent to the given criterion.

Below are some basic examples of boundary conditions that guarantee (4.7) and therefore determine domains over which regular Sturm–Liouville differential operators are symmetric.

(1) **Separate boundary conditions:** One at the point a and another at the point b:

$$\begin{cases} \alpha_1 u + \beta_1 \dfrac{du}{dx} \bigg|_{x=a} = 0, \\[2mm] \alpha_2 u + \beta_2 \dfrac{du}{dx} \bigg|_{x=b} = 0, \end{cases}$$

with
$$\alpha_1 \beta_2 \neq \alpha_2 \beta_1.$$

Two particularly relevant cases of these are the **Dirichlet** conditions from
$$\begin{cases} u(a) = 0, \\ u(b) = 0, \end{cases}$$

and the **Neumann** ones
$$\begin{cases} \dfrac{du}{dx}\bigg|_{x=a} = 0, \\ \dfrac{du}{dx}\bigg|_{x=b} = 0. \end{cases}$$

Another common type of boundary conditions is the **Robin** conditions
$$\begin{cases} \alpha u + \dfrac{du}{dx}\bigg|_{x=a} = 0, \\ \beta u + \dfrac{du}{dx}\bigg|_{x=b} = 0, \end{cases}$$

with α, β nonzero real numbers. Moreover, we could have mixed boundary conditions, mixing these types of boundary conditions.

(2) **Periodic boundary conditions:**
$$\begin{cases} u(a) = u(b), \\ p(a)\dfrac{du}{dx}\bigg|_{x=a} = p(b)\dfrac{du}{dx}\bigg|_{x=b}. \end{cases}$$

Throughout the book, the various boundary conditions mentioned above, Dirichlet, Neumann, Robin, and periodic boundary conditions, will be encountered in different contexts and applications. These conditions model distinct physical scenarios, each tailored to specific situations.

The importance of Sturm–Liouville differential operators is twofold. First, these operators give rise to a special class of eigenfunctions that form complete and orthogonal sets. This property means that, under suitable conditions, these eigenfunctions provide a powerful basis for expanding functions in the Hilbert space L^2.

Second, the completeness and orthogonality of these eigenfunction sets make them an indispensable tool for solving a wide range of boundary value problems for partial differential equations. Specifically, they play a crucial role in the method of eigenfunction expansion, enabling the separation of variables in complex mathematical problems. This technique has numerous applications, ranging from solving problems in mathematical physics to signal processing and data analysis, making Sturm–Liouville theory a cornerstone of modern mathematical analysis and its practical applications.

Regular Sturm–Liouville Operators I

If L is a symmetric regular Sturm–Liouville differential operator on a domain

$$\mathfrak{D} \subset L^2([a,b], \rho \, dx),$$

then there exists a complete orthogonal set of eigenfunctions

$$\{u_n\}_{n=1}^{\infty} \subset \mathfrak{D} \subset L^2([a,b], \rho \, dx).$$

Additionally, for any $u \in \mathfrak{D}$, the corresponding expansion

$$u = \sum_{n=1}^{\infty} c_n u_n,$$

that converges strongly in $L^2([a,b], \rho \, dx)$, also converges uniformly to u on the closed interval $[a,b]$.

Theorem

For the uniform convergence result, see [52, section 5.2, theorem 1]. Note that this is a sufficient condition, but not necessary as we will see when discussing Fourier series in the next chapter.

Regular Sturm–Liouville Operators II

If L is a regular Sturm–Liouville operator with separate boundary conditions in (a,b), then:

(1) The eigenvalues are simple.

(2) The eigenfunction u_n corresponding to the n-th eigenvalue λ_n has $n-1$ zeros in the open interval (a,b).

(3) Whenever $p(x) > 0$ in $[a,b]$, the eigenvalues can be ordered as follows:

$$\lambda_1 < \lambda_2 < \cdots,$$

with

$$\lim_{n \to \infty} \lambda_n = \infty.$$

If $p(x) < 0$ in $[a,b]$, then

$$\lambda_1 > \lambda_2 > \cdots,$$

and

$$\lim_{n \to \infty} \lambda_n = -\infty.$$

(4) Sturm oscillation: The eigenfunction u_{n+1} interlaces the eigenfunction u_n. That is, between two consecutive zeros of u_{n+1}, we have one and only one zero of u_n.

Theorem

In simpler terms, the Sturm oscillation theorem provides a way to determine the number of zeros that a solution to a second-order linear differential equation has within a specified interval by examining the sign changes in a sequence of functions associated with the equation.

This theorem is valuable in various areas of mathematics, physics, and engineering where it is used to solve boundary value problems and study the behavior of solutions to differential equations.

§4.5.2. Singular Sturm–Liouville Operators

For symmetric singular type Sturm–Liouville differential operators, there are generalizations of the previous results in box I; although their description requires introducing new concepts such as continuous spectrum and generalized eigenfunctions.

Singular Sturm–Liouville operators are essential mathematical tools with broad applications in modeling physical systems, analyzing differential equations with singularities, and solving boundary value problems. Their relevance spans various disciplines, making them a valuable subject of study in both theoretical mathematics and practical science and engineering. In control theory and optimal control problems, singular Sturm–Liouville operators can arise in the formulation of cost functionals and system dynamics. Analyzing these operators can provide insights into the controllability and stability of control systems.

A Sturm–Liouville differential operator is said to be singular when it is not regular, which means at least one of the following conditions is satisfied:

- Either $a = -\infty$ or $b = +\infty$.
- At least one of the functions ρ, p, q is singular at a or b.
- At least one of the functions ρ, p becomes zero at a or b.

> ### Singular Sturm–Liouville Operators
>
> If $[a, b]$ is bounded, a singular Sturm–Liouville differential operator with a domain $\mathfrak{D}$ in $L^2([a, b], \rho\,dx)$ is symmetric if and only if, for all $u, v \in \mathfrak{D}$, the following condition is satisfied:
>
> $$(4.9) \quad \lim_{x \to a} \left(p(x) \begin{vmatrix} \bar{u}(x) & v(x) \\ \bar{u}_x(x) & v_x(x) \end{vmatrix} \right) = \lim_{x \to b} \left(p(x) \begin{vmatrix} \bar{u}(x) & v(x) \\ \bar{u}_x(x) & v_x(x) \end{vmatrix} \right).$$
>
> **Theorem**

When the operator is symmetric, the properties of the general theorem for eigenvalues and eigenfunctions of a symmetric operator hold. However, the primary issue for singular operators is that their eigenfunctions $u \in \mathfrak{D}$ do not generally form a complete set. In such cases, we must consider solutions u of the equation $Lu = \lambda u$ not only outside the domain $\mathfrak{D}$ of the operator, but also outside of $L^2([a, b], \rho\,dx)$ (generalized eigenfunctions).

Let us see some relevant examples of singular Sturm–Liouville differential operators. In the examples $x \in (0, b]$ for $b > 0$.

- The Airy equation is a classic example of a singular Sturm–Liouville equation that arises in the study of various physical phenomena such as wave propagation and quantum mechanics.
  The Airy equation is

$$u'' - xu = 0.$$

  Here

$$p(x) = 1, \quad q(x) = x, \quad \rho(x) = 1.$$

  The singularity occurs at $x = 0$, where the coefficient $p(x) = 1$ is regular, but $q(x) = x$ is nonstandard, leading to an irregular behavior in the differential equation. This equation has solutions known as Airy functions, which are oscillatory and exponentially decaying functions.

- The Schrödinger equation for a particle under the influence of a Coulomb potential is another example of a singular Sturm–Liouville equation:

$$-\frac{\hbar^2}{2m} \frac{d^2 u}{dx^2} + \frac{k}{x} u = Eu.$$

  Here,

$$p(x) = \frac{\hbar^2}{2m}, \quad q(x) = \frac{k}{x}, \quad \rho(x) = 1.$$

  The singularity occurs at $x = 0$, where the coefficient $q(x) = \frac{k}{x} - E$ becomes infinite. This equation describes the behavior of a quantum particle in a Coulomb potential field and has solutions that involve special functions like the Coulomb wave functions.

For a further example, see the case of the Bessel equation in Chapter 7.

These examples demonstrate the variety of differential equations involving singular Sturm–Liouville operators and how special functions like Airy functions and Bessel functions play a crucial role in solving them.

§4.5.3.   Three-Dimensional Sturm–Liouville Operators

Sturm–Liouville operators can be extended to three dimensions to describe physical phenomena involving multiple spatial dimensions. The three-dimensional Sturm–Liouville operators are used to study partial differential equations in three-dimensional space and find applications in quantum mechanics, electromagnetism, fluid dynamics, and other areas of physics and engineering.

Three-Dimensional Sturm–Liouville Operators

A second-order differential operator defined on a domain

$$\mathfrak{D} \subset L^2(\overline{\Omega}, \rho\, d^3x)$$

with $\Omega \subset \mathbb{R}^3$ is of the Sturm–Liouville type if it is of the form

$$Lu = \frac{1}{\rho}\left(-\operatorname{div}(p\nabla u) + qu\right),$$

where ρ, p, q are real functions in $C^\infty(\overline{\Omega})$ such that

$$\rho, p > 0$$

on Ω.

Definition

To determine domains where these operators are symmetric, let's suppose that Ω is a bounded set in $\mathbb{R}^3$, and ρ, p, q are functions in $C^\infty(\overline{\Omega})$. Then, using the expression of the scalar product in $L^2(\overline{\Omega}, \rho\, d^3x)$ and applying the divergence theorem, we obtain **Lagrange's identity**:

$$(Lv, w) - (v, Lw) = \int_{\overline{\Omega}} \left(\bar{v}(x)\Big(\nabla \cdot \big(p(x)\nabla w(x) \big)\Big) \right.$$

$$\left. - \Big(\nabla \cdot \big(p(x)\nabla \bar{v}(x)\big)\Big) w(x) \right) d^3x$$

$$= -\int_{\overline{\Omega}} \nabla \cdot \left(p\Big((\nabla\bar{v})w - \bar{v}(\nabla w)\Big) \right) d^3x$$

$$= -\int_{\partial\Omega} \left(p\Big((\nabla\bar{v})w - \bar{v}(\nabla w)\Big) \right) \cdot dS$$

$$= -\int_{\partial\Omega} \left(p\Big(\frac{\partial\bar{v}}{\partial \boldsymbol{n}}w - \bar{v}\frac{\partial w}{\partial \boldsymbol{n}}\Big) \right) dS.$$

For $L = -\Delta$ this Lagrange identity is also known as the second Green's identity.

Therefore, we are led to:

Domains for Symmetric Three-Dimensional Sturm–Liouville Operators

Assume that Ω is a bounded domain in $\mathbb{R}^3$ and that ρ, p, q are functions in $C^\infty(\overline{\Omega})$. The Sturm–Liouville operator L is symmetric on a domain $\mathfrak{D}$ in $L^2(\overline{\Omega}, \rho\, d^3x)$ if and only if for all $v, w \in \mathfrak{D}$,

$$\int_{\partial\Omega} p\left(\frac{\partial\bar{v}}{\partial \boldsymbol{n}}w - \bar{v}\frac{\partial w}{\partial \boldsymbol{n}}\right) dS = 0.$$

Theorem

As an immediate consequence, we deduce the following domains on which the Sturm–Liouville operator is symmetric:

(1) Homogeneous Dirichlet conditions domain

$$\mathfrak{D} := \left\{ u \in C^{\infty}(\overline{\Omega}) : u|_{\partial\Omega} = 0 \right\}.$$

(2) Homogeneous Neumann conditions domain

$$\mathfrak{D} := \left\{ u \in C^{\infty}(\overline{\Omega}) : \frac{\partial u}{\partial \boldsymbol{n}}\bigg|_{\partial\Omega} = 0 \right\}.$$

Relevant examples of Sturm–Liouville operators in three dimensions are as follows:

(1) **Laplacian operator.** If we take $\rho(x) = p(x) \equiv 1$ we get the Laplacian operator with a negative sign

$$L = -\Delta.$$

The sign choice is made to ensure that $E(u) := (u, Lu) > 0$ holds for u such that $|\nabla u|^2$ is in $L^2(\mathbb{R}^n, d^n x)$. In mathematical terms, $E(u)$ represents the energy of the solutions. Hence, the objective is to guarantee that the energies are consistently positive.

(2) **Schrödinger's Hamiltonian.** If we take $\rho = 1$ and $p = \dfrac{\hbar^2}{2m}$, then the corresponding operator is the familiar Schrödinger's Hamiltonian operator in quantum mechanics of a particle in a force field with a potential $q = V$:

$$Lu = -\frac{\hbar^2}{2m}\Delta u + V(x, y, z)u.$$

Further reading: For Hilbert space and functional analysis, we recommend [58] and [67]. You may also find valuable insights in the classical texts [1], [61], and [52]. For discussions on symmetric operators and self-adjointness, please refer to [59] and [67]. Additionally, consider the concise treatment provided in the booklet [36] and the recent book [10] on spectral theory.

In connection with the mathematical theory of linear partial differential equations, we offer the authoritative monograph by Brezis [16] and the comprehensive four-volume treatment of Linear Partial Differential Operators by Hörmander [38]. Both are highly advanced and respected sources. A similar resource centered in linear PDEs with constant coefficients is the book [73]. For a classical treatment of Sturm–Liouville operators, refer to [39], we also recommend [37]. For an elementary approach, see [2].

For a detailed exploration of Green functions, their mathematical foundations, and their practical significance, turn to the work by Duffy [23]. This source provides valuable insights and examples that will enrich your knowledge in this area. For an encyclopedic reference on Green functions and related topics, we suggest consulting the classic text by Morse and Fesbach [51]. This renowned work offers an extensive and in-depth treatment of the subject, making it an invaluable resource for those looking to delve even further into Green functions and their diverse applications.

Undoubtedly, the most authoritative and highly regarded reference, which has seen three invaluable editions, is the work by Stakgold on Green functions and boundary problems [69]. Stakgold's comprehensive treatment of this subject has established itself as a timeless and indispensable resource in the field.

§4.6. Remarkable Lives and Achievements

Who was Hilbert? David Hilbert was a prominent German mathematician who made substantial contributions to various branches of mathematics and beyond. He was born on January 23, 1862, in Königsberg, Prussia (now Kaliningrad, Russia), and he passed away on February 14, 1943, in Göttingen, Germany.

One of Hilbert's most influential contributions was his formulation of 23 unsolved problems in mathematics at the 1900 International Congress of Mathematicians in Paris. These problems have had a profound impact on the development of various mathematical fields and continue to inspire research even today. Hilbert played a crucial role in the foundation of mathematics, particularly in the area of mathematical logic and formalization. He proposed the axiomatic method and contributed to the development of set theory, aiming to establish a solid logical foundation for all of mathematics.

Hilbert made relevant achievements to algebra and number theory. He worked on invariant theory, studying the properties of polynomial functions that remain unchanged under certain transformations. His work in number theory included studies on algebraic number fields and the solvability of equations. In the field of functional analysis, Hilbert's pioneering work in the theory of Hilbert spaces is fundamental to modern quantum mechanics and various other areas of mathematics and physics. Additionally, he worked on integral equations, developing methods to solve a wide range of problems arising in various mathematical disciplines.

Hilbert's legacy as one of the greatest mathematicians of all time is marked by his profound influence on mathematics and his role in shaping the course of 20th-century mathematics. His work continues to be studied and admired by mathematicians and scientists worldwide.

Who was Green? George Green was a remarkable British mathematician who left an enduring legacy in mathematics and physics. His work significantly impacted the fields of electricity and magnetism, particularly through the introduction of Green's functions and Green's theorem.

Born on July 14, 1793, in Sneinton, Nottingham, England, Green's journey is made even more remarkable by his humble origins and lack of formal university education.

Despite his limited formal education, Green displayed an extraordinary aptitude for mathematics and physics. He embarked on a journey of self-education, making use of the few mathematics books available to him. His passion for these subjects began to flourish from an early age.

In 1828, Green authored a groundbreaking essay titled "An Essay on the Application of Mathematical Analysis to the Theories of Electricity and Magnetism." Within this work, he introduced what is now known as Green's theorem, a cornerstone result in vector calculus. This theorem establishes a profound connection between line integrals around closed curves and double integrals over the plane regions enclosed by those curves. Green's theorem laid the groundwork for the application of Green's functions in solving partial differential equations.

Green's pioneering work on Green's functions provided a novel approach to solving differential equations, particularly those encountered in the study of electricity and magnetism. These functions have since become indispensable tools in various scientific and engineering disciplines, facilitating the resolution of problems in areas such as heat conduction, fluid dynamics, and quantum mechanics.

Although George Green's contributions did not receive widespread recognition during his lifetime, his work gradually gained acclaim in the years following his untimely passing on May 31, 1841, at the age of 47. Prominent mathematicians and physicists, including Lord Kelvin (William Thomson) and Lord Rayleigh (John Strutt), acknowledged the profound significance of his work. For more information, see [33].

Who was Sturm? Jacques Charles François Sturm was a French mathematician, physicist, and engineer who was born on September 29, 1803, in Geneva, Switzerland, and passed away on December 18, 1855, in Paris, France.

Sturm is best known for his work in mathematics, particularly his contributions to the theory of differential equations and the Sturm–Liouville theory. He collaborated with the prominent French mathematician Augustin-Louis Cauchy and made significant advancements in the study of second-order linear differential equations with boundary value problems.

The Sturm–Liouville theory, as developed by Sturm and Cauchy, provided a systematic approach to studying the eigenvalues and eigenfunctions of a class of differential operators now known as Sturm–Liouville operators. This theory played a crucial role in the analysis of differential equations and has applications in various areas of physics, engineering, and mathematical analysis.

Sturm also made important contributions to the theory of vibrations in elastic rods and bars, known as Sturm oscillations. His work on the vibrations of elastic systems contributed to the understanding of wave phenomena and mechanical engineering.

In terms of honors, Sturm did not receive as much recognition during his lifetime as some of his contemporaries. However, his work in mathematics and physics was well-regarded, and the Sturm–Liouville theory that he co-developed with Cauchy remains a significant and enduring contribution to mathematical analysis.

Despite not receiving the same level of honors and awards as some other mathematicians of his time, Sturm's legacy lives on through his important mathematical contributions, and his name is associated with the fundamental Sturm–Liouville theory.

Who was Liouville? Joseph Liouville was a French mathematician who made significant contributions to various areas of mathematics, especially in the branches of number theory, complex analysis, and mathematical physics. He was born on March 24, 1809, in Saint-Omer, France, and passed away on September 8, 1882, in Paris, France.

Liouville is best known for his work in number theory and his discovery of transcendental numbers. He proved the existence of transcendental numbers, which are numbers that are not the root of any nonzero polynomial equation with integer coefficients. His groundbreaking work on transcendental numbers had a profound impact on the study of real and complex numbers.

In the field of complex analysis, Liouville made important contributions to the theory of elliptic functions and the Riemann mapping theorem. He also found that any bounded entire function in the complex plane must be constant. This theorem is a fundamental result in complex analysis and has various applications in mathematics and physics. In this context it is known as Liouville theorem.

Liouville also worked on the theory of differential equations and partial differential equations, particularly in the study of solutions with certain types of singularities. His investigations in this area led to the development of the theory of Surm–Liouville equations.

Liouville's work in mathematics and his contributions to various fields have had a lasting impact on the development of mathematics. His name is remembered through concepts like Liouville's theorem and the Liouville equation, which continue to be studied and applied in modern mathematics and mathematical physics. For more information, see [11].

§4.7. Exercises

§4.7.1. Exercises with Solutions

(1) Let us consider the operator $L = -D^2$ on the set of smooth functions u in the interval $[a, b]$ with weight $\rho(x) = 1$. We need to determine which of the following boundary conditions define domains where the operator is symmetric.

(a) $u(a) = 0$, $u'(a) = 0$.
(b) $2u(a) - u'(a) = 0$, $3u(a) + 2u(b) = 0$.
(c) $u(a) + u'(a) = 0$, $2\,i\,u(a) + u(b) = 0$.
(d) $u(a) - 2u'(b) = 0$, $u(b) + 2u'(a) = 0$.
(e) $u(a) + u(b) = 0$, $u'(a) - u'(b) = 1$.

Solution: The boundary conditions must be homogeneous and have real coefficients, which eliminates options (c) and (e). To determine the correct answer among the three remaining options, we recall that the operator $-D^2$ is symmetric if for all pairs of functions u and v in the domain, the condition $\bar{u}(a)v'(a) - \bar{u}'(a)v(a) = \bar{u}(b)v'(b) - \bar{u}'(b)v(b)$ holds.

Option (a) is immediately discarded since generically

$$0 \neq \bar{u}(b)v'(b) - \bar{u}'(b)v(b).$$

In option (b), we can express $u'(a) = 2u(a)$ and $u(b) = -3/2u(a)$ in terms of $u(a)$. Thus, we have

$$\bar{u}(a)v'(a) - \bar{u}'(a)v(a) = 0,$$
$$\bar{u}(b)v'(b) - \bar{u}'(b)v(b) = -3/2(\bar{u}(a)v'(b) - v(a)\bar{u}'(b)).$$

Therefore, the operator is not symmetric.

However, option (d) is correct. Now we have $u'(b) = u(a)/2$ and $u'(a) = -u(b)/2$, which leads to $\bar{u}(a)v'(a) - \bar{u}'(a)v(a) = -(\bar{u}(a)v(b) - \bar{u}(b)v(a))/2$ and $\bar{u}(b)v'(b) - \bar{u}'(b)v(b) = (\bar{u}(b)v(a) - \bar{u}(a)v(b))/2$, and equality is satisfied.

(2) Determine the eigenvalues of the differential operator:

$$Lu = -D^2u,$$

where u is a smooth function defined in the domain $[0, 1]$. The boundary conditions that must be satisfied are:

$$u_x(0) = 0,$$
$$u(1) = 0.$$

Solution: First, we observe that $\lambda = 0$ is not an eigenvalue because if it were the case, then the solutions to the differential equation

$$-u_{xx} = \lambda u$$

would be of the form
$$u(x) = A + Bx,$$
where A and B are constants. The given boundary conditions lead to $A = B = 0$, which implies the trivial solution, and it does not provide any nontrivial eigenvalues.

For the nontrivial case where $\lambda \neq 0$, the solutions to the eigenvalue problem
$$-u_{xx} = \lambda u$$
are of the form $u = A\,e^{ikx} + B\,e^{-ikx}$, with $\lambda = k^2$.

When imposing the boundary conditions, we obtain the following linear system:
$$\begin{cases} ik(A - B) = 0, \\ e^{ik}\,A + e^{-ik}\,B = 0. \end{cases}$$

This system has nonzero solutions, represented by $(A, B) \neq (0, 0)$, as long as the determinant of the coefficient matrix is zero:
$$\begin{vmatrix} ik & -ik \\ e^{ik} & e^{-ik} \end{vmatrix} = 0.$$

Solving this determinant leads to $\cos k = 0 \Leftrightarrow k = \left(n + \tfrac{1}{2}\right)\pi$, where $n \in \mathbb{N}_0$. Thus, the eigenvalues are given by $\lambda_n = \left(n + \tfrac{1}{2}\right)^2 \pi^2$ for $n \in \mathbb{N}_0$. These are the values that satisfy the differential equation and the given boundary conditions.

(3) Given the differential operator
$$Lu := -D(e^x Du),$$
on the smooth functions in $[0, 1]$, determine for which value of c the following boundary conditions
$$\begin{cases} u(0) + cu(1) = 0, \\ u'(0) + cu'(1) = 0, \end{cases}$$
define a domain of $L^2[0, 1]$ in which the operator is symmetric.

Solution: The operator L is of the Sturm–Liouville type with $p(x) = e^x$ and $q(x) = 0$ over the closed interval $[0, 1]$. Therefore, the condition for it to be symmetric is given by:
$$\begin{vmatrix} \bar{v}(0) & w(0) \\ \bar{v}_x(0) & w_x(0) \end{vmatrix} = e \begin{vmatrix} \bar{v}(1) & w(1) \\ \bar{v}_x(1) & w_x(1) \end{vmatrix},$$
for all pairs of functions $v(x)$ and $w(x)$ in the domain. By applying the boundary conditions, we find that the first determinant is:
$$\begin{vmatrix} \bar{v}(0) & w(0) \\ \bar{v}_x(0) & w_x(0) \end{vmatrix} = c^2 \begin{vmatrix} \bar{v}(1) & w(1) \\ \bar{v}_x(1) & w_x(1) \end{vmatrix},$$
leading to the conclusion that L is a symmetric operator whenever $c^2 = e$.

(4) Let us consider the differential operator

$$Lu = -D^2 u,$$

acting on the domain of smooth functions in $[0, 1]$ that fulfill the boundary conditions

$$\begin{cases} u'(0) = 0, \\ u(1) + u'(1) = 0. \end{cases}$$

Find the relationship that determines the eigenvalues $\lambda = k^2$ of L.
Solution: For $\lambda = 0$ the sought solutions are

$$u(x) = a + bx,$$

which, once the given boundary conditions are applied, leads to

$$u = 0.$$

Therefore, 0 is not an eigenvalue.
The possible eigenfunctions of L must be in the form of

$$u(x) = a \cos kx + b \sin kx,$$

for

$$a, b \in \mathbb{R}$$

and with nonzero eigenvalue

$$\lambda = k^2.$$

The derivative is

$$u'(x) = -ak \sin kx + bk \cos kx.$$

Therefore, the existence of nontrivial solutions leads to

$$\begin{vmatrix} 0 & k \\ \cos k - k \sin k & \sin k + k \cos k \end{vmatrix} = 0$$

implying

$$k(\cos k - k \sin k) = 0,$$

and therefore the spectral relationship is

$$\boxed{k \tan k = 1.}$$

A table of the initial 9 zeros and a graph of the corresponding functions follow.

Initial zeros of $k \tan k = 1$								
0.86	3.43	6.44	9.53	12.64	15.77	18.90	22.04	25.17

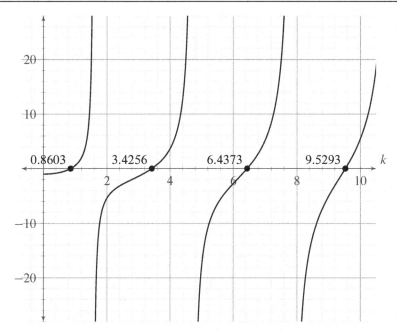

Graph of $f(k) = k \tan(k) - 1$ and the initial four zeros

(5) Compute the eigenvalues of the boundary value problem

$$\begin{cases} -(1+x)\dfrac{d}{dx}\left((1+x)\dfrac{du}{dx}\right) = \lambda u, & 0 < x < 1, \\[2mm] \begin{cases} u(0) = 0, \\ u(1) = 0. \end{cases} \end{cases}$$

Solution: Using the variable $y := x+1$, the eigenvalue problem with $\lambda = k^2$, $k \in \mathbb{C}$, transforms into an Euler equation

$$y^2 u_{yy} + y u_y + k^2 u = 0,$$

whose general solution is given by

$$u(y) = a\mathrm{e}^{ik\log y} + b\mathrm{e}^{-ik\log y}, \qquad\qquad a, b \in \mathbb{C}.$$

Note that since $y \in (1,2)$, the function $\log y$ can be considered as a real function, avoiding the need for a discussion on determinations. In terms of the variable x, the solution is written as follows

$$u(x) = a\mathrm{e}^{ik\log(x+1)} + b\mathrm{e}^{-ik\log(x+1)}, \qquad\qquad a, b \in \mathbb{C}.$$

Now, let's impose the boundary conditions, i.e.,

$$a + b = 0,$$

$$a\mathrm{e}^{ik\log 2} + b\mathrm{e}^{-ik\log 2} = 0.$$

Therefore, the existence of nontrivial solutions requires

$$\mathrm{e}^{i2k\log 2} = 1,$$

which leads to $k_n = \dfrac{\pi}{\log 2} n$, $n \in \mathbb{N}$, and the eigenvalues are

$$\boxed{\lambda_n = \frac{\pi^2}{(\log 2)^2} n^2}$$

for $n \in \mathbb{N}$.

(6) The differential operator is defined as

$$L := -x \frac{\mathrm{d}^2}{\mathrm{d}x^2},$$

and it acts on the set of smooth functions in the interval $[1, \mathrm{e}]$ that satisfy the following boundary conditions:

$$\begin{cases} u(1) = 0, \\ u(\mathrm{e}) + c u'(\mathrm{e}) = 0. \end{cases}$$

We need to determine which of the following values of the constant c defines a domain over which the operator L is symmetric:

(a) c arbitrary real.

(b) c arbitrary pure imaginary.

(c) $c = 1 + \mathrm{i}$.

(d) There is no c for which the operator is symmetric.

(e) $c = 1 - \mathrm{i}$.

 Solution: The operator is of the Sturm–Liouville type. Moreover, if $c \in \mathbb{R}$, then L is regular with separate boundary conditions and, therefore, symmetric. It can also be verified that this is a necessary condition. The correct answer is (a).

(7) Given $h, l > 0$ find the spectral equation that determines the eigenvalues $\lambda = k^2$ of the boundary value problem

$$\begin{cases} -u_{xx} = \lambda u, \quad 0 < x < l, \\ \begin{cases} h u(0) - u_x(0) = 0, \\ h u(l) + u_x(l) = 0. \end{cases} \end{cases}$$

 Solution: We encounter separate boundary conditions and a regular Sturm–Liouville operator, which makes the operator symmetric, resulting in simple and increasing eigenvalues. First, we consider the case $\lambda = 0$, where the solution $u = c_1 + c_2 x$ satisfy the boundary conditions:

$$\begin{bmatrix} h & -1 \\ h & hl + 1 \end{bmatrix} \begin{bmatrix} c_1 \\ c_2 \end{bmatrix} = \begin{bmatrix} 0 \\ 0 \end{bmatrix}.$$

However, this system has only the trivial solution since leads to $h(hl + 1) = -h$, which is not possible given $h, l > 0$. Hence, $\lambda = 0$ is not an eigenvalue.

Next, we consider the case $\lambda \neq 0$ and impose the boundary conditions on the general solution

$$u(x) = c_1 \cos kx + c_2 \sin kx.$$

The derivative of u is $u_x(x) = -kc_1 \sin kx + kc_2 \cos kx$, and we obtain the following system of equations:

$$\begin{bmatrix} h & -k \\ h \cos kl - k \sin kl & h \sin kl + k \cos kl \end{bmatrix} \begin{bmatrix} c_1 \\ c_2 \end{bmatrix} = \begin{bmatrix} 0 \\ 0 \end{bmatrix}.$$

To have nontrivial solutions, we require that

$$\begin{vmatrix} h & -k \\ h \cos kl - k \sin kl & h \sin kl + k \cos kl \end{vmatrix} = 0,$$

which simplifies to:

$$(h^2 - k^2) \sin kl + 2kh \cos kl = 0.$$

The spectral relationship is given by:

$$\tan kl = \frac{2hk}{k^2 - h^2}.$$

Therefore, the eigenvalues λ are determined by this spectral equation for k, where $\lambda = k^2$, and k satisfies $\tan kl = \frac{2hk}{k^2 - h^2}$.

(8) Let us consider the spectral problem

$$\begin{cases} -u_{xx} = \lambda u, \quad 0 < x < 1, \\ u_x(0) + u(0) = 0, \\ \qquad\quad u(1) = 0. \end{cases}$$

(a) Find the equation that determines eigenvalues and prove it has infinite solutions.
(b) Calculate the corresponding eigenfunctions. Do they form a complete orthogonal set in $L^2[0,1]$?
(c) Expand the function $u(x) = x \sin \pi x$ in a series of such eigenfunctions.

Hints:

$$\int_0^1 x \sin \pi x \sin k_n x \, dx = \pi \frac{\pi^2 \sin k - 2k_n \cos k_n - k_n^2 \sin k_n - 2k_n}{(\pi^2 - k_n^2)^2},$$

$$\int_0^1 x \sin \pi x \cos k_n x \, dx = \pi \frac{\pi^2 \cos k - 2k_n \sin k_n - k_n^2 \cos k_n}{(\pi^2 - k_n^2)^2},$$

$$\int_0^1 \sin^2 k_n x \, dx = \frac{1}{2} \left(1 - \frac{\sin k_n \cos k_n}{k_n} \right),$$

$$\int_0^1 x \sin k_n x \cos k_n x \, dx = \frac{\sin^2 k_n}{2k_n},$$

$$\int_0^1 \cos^2 k_n x \, dx = \frac{1}{2} \left(1 + \frac{\sin k_n \cos k_n}{2k_n} \right).$$

Solution:

(a) The given Sturm–Liouville problem is regular with separate boundary conditions, which ensures that its eigenvalues are real, simple, and form an increasing sequence: $\lambda_0 < \lambda_1 < \ldots$, with $\lambda_n \xrightarrow[n\to\infty]{} \infty$. Additionally, the corresponding set of eigenfunctions $\{u_n\}_{n=1}^\infty$ is an orthogonal basis for $L^2([0,1]dx)$.

Let's denote $\lambda = k^2$. The function $u(x) = x \sin \pi x$ satisfies the boundary conditions:

$$u(1) = (x \sin \pi x)\big|_{x=1} = \sin \pi = 0,$$
$$u'(0) + u(0) = (\sin \pi x + x\pi \cos \pi x)\big|_{x=0} + (x \sin \pi x)\big|_{x=0} = 0.$$

Thus, the solution

$$u(x) = x \sin \pi x$$

satisfies the boundary conditions and is therefore in the domain of the differential operator. As a result, we have uniform convergence of this eigenfunction series.

If $k = 0$, the eigenfunction has the form

$$u(x) = a + bx,$$

and the boundary conditions are satisfied if and only if

$$a + b = 0,$$

leading to 0 being an eigenvalue with the corresponding eigenfunction given by

$$u_0(x) = 1 - x.$$

On the other hand, if $k \neq 0$, the boundary conditions on the possible solution

$$a \sin kx + b \cos kx$$

imply

$$\begin{vmatrix} k & 1 \\ \sin k & \cos k \end{vmatrix} = 0,$$

which yields the spectral equation:

$$\boxed{\tan k = k.}$$

Therefore, considering the parity of the spectral equation, we only deal with the solutions with $k > 0$.

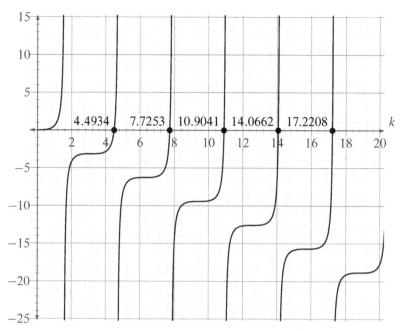

Graph of $f(k) = \tan(k) - k$ and the initial five zeros

Initial zeros of $\tan k = k$

4.4934	7.7253	10.9041	14.0662	17.2208	20.4204

An asymptotic analysis of the tangent near its poles leads to zeros k_n satisfying

$$k_n = \left(n + \frac{1}{2}\right)\pi - \frac{1}{\left(n + \frac{1}{2}\right)\pi} + O\left(\frac{1}{n^2}\right)$$

as $n \to \infty$. Specifically, from the addition relations for trigonometric functions, we know that

$$\tan\left(\left(n + \frac{1}{2}\right)\pi + \epsilon\right) = -\cot\epsilon.$$

Thus, the relationship

$$\tan k = k$$

for

$$k = \left(n + \frac{1}{2}\right)\pi + \epsilon,$$

with $\epsilon > 0$ small, leads to

$$-\frac{\cos\epsilon}{\sin\epsilon} = \left(n + \frac{1}{2}\right)\pi + \epsilon.$$

For $\epsilon \sim 0$ small, we have

$$\cos \epsilon \sim 1,$$

$$\sin \epsilon \sim \epsilon,$$

which leads to

$$-\frac{1}{\epsilon} \sim \left(n + \frac{1}{2}\right)\pi$$

as $n \to \infty$.

(b) Having discussed the eigenvalues

$$\lambda_n = k_n^2,$$

let's proceed to consider the eigenfunctions. Given the spectral equation, the eigenfunctions can be chosen in the form

$$\boxed{u_n(x) = \sin k_n x - k_n \cos k_n x,}$$

for $n \in \mathbb{N}$.

(c) These eigenfunctions

$$\{u_n(x)\}_{n=0}^{\infty}$$

form an orthogonal basis, allowing us to expand

$$x \sin \pi x$$

as a series of such eigenfunctions:

$$x \sin \pi x = \sum_{n=0}^{\infty} c_n u_n(x),$$

where the series coefficients are given by the following formulas:

$$c_n = \begin{cases} \dfrac{\int_0^1 (1-x)x \sin \pi x \, dx}{\int_0^1 (1-x)^2 dx}, & n = 0, \\[3ex] \dfrac{\int_0^1 (\sin k_n x - k_n \cos k_n x)x \sin \pi x \, dx}{\int_0^1 (\sin k_n x - k_n \cos k_n x)^2 dx}, & n \in \mathbb{N}. \end{cases}$$

So, for $n = 0$, we get $c_0 = \frac{12}{\pi^3}$. On the other hand, using the hint, we can write

$$H_n := \int_0^1 (\sin k_n x - k_n \cos k_n x)^2 dx$$

$$= \int_0^1 \sin^2 k_n x \, dx + k_n^2 \int_0^1 \cos^2 k_n x \, dx - 2k_n \int_0^1 \sin k_n x \cos k_n x \, dx$$

$$= \frac{1}{2}\left(1 - \frac{\sin k_n \cos k_n}{k_n}\right) + k_n^2 \frac{1}{2}\left(1 + \frac{\sin k_n \cos k_n}{k_n}\right) - 2k_n \frac{\sin^2 k_n}{2k_n}$$

$$= \frac{1 + k_n^2}{2} - \frac{1 - k_n^2}{2}\frac{\sin k_n \cos k_n}{k_n} - \sin^2 k_n,$$

and if we now remember that $\sin k_n = k_n \cos k_n$, we have

$$
\begin{aligned}
H_n &= \frac{1 + k_n^2}{2} - \frac{1 - k_n^2}{2} \cos^2 k_n - k_n^2 \cos^2 k_n \\
&= \frac{1 + k_n^2}{2} - \frac{1 + k_n^2}{2} \cos^2 k_n \\
&= \frac{1 + k_n^2}{2} \sin^2 k_n.
\end{aligned}
$$

On the other hand, from the hints, we deduce that

$$
\begin{aligned}
I_n &:= \int_0^1 (\sin k_n x - k_n \cos k_n x) x \sin \pi x \, dx \\
&= \frac{\pi}{(\pi^2 - k_n^2)^2} (A_n - k_n B_n),
\end{aligned}
$$

where

$$
\begin{aligned}
A_n &:= \pi^2 \sin k_n - 2k_n \cos k_n - k_n^2 \sin k_n - 2k_n, \\
B_n &:= \pi^2 \cos k_n + 2k_n \sin k_n - k_n^2 \cos k_n.
\end{aligned}
$$

But, from the spectral relation $\sin k_n = k_n \cos k_n$, it follows that

$$
\begin{aligned}
A_n - k_n B_n &= \pi^2 \sin k_n - 2k_n \cos k_n - k_n^2 \sin k_n - 2k_n \\
&\quad - k_n(\pi^2 \cos k_n + 2k_n \sin k_n - k_n^2 \cos k_n) \\
&= -2(1 + k_n^2) \sin k_n - 2k_n.
\end{aligned}
$$

Finally, remembering the spectral relation, we can write

$$
\begin{aligned}
c_n &= -\frac{4\pi}{(\pi^2 - k_n^2)^2} \frac{k_n + (1 + k_n^2) \sin k_n}{(1 + k_n^2) \sin^2 k_n} \\
&= -\frac{4\pi}{(\pi^2 - k_n^2)^2} \left(\frac{\frac{\sin k_n}{\cos k_n}}{1 + \frac{\sin^2 k_n}{\cos^2 k_n}} \frac{1}{\sin^2 k_n} + \frac{1}{\sin k_n} \right) \\
&= -\frac{4\pi}{(\pi^2 - k_n^2)^2} \left(\frac{\cos k_n}{\sin k_n} + \frac{1}{\sin k_n} \right),
\end{aligned}
$$

which leads to

$$
\boxed{ c_n = -\frac{4\pi}{(\pi^2 - k_n^2)^2} \left(\frac{1}{\sin k_n} + \frac{1}{k_n} \right). }
$$

From where we get the table:

Initial coefficients c_n					
0.39	0.09	−0.006	9.6×10^{-4}	-3.82×10^{-4}	-7.95×10^{-5}

Therefore, the series expansion requested in $[0, 1]$ will be

$$x \sin \pi x = \frac{12}{\pi^3}(1 - x) - \sum_{n=1}^{\infty} \frac{4\pi}{(\pi^2 - k_n^2)^2}\left(\frac{1}{\sin k_n} + \frac{1}{k_n}\right)(\sin k_n x - k_n \cos k_n x).$$

In particular, the truncation to three terms is

$$x \sin \pi x \sim 0.3870(1 - x) + 0.0946 \sin 4.4934x - 0.4251 \cos 4.4934x$$
$$- 0.0058 \sin 7.7253x + 0.0448 \cos 7.7253x,$$

and we compare its graph with that of $u(x) = x \sin \pi x$ in the following figure:

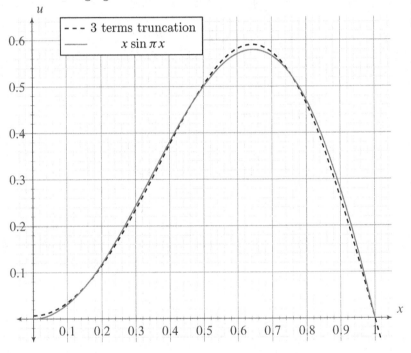

§4.7.2. Exercises

(1) Prove that if L_1 and L_2 are symmetric operators on a certain domain $\mathfrak{D}$, then all the linear combinations $\lambda_1 L_1 + \lambda_2 L_2$, with real coefficients λ_1 and λ_2, are also symmetric on that domain.

(2) Prove that the operator $L = -\dfrac{\mathrm{d}^2}{\mathrm{d}x^2}$ is symmetric over the domain $\mathfrak{D}$ of class C^∞ functions in $[a, b]$ such that they satisfy one of the following types of boundary conditions:

$$u(b) = \alpha u(a) + \beta u'(a),$$
$$u'(b) = \gamma u(a) + \delta u'(a),$$

being $\alpha, \beta, \gamma, \delta$ real numbers such that:

$$\alpha\delta - \beta\gamma = 1.$$

(3) Let us consider the operator $L = -\mathrm{i}\dfrac{\mathrm{d}}{\mathrm{d}x}$ on the domain $\mathfrak{D}$ of functions of class C^∞ in $[a, b]$ such that they satisfy the boundary value condition

$$\alpha u(a) + \beta u(b) = 0,$$

being α and β given complex numbers. Prove that L is symmetric if and only if

$$|\alpha| = |\beta|.$$

Determine in such a case the spectrum and eigenfunctions of L.

(4) Let us consider the operator $Lu = \dfrac{\mathrm{d}^2 u}{\mathrm{d}x^2} - 2u$ on the domain $\mathfrak{D}$ of functions of class C^2 in $[0, 1]$ such that

$$u(0) = 0, \quad u(1) = 0.$$

Determine the spectrum and the eigenfunctions.

(5) Prove that an operator of Sturm–Liouville regular on an interval $[a, b]$ of the form

$$Lu = \frac{1}{\rho}\big(- D(pDu) + qu\big),$$

is symmetric over domains corresponding to the following boundary conditions

(a) Separate:

$$\begin{cases} u(a) + \alpha'u'(a) = 0, \\ \beta u(b) + \beta'u'(b) = 0, \end{cases}$$

with $\alpha, \alpha', \beta, \beta' \in \mathbb{R}$.

(b) Periodic:

$$\begin{cases} u(a) = u(b), \\ p(a)u'(a) = p(b)u'(b). \end{cases}$$

5. Fourier Analysis

Contents

FOURIER analysis, encompassing series and transformations, is the topic we are about to introduce in this chapter. We will divide it into two parts. Firstly, we will study Fourier series expansions, which can involve imaginary exponentials, sine and cosines, or exclusively sine or cosine functions. Secondly, we will delve into the continuous version of these expansions, which arises when the domains of the expanded functions are unbounded. This body of knowledge, known as Fourier analysis, is a fundamental tool for the comprehension natural phenomena. It was first introduced around 200 hundred years ago by Joseph Fourier in the study of heat transmission.

§5.1. Fourier Trigonometric Bases

EIGENVALUE problems associated with two particularly simple differential operators will be considered. The corresponding expansions in eigenfunctions constitute the so-called **Fourier trigonometric series**. The operator (which in quantum mechanics is the linear momentum operator)

$$L = -i\frac{d}{dx}.$$

is symmetric in a domain built up with smooth functions u in a bounded interval $[a, b]$ such that $u(a) = u(b)$, i.e., functions that satisfy periodic conditions.

The problem of eigenvalues

$$\begin{cases} -i\dfrac{du}{dx} = \lambda u, \\ u(a) = u(b), \end{cases}$$

has the solution

$$u(x) = ce^{i\lambda x},$$

and the boundary value condition is satisfied if and only if

$$e^{i\lambda(b-a)} = 1.$$

This condition determines the possible elements of the spectrum

$$\lambda_n = n\omega, \quad \omega := \frac{2\pi}{b-a},$$

for $n \in \mathbb{Z}$, with the following associated eigenfunctions

$$u_n(x) = e^{in\omega x}.$$

The second operator we consider is

$$L = -\frac{d^2}{dx^2}.$$

We will study domains whose elements are smooth functions associated with three types of boundary conditions.

(1) **Periodic.** The problem of eigenvalues is

$$\begin{cases} -\dfrac{d^2u}{dx^2} = \lambda u, \\ u(a) = u(b), \\ \dfrac{du}{dx}(a) = \dfrac{du}{dx}(b). \end{cases}$$

When $\lambda \neq 0$, the solution is

$$u(x) = c_1 \cos\sqrt{\lambda}x + c_2 \sin\sqrt{\lambda}x,$$

which satisfies the boundary conditions if and only if the following homogeneous linear system for (c_1, c_2) is fulfilled:

(5.1)
$$\left(\cos\sqrt{\lambda}a - \cos\sqrt{\lambda}b\right)c_1 + \left(\sin\sqrt{\lambda}a - \sin\sqrt{\lambda}b\right)c_2 = 0,$$
$$\left(\sin\sqrt{\lambda}a - \sin\sqrt{\lambda}b\right)c_1 - \left(\cos\sqrt{\lambda}a - \cos\sqrt{\lambda}b\right)c_2 = 0.$$

Nontrivial solutions $(c_1, c_2) \neq (0,0)$ of this system exist if and only if the determinant of the matrix of coefficients is zero:

$$\begin{vmatrix} \cos\sqrt{\lambda}a - \cos\sqrt{\lambda}b & \sin\sqrt{\lambda}a - \sin\sqrt{\lambda}b \\ -\sin\sqrt{\lambda}a + \sin\sqrt{\lambda}b & \cos\sqrt{\lambda}a - \cos\sqrt{\lambda}b \end{vmatrix} = 0,$$

which simplifies to

$$\left(\cos\sqrt{\lambda}a - \cos\sqrt{\lambda}b\right)^2 + \left(\sin\sqrt{\lambda}a - \sin\sqrt{\lambda}b\right)^2 = 0.$$

Expanding the squares and using trigonometric formulas, we get

$$1 - \cos\sqrt{\lambda}(b-a) = 0.$$

Thus, nonzero eigenvalues are given by

$$\lambda_n = \omega^2 n^2, \qquad\qquad \omega := \frac{2\pi}{b-a},$$

for $n \in \mathbb{N}$. These eigenvalues are not simple (note that this is a homogeneous Sturm–Liouville problem with periodic boundary conditions), as for any of them, the linear system (5.1) reduces to a system in which the matrix of coefficients is the zero matrix, allowing all values of (c_1, c_2) as solutions. In this way,

$$\mathfrak{D}_{\lambda_n} = \mathbb{C}\{\sin n\omega x, \cos n\omega x\}.$$

On the other hand, $\lambda = 0$ is also an eigenvalue, and an eigenfunction can be taken as the constant 1 (the solutions of $-u'' = 0$ are $u(x) = c_1 + c_2 x$, but the boundary conditions impose $c_2 = 0$). Therefore, the set of eigenvalues is

$$\boxed{\lambda_n = \omega^2 n^2, \quad \omega := \frac{2\pi}{b-a},}$$

for $n \in \mathbb{N}_0$ and the set of eigenfunctions is

$$\{1, \cos n\omega x, \sin n\omega x\}_{n=1}^\infty,$$

constituting a complete orthogonal set in $L^2([a,b], \mathrm{d}x)$.

(2) **Homogeneous Dirichlet**. The problem of eigenvalues is given by:

$$\boxed{\begin{cases} -\dfrac{\mathrm{d}^2 u}{\mathrm{d}x^2} = \lambda u, \\ u(a) = 0, \\ u(b) = 0. \end{cases}}$$

The value $\lambda = 0$ is not a part of the spectrum because if it were, the corresponding eigenfunction would be $u(x) = c_1 + c_2 x$, and the boundary conditions would imply that $c_1 + c_2 a = c_1 + c_2 b = 0$. However, since $b - a \neq 0$, we have $c_1 = c_2 = 0$. For $\lambda \neq 0$, the solutions to the ordinary differential equation have the form:

$$u(x) = c_1 \cos \sqrt{\lambda} x + c_2 \sin \sqrt{\lambda} x,$$

and the boundary conditions are satisfied if and only if:

$$c_1 \cos \sqrt{\lambda} a + c_2 \sin \sqrt{\lambda} a = 0,$$
$$c_1 \cos \sqrt{\lambda} b + c_2 \sin \sqrt{\lambda} b = 0.$$

Nontrivial solutions $(c_1, c_2) \neq (0, 0)$ exist if and only if:

$$\begin{vmatrix} \cos \sqrt{\lambda} a & \sin \sqrt{\lambda} a \\ \cos \sqrt{\lambda} b & \sin \sqrt{\lambda} b \end{vmatrix} = 0,$$

which simplifies to:

$$\sin\left(\sqrt{\lambda}(b-a)\right) = 0.$$

As a result, the eigenvalues are given by:

$$\lambda_n = \frac{\omega^2}{4}n^2, \quad \omega := \frac{2\pi}{b-a},$$

for $n \in \mathbb{N}$ and the corresponding eigenfunctions are:

$$u_n(x) = -\sin n\frac{\omega}{2}a \, \cos n\frac{\omega}{2}x + \cos n\frac{\omega}{2}a \, \sin n\frac{\omega}{2}x = \sin n\frac{\omega}{2}(x-a).$$

(3) **Homogeneous Neumann.** Now, we have:

$$\begin{cases} -\dfrac{\mathrm{d}^2 u}{\mathrm{d}x^2} = \lambda u, \\[2mm] \begin{cases} \dfrac{\mathrm{d}u}{\mathrm{d}x}(a) = 0, \\[2mm] \dfrac{\mathrm{d}u}{\mathrm{d}x}(b) = 0. \end{cases} \end{cases}$$

In this case, $\lambda = 0$ is in the spectrum: the corresponding solution is $u(x) = c_1 + c_2 x$. The boundary conditions imply $c_2 = 0$, and the corresponding eigenfunction can be taken as

$$u_0(x) = 1.$$

For $\lambda \neq 0$, the solutions have the form:

$$u(x) = c_1 \cos\sqrt{\lambda}x + c_2 \sin\sqrt{\lambda}x,$$

and the boundary conditions are satisfied if and only if:

$$-c_1\sqrt{\lambda}\sin\sqrt{\lambda}a + c_2\sqrt{\lambda}\cos\sqrt{\lambda}a = 0,$$
$$-c_1\sqrt{\lambda}\sin\sqrt{\lambda}b + c_2\sqrt{\lambda}\cos\sqrt{\lambda}b = 0.$$

Nontrivial solutions $(c_1, c_2) \neq (0, 0)$ exist if and only if:

$$\sqrt{\lambda}\begin{vmatrix} -\sin\sqrt{\lambda}a & \cos\sqrt{\lambda}a \\ -\sin\sqrt{\lambda}b & \cos\sqrt{\lambda}b \end{vmatrix} = 0.$$

This simplifies to:

$$\sin\left(\sqrt{\lambda}(b-a)\right) = 0.$$

Thus, the eigenvalues are given by:

$$\lambda_n = \frac{\omega^2}{4}n^2, \quad \omega := \frac{2\pi}{b-a},$$

for $n \in \mathbb{N}_0$, and the eigenfunctions are:

$$u_n(x) = \cos n\frac{\omega}{2}a \, \cos n\frac{\omega}{2}x + \sin n\frac{\omega}{2}a \, \sin n\frac{\omega}{2}x = \cos n\frac{\omega}{2}(x-a).$$

§5.2.   Fourier Series

OURIER series stands as a fundamental mathematical concept with far-reaching applications spanning diverse fields, including mathematics, physics, engineering, and signal processing. Its primary significance lies in its extraordinary capacity to represent complex, periodic functions through combinations of more elementary trigonometric functions, namely sines and cosines.

In the realm of signal processing, the Fourier series assumes paramount importance, serving as a bedrock for analyzing and deconstructing signals into their constituent frequencies. It plays a pivotal role in applications as varied as audio processing, image manipulation, and data compression.

In the domain of electrical engineering, the Fourier series takes center stage in the analysis of AC circuits, commonly encountered in power distribution and electronics. It provides valuable insights into the intricate voltage and current waveforms, serving as a cornerstone for designing critical components such as filters and transformers.

Furthermore, within mechanical engineering, the Fourier series emerges as an indispensable tool for the analysis and depiction of vibrations and oscillations in mechanical systems. Engineers rely on it to unravel the dynamic behavior of structures, machinery, and vehicles subject to periodic forces or disturbances.

Expanding its horizons, as we will discuss, the Fourier series finds itself entwined with the solution of problems related to heat conduction, wave propagation, and diffusion. In the realm of quantum mechanics, it plays a pivotal role, as wave functions can often be expressed as superpositions of energy eigenstates, akin to Fourier series representations. This aids physicists in comprehending the intricate behaviors of quantum systems and predicting diverse outcomes.

In the world of music, the Fourier series lends its insights to the analysis of musical tones and harmonics, offering a window into the composition of intricate musical notes and guiding the design of musical instruments and sound systems.

In the sphere of control theory, engineers harness the Fourier series to dissect the frequency response of control systems, a critical endeavor for the development of stable and efficient systems, with applications ranging from robotics to aerospace.

Furthermore, telecommunications owes a debt to the Fourier series, as it is employed in the modulation and demodulation of signals for transmission and reception. Its applications extend across wireless communication, digital signal processing, and data transmission.

Expansions in the eigenfunctions of the symmetric differential operators discussed in the previous section constitute the so-called Fourier trigonometric series expansions. The operator $L = -\mathrm{i}\dfrac{\mathrm{d}}{\mathrm{d}x}$ is symmetric over the domain $\mathfrak{D}$ of smooth functions $u = u(x)$ in the interval $[a, b]$ such that $u(a) = u(b)$.

Exponential Series Expansions

Any function $u \in L^2([a, b], \mathrm{d}x)$ can be expanded in the form

$$u = \sum_{n=-\infty}^{\infty} c_n \mathrm{e}^{\mathrm{i}n\omega x}, \quad \omega := \frac{2\pi}{b - a},$$

with coefficients given by

$$c_n = \frac{1}{b - a} \int_a^b \mathrm{e}^{-\mathrm{i}n\omega x} u(x)\mathrm{d}x.$$

In general, the series converges in square mean to the function u; however, if u is in $\mathfrak{D}$, the series converges uniformly.

Theorem

Note that in the previous calculation of the coefficients c_n, it has been taken into account that $\|u_n\|^2 = \int_a^b \mathrm{d}x = b - a$.

The operator $L = -\dfrac{\mathrm{d}^2}{\mathrm{d}x^2}$ is of the Sturm–Liouville regular and symmetric type, provided that $[a, b]$ is a finite interval and any of the three boundary conditions studied occur: periodic, Dirichlet, and Neumann. Therefore, the associated sets of eigenfunctions are orthogonal bases in $L^2([a, b], \mathrm{d}x)$. Thus, all the series that we are going to deal with converge in L^2 mean to the function being expanded, and if, in addition, this function belongs to the corresponding domain $\mathfrak{D}$, the convergence is uniform.

Sine and Cosine Series (Periodic Conditions)

All functions $u \in L^2([a, b], \mathrm{d}x)$ can be expanded in the form

$$u = \frac{a_0}{2} + \sum_{n=1}^{\infty} (a_n \cos n\omega x + b_n \sin n\omega x), \quad \omega := \frac{2\pi}{b - a},$$

where the coefficients are determined by the so-called **Euler formulas**

$$a_n = \frac{2}{b - a} \int_a^b \cos n\omega x \, u(x)\mathrm{d}x, \quad n \geq 0,$$

$$b_n = \frac{2}{b - a} \int_a^b \sin n\omega x \, u(x)\mathrm{d}x, \quad n \geq 1.$$

Theorem

The eigenfunction basis used is

$$\left\{u_0 = \frac{1}{2}, u_n = \cos n\omega x, v_n = \sin n\omega x\right\}_{n=1}^{\infty},$$

and it has been taken into account that $\|u_0\|^2 = \frac{1}{4}\int_a^b dx = \frac{b-a}{4}$, and for $n \in \mathbb{N}$ that

$$\|u_n\|^2 = \int_a^b \cos^2 n\omega x\, dx = \frac{1}{2}\int_a^b (1 + \cos 2n\omega x)dx = \frac{b-a}{2},$$

$$\|v_n\|^2 = \int_a^b \sin^2 n\omega x\, dx = \int_a^b (1 - \cos^2 n\omega x)dx = \frac{b-a}{2}.$$

It is clear that there is a close relationship between the Fourier series of exponentials and those of sines and cosines. We can go from one to another by using the formulas

$$e^{in\omega x} = \cos n\omega x + i\sin n\omega x,$$

$$\cos n\omega x = \frac{e^{in\omega x} + e^{-in\omega x}}{2}, \quad \sin n\omega x = \frac{e^{in\omega x} - e^{-in\omega x}}{2i}.$$

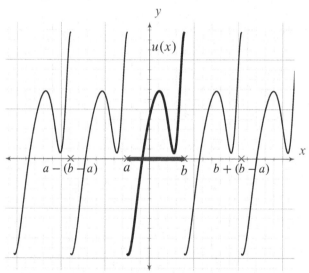

Periodic extension

Although the function u is, in principle, defined only on the interval $[a, b]$, the two types of Fourier expansions that we have discussed provide a series of periodic functions with period $(b - a)$. Thus, the functions defined by these series when they are considered on the domain $\mathbb{R}$ are periodic functions with period $(b - a)$. Therefore, these Fourier series lead to the extension of u defined as follows.

Given $x \in \mathbb{R}$, there is only one integer $m \in \mathbb{Z}$ such that $x \in [a + m(b - a), b + m(b - a)]$, and the periodic extension is $u_{\text{per}}(x) := u(x - m(b - a))$. That is, the graph of $u_{\text{per}}(x)$ is constructed by simply pasting copies, one after the other, of the graph of u over $[a, b]$. Expansions in exponential series or cosines and sines can be applied to either u or the corresponding periodic extension. To conclude the construction, we need to handle the matching points of this copy and paste process. We take $\frac{u(a)+u(b)}{2}$ as the value of the extended function at the matching points. Observe that the extended function $u_{\text{per}}(x)$ is continuous on $\mathbb{R}$ if and only if the function u is periodic on $[a, b]$, i.e., $u(a) = u(b)$. In other words, the continuity of $u_{\text{per}}(x)$ on $\mathbb{R}$ is equivalent to stating that u belongs to the domain $\mathfrak{D}$.

We now transition to the sine Fourier series and cosine Fourier series. These expansions are associated with the so-called odd periodic and even periodic extensions of a function defined within a given interval $[a, b]$. In the following illustrations, both extensions are depicted. The construction will be explained immediately.

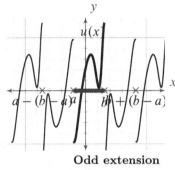

Odd extension

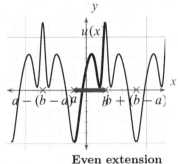

Even extension

Sine Series (Dirichlet Conditions)

All functions $u \in L^2([a, b], \mathrm{d}x)$ can be expanded as

$$u = \sum_{n=1}^{\infty} b_n \sin \frac{n\omega}{2}(x - a), \quad \omega := \frac{2\pi}{b - a},$$

with the coefficients of the series given by

$$b_n = \frac{2}{b - a} \int_a^b \sin \frac{n\omega}{2}(x - a)\, u(x)\mathrm{d}x, \quad n \in \mathbb{N}.$$

Theorem

The orthogonal basis of eigenfunctions is given by

$$\left\{ v_n = \sin \frac{n\omega}{2}(x - a) \right\}_{n=1}^{\infty},$$

with their squared norms being:

$$\|v_n\|^2 = \int_a^b \sin^2 \frac{n\omega}{2}(x - a)\mathrm{d}x = \int_a^b \frac{1}{2}\left(1 - \cos n\frac{2\pi}{b - a}(x - a)\right)\mathrm{d}x = \frac{b - a}{2}.$$

The Fourier sine series of u makes sense not only in the interval $[a, b]$ but also throughout the entire real line $\mathbb{R}$. The result is known as the odd periodic extension of the function u. This odd extension of u to the entire real line is determined in two stages: first, the function is extended to the interval

$$[a - (b - a), b] = [2a - b, a] \cup [a, b]$$

as an odd function. That is, if $x \in [a, b]$, we define

$$u_{\text{odd}}(a - (x - a)) = -u(x),$$

and then a periodic extension is made from the interval $[2a - b, a]$ to the entire real line. This construction is justified by observing that the functions

$$\frac{\omega}{2}(x - a)$$

that appear in the series, considered as functions in $\mathbb{R}$, fulfill the following property:

$$\sin \frac{\omega}{2}(X - a)\Big|_{X = 2a - x} = -\sin \frac{\omega}{2}(x - a),$$

which leads to the odd extension that is also periodic with a period of $2(b - a)$. Again, to complete the construction, we need to address the matching points of this copy and paste procedure. For the odd extension, $u_{\text{odd}}(x)$, we take 0 as the value of the extended function at the matching points.

Then, at the boundaries of the doubled interval, we have opposite values for the odd extension, so we again take 0 as the value at all the other matching points. Note that the extended function is continuous on $\mathbb{R}$ if and only if the function satisfies the Dirichlet conditions, i.e., $u \in \mathfrak{D}$.

Cosine Series (Neumann conditions)

All functions $u \in L^2([a, b], dx)$ can be expanded as

$$u = \frac{a_0}{2} + \sum_{n=1}^{\infty} a_n \cos n \frac{\omega}{2}(x - a), \quad \omega := \frac{2\pi}{b - a},$$

where the coefficients of the series are given by

$$a_n = \frac{2}{b - a} \int_a^b \cos n \frac{\omega}{2}(x - a)\, u(x) dx, \quad n \in \mathbb{N}_0.$$

Theorem

The eigenfunction basis used is

$$\left\{ u_0 = \frac{1}{2}, u_n = \cos\left(n \frac{\omega}{2}(x - a)\right) \right\}_{n \geq 1},$$

with squared norms

$$\|u_0\|^2 = \frac{1}{4} \int_a^b \mathrm{d}x$$

$$= \frac{b-a}{4},$$

$$\|u_n\|^2 = \int_a^b \cos^2 n \frac{\omega}{2} (x-a) \mathrm{d}x$$

$$= \frac{1}{2} \int_a^b \left(1 + \cos n \frac{2\pi}{b-a} (x-a) \right) \mathrm{d}x$$

$$= \frac{b-a}{2}.$$

The Fourier cosine series is a function defined on $\mathbb{R}$ which is the even extension of u defined as follows: we extend the function u to the interval

$$[a - (b-a), b] = [2a - b, a] \cup [a, b]$$

evenly, that is, if $x \in [a, b]$ we define

$$u_{\mathrm{even}}(a - (x - a)) = u(x),$$

and then the periodic extension is made from $[2a - b, b]$ to the whole real line. The justification for this is analogous to the sine Fourier series. We observe that the functions

$$\cos n \frac{\omega}{2} (x - a)$$

considered as functions in $\mathbb{R}$ fulfill that

$$\cos \frac{n\omega}{2} (X - a) \Big|_{X = 2a - x} = \cos \frac{n\omega}{2} (x - a),$$

which leads to even extension and is also periodic with period $2(b - a)$. In this even situation there is no need to pay special attention to the matching points, as the laterals limits are equal. Therefore, the even extension is always continuous even though u does not satisfy the Neumann conditions. If $u \in \mathfrak{D}$ then

$$u_{\mathrm{even}} \in C^1(\mathbb{R}).$$

Examples: Consider the function

$$u(x) = e^x$$

in the interval $[0, 1]$. Let's analyze the three different Fourier trigonometric expansions. First, we consider the sine and cosine series expansion and their periodic extension. Then, we analyze series expansions in cosine only and in sines only, and their even and odd extensions, respectively.

(1) The Fourier sine and cosine series of e^x in the interval $[0, 1]$ is given by:

$$e^x = \frac{a_0}{2} + \sum_{n=1}^{\infty}(a_n \cos 2\pi nx + b_n \sin 2\pi nx),$$

where the coefficients are as follows:

$$a_n = 2\int_0^1 e^x \cos 2\pi nx\,dx$$

$$= 2\,\mathrm{Re}\int_0^1 e^x e^{2i\pi nx}\,dx$$

$$= 2\,\mathrm{Re}\int_0^1 e^{(1+2i\pi n)x}\,dx$$

$$= 2\,\mathrm{Re}\left[\frac{e^{(1+2i\pi n)x}}{1+2i\pi n}\right]_0^1$$

$$= 2\frac{e-1}{1+4\pi^2 n^2},$$

$$b_n = 2\int_0^1 e^x \sin 2\pi nx\,dx$$

$$= 2\,\mathrm{Im}\int_0^1 e^x e^{2i\pi nx}\,dx$$

$$= 2\,\mathrm{Im}\left[\frac{e^{(1+2i\pi n)x}}{1+2i\pi n}\right]_0^1$$

$$= -2\pi n a_n.$$

The first terms of the series in sines and cosines are as follows:

$$[e^x]_{\mathrm{per}} = 2(e-1)\left(\frac{1}{2} + \frac{1}{1+4\pi^2}\cos 2\pi x + \frac{1}{1+16\pi^2}\cos 4\pi x\right.$$

$$+ \frac{1}{1+36\pi^2}\cos 6\pi x + \frac{1}{1+64\pi^2}\cos 8\pi x + \cdots\Big)$$

$$- 4\pi(e-1)\left(\frac{1}{1+4\pi^2}\sin 2\pi x + \frac{2}{1+16\pi^2}\sin 4\pi x\right.$$

$$+ \frac{3}{1+36\pi^2}\sin 6\pi x + \frac{4}{1+64\pi^2}\sin 8\pi x + \cdots\Big).$$

Below, we have the graph of the periodic extension of the exponential function and its Fourier trigonometric series truncated to 7 terms:

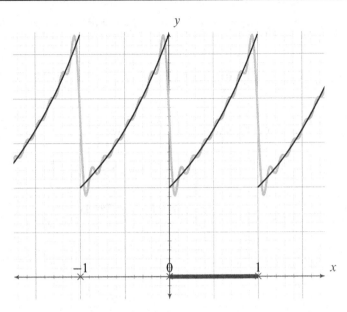

Sines & cosines: 14 terms

Observe that at the discontinuities of the function, the series tends to the mean value of the left and right values of the given function. Additionally, we can see that near these discontinuities, the truncated series oscillates, exhibiting the so-called Gibbs phenomenon. This phenomenon persists even when we take more and more terms in the series.

(2) We now consider the series with only cosines and only sines, respectively,

$$e^x = \frac{a_0}{2} + \sum_{n=1}^{\infty} a_n \cos \pi n x, \quad e^x = \sum_{n=1}^{\infty} b_n \sin \pi n x.$$

We have the following coefficients:

$$a_n = 2 \int_0^1 e^x \cos \pi n x \, dx = 2 \operatorname{Re} \int_0^1 e^x e^{i\pi n x} dx$$

$$= 2 \operatorname{Re} \int_0^1 e^{(1+i\pi n)x} dx = 2 \operatorname{Re} \left[\frac{e^{(1+i\pi n)x}}{1+i\pi n} \right]_0^1$$

$$= \frac{2[(-1)^n e - 1]}{1 + \pi^2 n^2},$$

$$b_n = 2 \int_0^1 e^x \sin \pi n x \, dx = 2 \operatorname{Im} \int_0^1 e^x e^{i\pi n x} dx$$

$$= 2 \operatorname{Im} \left[\frac{e^{(1+i\pi n)x}}{1+i\pi n} \right]_0^1$$

$$= -n\pi a_n$$

The corresponding Fourier series expansions are

$$[e^x]_{\text{even}} = e - 1 + 2\Big(-\frac{e+1}{1+\pi^2}\cos\pi x$$
$$+ \frac{e-1}{1+4\pi^2}\cos 2\pi x - \frac{e+1}{1+9\pi^2}\cos 3\pi x$$
$$+ \frac{e-1}{1+16\pi^2}\cos 4\pi x + \cdots\Big),$$

$$[e^x]_{\text{odd}} = 2\pi\Big(\frac{e+1}{1+\pi^2}\sin\pi x$$
$$- 2\frac{e-1}{1+4\pi^2}\sin 2\pi x + 3\frac{e+1}{1+9\pi^2}\sin 3\pi x$$
$$- 4\frac{e-1}{1+16\pi^2}\sin 4\pi x + \cdots\Big).$$

Next, we present the graphs of the extended functions and their truncated series. For $[e^x]_{\text{even}}$, we have:

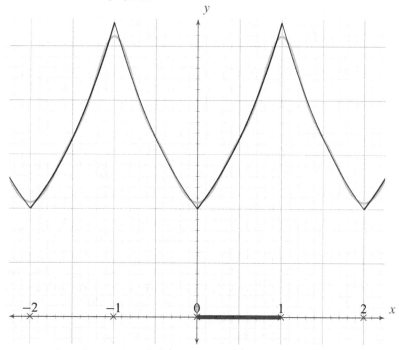

Cosine series: 4 terms

We observe that it is a very good approximation. Furthermore, we notice that at the points $x \in \mathbb{Z}$, the tangent to the truncated series has a null slope, which means that the derivative is zero (as it should be, since the derivative only has sines that vanish at those points), in contrast to the original function where the derivative jumps from -1 to 1.

As for $[e^x]_{odd}$, we have:

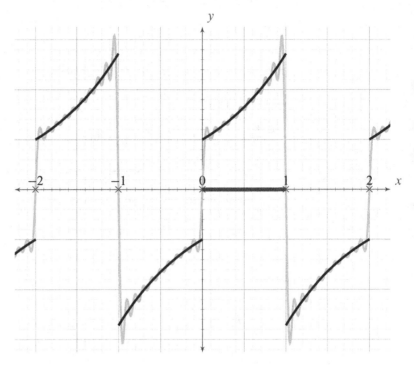

Sine series: 20 terms

The oscillating line represents the Fourier series of sines truncated to the sixth order. We should note that in the problematic points $0, \pm1, \pm2, \ldots$, the discrepancy between the function and the truncated series is remarkable. For example, at $x = 0$, the original function has a discontinuous jump of 2 units, while the series at that point is zero, as it should be. These phenomena are due to the fact that the chosen function does not comply with the boundary requirements, and its odd extension is discontinuous at the mentioned points. Nevertheless, in the interior, the Fourier series converges to the given extension.

Below, we show the approximation with 60 terms of the Fourier sine series, which starts to approximate the original graph quite well. However, when compared to the previous expansions that approximated even better with a much smaller number of terms, the **Gibbs phenomenon** becomes clearly visible. The partial sum near the discontinuity point, with a jump of Δ, deviates from the expected jump by approximately 9% of the total jump of the function approaching at that point, $\delta = 0.089489872236\Delta$. The lower the jump, the lower the rebound due to the Gibbs phenomenon.

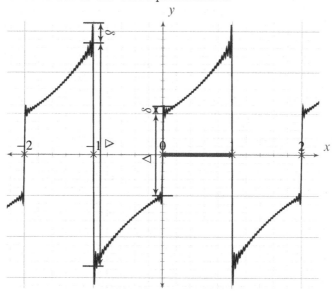

Gibbs phenomenon: 60 terms

§5.3. Convergence of Fourier Series

CONVERGENCE of the Fourier series of a function $u(x)$ is the subject we will explore next. As we already know, *for any function $u \in L^2([a, b], \mathrm{d}x)$, their exponential Fourier series, sines and cosines series, and series with only sines or only cosines converge strongly.* Moreover, if the function is smooth and satisfies suitable boundary conditions (periodic, homogeneous Dirichlet, or homogeneous Neumann), the corresponding series converges uniformly (and therefore also pointwise) to the function u. A fundamental question is to determine when the series of a general function $u \in L^2([a, b], \mathrm{d}x)$ converge pointwise and what the relationship is between the series and the function u. A class of functions for which we can provide precise answers to these questions is as follows:

C^1 Piecewise Functions

A function $u = u(x)$ is considered **C^1 piecewise** on an interval $[a, b]$ if there exists a partition $a = c_1 < c_2 < \ldots < c_M = b$ of $[a, b]$ such that for all $i = 1, \ldots, M - 1$, both u and its first derivative u' are continuous in the subintervals (c_i, c_{i+1}) and have finite lateral limits at the edges c_i and c_{i+1}.

Definition

For example, to investigate the pointwise convergence of the series of sines and cosines of a function $u \in L^2([a,b], dx)$, one needs to determine for which points $x \in \mathbb{R}$ the limit $\lim_{N\to\infty} S_N(u,x)$, exists, where $S_N(u,x)$ represents the N-th partial sum of the series of sines and cosines as follows:

$$S_N(u,x) := \frac{a_0}{2} + \sum_{n=1}^{N} (a_n \cos n\omega x + b_n \sin n\omega x),$$

and understand its relationship to the value of u at x. One fundamental result that addresses this question is the following theorem:

Dirichlet Theorem (1829)

If $u(x)$ is a C^1 piecewise function on $[a,b]$, then its Fourier series converges pointwise at all points $x \in \mathbb{R}$, and the limit is equal to

$$\frac{u_{\text{ext}}(x+0^+) + u_{\text{ext}}(x+0^-)}{2},$$

where u_{ext} represents the corresponding extension (periodic, even, or odd) of u to the entire real line $\mathbb{R}$, and

$$u_{\text{ext}}(x+0^{\pm}) := \lim_{\varepsilon\to 0} u_{\text{ext}}(x \pm |\varepsilon|).$$

Theorem

This result completely solves the problem of the pointwise convergence of the Fourier series discussed above. Essentially, we only need to know the extension of the function $u = u(x)$ to $\mathbb{R}$ and apply the following consequences of Dirichlet's theorem:

(1) If x is a point of continuity of u_{ext}, then both limits

$$u_{\text{ext}}(x+0^{\pm})$$

match $u_{\text{ext}}(x)$, and consequently, the corresponding Fourier series pointwise converges to u_{ext}.

(2) If x is a discontinuity point of u_{ext}, then the two limits

$$u_{\text{ext}}(x+0^{\pm})$$

are different and represent two equally attractive *candidates* for the limit of the Fourier series. In this case, the series makes a *Solomonic* choice and decides to converge to the semi-sum of these two limits.

Example: Let's consider the function:

$$u(x) = \begin{cases} \dfrac{\pi}{4}, & 0 \le x \le \pi, \\ -\dfrac{\pi}{4}, & -\pi \le x < 0, \end{cases}$$

and consider its corresponding Fourier expansions in exponentials. Bearing in mind that in this case $b - a = 2\pi$, and $\omega = 1$, we have:

$$u(x) = \sum_{n=-\infty}^{\infty} c_n e^{inx},$$

$$c_n = \frac{1}{2\pi} \int_{-\pi}^{\pi} u(x) e^{-inx} dx = \begin{cases} 0, & \text{if } n \text{ is even,} \\ \dfrac{1}{2in}, & \text{if } n \text{ is odd.} \end{cases}$$

Therefore,

$$u(x) = \frac{1}{2i}\left(e^{ix} + \frac{1}{3}e^{i3x} + \cdots\right) + \frac{1}{2i}\left(-e^{-ix} - \frac{1}{3}e^{-i3x} - \cdots\right),$$

that after grouping the terms, it can be expressed as:

$$u(x) = \sin x + \frac{1}{3}\sin 3x + \cdots$$

$$= \sum_{n=0}^{\infty} \frac{\sin(2n+1)x}{2n+1}.$$

This is the Fourier sine series of u for x in the interval $[-\pi, \pi]$ (the terms in the cosines have zero coefficient).

The function u is piecewise C^1 in $[-\pi, \pi]$, and thus its Fourier series must satisfy the properties required by Dirichlet's theorem. We can directly check some of these properties. Let's consider the points $x = 0, \pm\pi$. At these points, all the sine terms of the series cancel out, leading to a pointwise convergence of the sum to zero. This value matches precisely the value that the periodic extension takes at these points:

$$\frac{u_{\text{ext}}(x + 0^+) + u_{\text{ext}}(x + 0^-)}{2}.$$

Furthermore, note that the pointwise convergence for $x = \frac{\pi}{2}$ implies the following result:

$$\boxed{\frac{\pi}{4} = \sum_{n=0}^{\infty} \frac{(-1)^n}{2n+1}.}$$

Taking into consideration that

$$\frac{\pi}{4} = \arctan 1,$$

this formula can be derived from Gregory's series, which represents the inverse tangent function as an infinite Taylor series expansion centered at the origin:

$$\arctan x = x - \frac{x^3}{3} + \frac{x^5}{5} - \frac{x^7}{7} + \cdots$$

$$= \sum_{k=0}^{\infty} \frac{(-1)^k x^{2k+1}}{2k+1}.$$

This power series converges within the unit disk $D(0,1)$.

Historical notes: Interestingly, the origin of this series dates back to the 14th century and is credited to Madhava of Sangamagrama (c. 1340 – c. 1425), as documented by his Kerala school follower Jyesthadeva (c. 1530). In more recent literature, it is occasionally referred to as the Madhava–Gregory series, acknowledging Madhava's precedence (also known as the Madhava series). Furthermore, the series was independently rediscovered by James Gregory in 1671 and Gottfried Leibniz in 1673, with the latter obtaining the formula for π as a special case.

The graph of the function and the partial sum of the Fourier series for $N = 40$ is

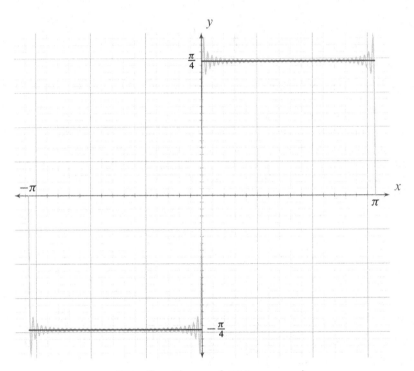

Step function and 40 terms series

§5.3.1. Advanced Results on Convergence

We will now explore more sophisticated results regarding the uniform convergence of the Fourier series and the continuity type properties of the

corresponding extension of the function being expanded. To do that, we first need to consider the concept of the variation of a given function.

Bounded Variation Functions

Given a function $u(x)$ in the closed interval $[a,b]$ and a partition $\mathcal{P} = \{x_i\}_{i=0}^{n}$ with $a = x_0 < x_1 < \cdots < x_{n-1} < x_n = b$, the **variation** of u associated with the partition $\mathcal{P}$ is given by

$$\operatorname{var}(u, \mathcal{P}) := \sum_{i=0}^{n-1} \left| u(x_{i+1}) - u(x_i) \right|.$$

The **total variation** of $u(x)$ is defined as

$$\operatorname{var} u := \sup_{\mathcal{P}} \operatorname{var}(u, \mathcal{P}).$$

Definition

Bounded monotonic functions are of bounded variation with $\operatorname{var} u = |u(b) - u(a)|$. If a function $u(x)$ is differentiable (in the sense that it has a derivative in some open set containing $[a,b]$ with an integrable derivative $u'(x)$ in $[a,b]$), its variation is given by $\operatorname{var} u = \int_a^b |u'(x)| \mathrm{d}x$. Jordan's decomposition of a function assures us that a function is of bounded variation if and only if it can be written as the difference of two nondecreasing differentiable functions $u(x) = u_1(x) - u_2(x)$. A function $u(x)$ belongs to the class of bounded variation $BV([a,b])$ if $\operatorname{var} u < +\infty$.

Example: The function

$$u(x) = \begin{cases} x^2, & x \in [-1,0], \\ -x^2, & x \in (0,1] \end{cases}$$

has derivative given by

$$u'(x) = \begin{cases} 2x, & x \in [-1,0], \\ -2x, & x \in (0,1] \end{cases}$$

and, consequently, its variation is

$$\operatorname{var} u = \int_{-1}^{1} |u'(x)| \mathrm{d}x = \int_{-1}^{0} |-2x| \mathrm{d}x + \int_{0}^{1} |2x| \mathrm{d}x$$
$$= -x^2 \big|_{-1}^{0} + x^2 \big|_{0}^{1} = 2 = u(b) - u(a).$$

Two additional important concepts are the extended idea of continuity pursued by Hölder and that of continuity modulus.

> ### Hölder Continuity
>
> It is said that $u(x)$ is of Hölder class, or **Hölder continuous**, provided that
> $$|u(x) - u(y)| \leq C|x - y|^{\alpha},$$
> where α and C are non-negative constants. If $\alpha = 1$, we say that the function is in the **Lipschitz class**.
>
> Definition

When $\alpha > 0$, the continuity is assured, and when $\alpha = 0$, it is equivalent to the boundedness of the function.

> ### Continuity Module
>
> A function $u(x)$ admits the function $\omega : \mathbb{R} \to \mathbb{R}_+$ as a **continuity modulus** in $[a, b]$ if:
> $$|u(x) - u(y)| \leq \omega(|x - y|).$$
>
> Definition

Example: One example of a continuity module is the function $\omega(x) = |x|^{\alpha}$ for some $\alpha > 0$. This function represents the modulus of continuity for Hölder continuous functions. In this case, α determines the degree of smoothness or regularity of the function. For instance, if $\alpha = 1$, then the continuity module is $\omega(x) = |x|$, and a function $f(x)$ is Lipschitz continuous if there exists a constant C such that $|f(x) - f(y)| \leq C|x - y|$ for all x, y in the domain of f. Similarly, for $\alpha = 1/2$, we have the function $\omega(x) = \sqrt{|x|}$, which corresponds to functions with square root continuity, and so on for other values of α. The continuity module $\omega(x)$ allows us to characterize and quantify the regularity of functions and plays an essential role in understanding the convergence properties of Fourier series and other integral transforms.

In terms of the continuity properties of the extension of the function u, we have the following properties for the coefficients of the Fourier series expansion.

> ### $u \in BV([a,b])$
>
> $$|c_n| \leq \frac{\text{var}\, u}{b - a} \frac{1}{|n|}.$$

$u \in C^p([a,b])$

$$|c_n| \leq \frac{\int_a^b |u^{(p)}(x)|\mathrm{d}x}{|n|^p}.$$

Continuity module

If $u \in C^p([a,b])$ with continuity module $\omega(x)$, then:
$$|c_n| \leq \omega\left(\frac{b-a}{n}\right) \frac{1}{|n|^p}.$$

Hölder

If $u(x)$ is α-Hölder continuous, then there is $K > 0$ such that
$$|c_n| \leq \frac{K}{|n|^\alpha}.$$

Regarding the uniform convergence of the Fourier series, we have the following results:

Uniform Convergence

Given a periodic function $u : [a,b] \to \mathbb{C}$, the Fourier series expansion in exponentials satisfy the following:

(1) If u is a Hölder function, then the series converges uniformly and
$$|u(x) - S_N(u,x)| \leq K\frac{\log N}{N^\alpha},$$
for some $K > 0$.

(2) If u has bounded variation, then the Fourier series converges uniformly.

(3) If $u \in C^p([a,b])$ and $u^{(p)}(x)$ has $\omega(x)$ as its continuity module, then the Fourier series converges uniformly, and there is $K > 0$ such that the following bound, due to Dunham Jackson is satisfied,
$$|u(x) - S_N(u,x)| \leq K\frac{\log N}{N^p}\omega\left(\frac{b-a}{N}\right).$$

Theorem

These results extend to the odd and even extensions.

Further reading: To gain deeper insights into Fourier Series, we recommend a selection of comprehensive resources that cover various aspects of this topic. For insightful discussions and analytical insights into Fourier Series, we suggest referring to the works of Körner [46] and Schwartz [65]. These texts offer valuable mathematical perspectives and techniques. Katznelson's book [45] provides a comprehensive treatment of Fourier series, making it an excellent resource for those seeking a thorough understanding of the subject. To explore the historical development of Fourier series, delve into the old writings of Jackson [42] and Byerly [15]. These sources provide valuable historical perspectives and context. Finally, for an authoritative and in-depth examination of Fourier Series, Simon's work [67] is highly recommended. This source is ideal for those looking to delve into the subject at an advanced level.

Historical notes: The Gibbs phenomenon was first observed and analyzed by Henry Wilbraham in 1848 in his paper [78]. However, it did not gain significant attention until 1914 when Heinrich Burkhardt mentioned it in his review of mathematical analysis in Klein's encyclopedia. In his paper, Wilbraham studied the behavior of the Fourier series of certain periodic functions, particularly the sawtooth wave function. He noticed a characteristic overshoot near the points of discontinuity in the partial sums of the Fourier series, which later became known as the "Gibbs overshoot" or the "Gibbs phenomenon." This phenomenon occurs when the partial sums of the Fourier series oscillate around the point of discontinuity and do not converge smoothly to the function being approximated.

In 1898, A. A. Michelson designed a device capable of computing and re-synthesizing the Fourier series. Despite a popular myth claiming that the machine's graphs oscillated at the discontinuities of a square wave input due to manufacturing errors, Michelson did not notice the Gibbs phenomenon in his work. The graphs produced by the machine were not clear enough to distinctly display the phenomenon.

The existence of the Gibbs phenomenon became more widely known through the correspondence in the scientific journal *Nature* between Michelson and A. E. H. Love, discussing the convergence of the Fourier series of the square wave function. In response, J. Willard Gibbs published a note in 1898, emphasizing the critical distinction between the limit of the graphs of partial sums of the Fourier series of a sawtooth wave and the graph of the actual limit of those partial sums. However, in his initial letter, Gibbs did not notice the Gibbs phenomenon and inaccurately described the limit of the graphs of partial sums. Nonetheless, in a correction published in 1899, he acknowledged the overshoot at the point of discontinuity [31].

In 1906, Maxime Bôcher conducted a detailed mathematical analysis of the overshoot, coining the term "Gibbs phenomenon" and popularizing its use. This analysis can be found in [9]. Subsequently, in 1925, Horatio Scott Carslaw noted that while the phenomenon could still be referred to as "Gibbs's phenomenon," it could no longer be exclusively credited to Gibbs as its discovery. Carslaw's discussion can be found in [17].

The sequence of events is inaccurately depicted in [46], while it is accurately narrated in [67].

§5.4. Fourier Transform

OURIER transform represents a potent mathematical tool renowned for its wide-ranging applicability in the realms of science, engineering, and technology. Its primary role revolves around the analysis and comprehension of the frequency characteristics embedded within signals and functions.

In the domain of signal processing, the Fourier transform empowers analysts to dissect complex signals, unraveling their inherent frequency components. This capability proves indispensable in a plethora of applications, spanning from audio processing and speech recognition to image manipulation. Furthermore, the world of communication systems draws upon the Fourier transform for both modulation and demodulation tasks, fostering efficient transmission of information across various mediums, including radio, television, and digital data channels.

Electrical engineers find help in the Fourier transform's prowess when scrutinizing the frequency responses of electrical circuits. This insight facilitates the design and optimization of vital components, such as filters and amplifiers. Additionally, it aids in the calculation of power spectral density, a critical metric in comprehending noise and interference within electronic systems. In the realm of mechanical engineering, the Fourier transform proves instrumental in the analysis of mechanical vibrations, shedding light on the frequency constituents underpinning oscillatory systems.

Quantum mechanics benefits from the Fourier transform's analytical capabilities, which are employed to scrutinize wave functions and probability distributions, offering profound insights into quantum system behavior. Crystallographers employ the Fourier transform as a cornerstone in x-ray crystallography, enabling the determination of the intricate three-dimensional structures of crystals. Within optics, the Fourier transform stands as a versatile instrument for the analysis of diffraction patterns and the recovery of images from holograms. Seismologists rely on Fourier transforms to decode seismic data, uncovering the distinctive frequency content and characteristics of seismic events, including earthquakes.

Medical imaging technologies, including magnetic resonance imaging and computed tomography, harness Fourier transformations to translate raw data into diagnostic images, thus revolutionizing the field of healthcare.

In the realm of chemical analysis, spectroscopic techniques, such as infrared and nuclear magnetic resonance spectroscopy, exploit Fourier transformations to unveil the intricate molecular structures and chemical compositions of substances.

Data compression algorithms leverage Fourier-based techniques like the discrete Fourier transform and fast Fourier transform to efficiently store and transmit data in diverse applications, ranging from multimedia to data storage. Machine learning practitioners find utility in Fourier analysis, particularly for feature extraction tasks. This proves especially valuable when dealing with time-series data and signal processing challenges. Control theorists delve into the dynamics of control systems, employing the Fourier transform to inform the design of controllers across an array of applications, spanning from robotics to aerospace.

In the realm of mathematics, the Fourier transform stands as a foundational concept in functional analysis, contributing substantially to our understanding of function spaces and distributions, underpinning a myriad of mathematical theories and applications.

Fourier transformation of a function $u = u(x)$ can be described as an expansion of u in eigenfunctions of the operator $Lu = -iDu$, where the domain of L is

$$\mathfrak{D} := \left\{ u \in L^2(\mathbb{R}, dx) \cap C^\infty(\mathbb{R}) : Lu \in L^2(\mathbb{R}, dx) \right\}.$$

The definition of $\mathfrak{D}$ is motivated by the fact that when considering functions u defined over the whole real line, it is not assured that $u \in L^2(\mathbb{R}, dx)$ when $u \in C^\infty(\mathbb{R})$, so we have to demand that u belong to both spaces. On the other hand, we must also demand that $Lu \in L^2(\mathbb{R}, dx)$ for L be an application $\mathfrak{D} \to L^2(\mathbb{R}, dx)$.

The special thing about this domain is that, even though L is symmetric in it, there are no eigenfunctions of L in $\mathfrak{D}$. This is clear since

$$Lu = \lambda u,$$

implies

$$u = c e^{i\lambda x}, \quad c \neq 0,$$

but such functions are not in $L^2(\mathbb{R}, dx)$ since

$$\int_{-\infty}^{+\infty} |u|^2 dx = |c|^2 \int_{-\infty}^{+\infty} e^{-2\,\mathrm{Im}(\lambda)x} dx = \infty.$$

However, we are going to show that a subset of these eigenfunctions

$$\mathscr{B} := \left\{ e_k(x) := e^{ikx}/\sqrt{2\pi} : k \in \mathbb{R} \right\},$$

allows us to expand any function $u \in L^2(\mathbb{R}, dx)$. Now, the expansion will not be a series, but an integral. In fact, the basis $\mathscr{B}$ of functions in the expansion is not a discrete set, as it has as many elements e_k, $k \in \mathbb{R}$, as the cardinality of the set of real numbers.

To introduce this new expansion, we start from the Fourier series of exponentials of a function $u \in L^2([a, b], dx)$

$$u(x) = \frac{1}{\sqrt{2\pi}} \sum_{n \in \mathbb{Z}} c_n e^{ik_n x}, \quad c_n = \frac{\sqrt{2\pi}}{b - a} \int_a^b e^{-ik_n x} u(x) dx,$$

where we have changed the expression that we used for the coefficients c_n by a factor of $\sqrt{2\pi}$, and we denote

$$k_n := \omega n, \quad \omega := \frac{2\pi}{b-a}.$$

The idea is to write the Fourier series in the form of Riemann's sum of an integral with respect to the variable k. In this sense, we have

$$u(x) = \frac{1}{\sqrt{2\pi}} \sum_{n \in \mathbb{Z}} c(k_n) e^{ik_n x} \Delta k$$

with

$$\Delta k := k_{n+1} - k_n = \omega,$$
$$c(k_n) := \frac{c_n}{\Delta k} = \frac{1}{\sqrt{2\pi}} \int_a^b e^{-ik_n x} u(x) \mathrm{d}x.$$

If we take the limit

$$a \to -\infty, \quad b \to +\infty,$$

observe that $\Delta k \to 0$, and that we have

$$u(x) = \frac{1}{\sqrt{2\pi}} \int_{\mathbb{R}} c(k) e^{ikx} \mathrm{d}k,$$
$$c(k) = \frac{1}{\sqrt{2\pi}} \int_{\mathbb{R}} e^{-ikx} u(x) \mathrm{d}x.$$

This is the expansion of u in the $e_k(x)$ eigenfunctions. The function $c = c(k)$ plays the role of *expansion coefficients*. All the information of the source function $u = u(x)$ is encoded in the new function $c = c(k)$. The usual nomenclature and notation for this expansion are:

$$c(k) =: \mathcal{F}(u), \quad \text{Fourier transform of } u,$$
$$u(x) =: \mathcal{F}^{-1}(c), \quad \text{inverse Fourier transform of } c.$$

The Fourier transform thus appears as a continuous limit of the Fourier series expansion concept.

We now consider the n-dimensional extension of the Fourier transform. We will denote

$$x = (x_0, x_1, \ldots, x_{n-1}) \in \mathbb{R}^n, \quad k = (k_0, k_1, \ldots, k_{n-1}) \in \mathbb{R}^n,$$
$$x \cdot k = x_0 k_0 + \cdots + x_{n-1} k_{n-1}.$$

> ### Fourier Transform
>
> The Fourier transform of a function $u(x)$ is defined by
>
> $$c(k) = \mathcal{F}(u) = \frac{1}{(2\pi)^{\frac{n}{2}}} \int_{\mathbb{R}^n} e^{-ik \cdot x} u(x) d^n x.$$
>
> **Definition**

For the following results, we refer the reader to [58]. A basic problem is knowing when the Fourier transform of a function does, in fact, exist.

> ### Fourier Transform in $L^1(\mathbb{R}^n, d^n x)$
>
> If the absolute value of $u(x)$ is integrable, that is, if
>
> $$\int_{\mathbb{R}^n} |u(x)| d^n x < \infty,$$
>
> then the Fourier transform $\mathcal{F}(u)$ **exists** and is a continuous function on $\mathbb{R}^n$.
>
> **Theorem**

However, the space of functions with integrable absolute value does not remain invariant under the Fourier transform. That is, **there are functions with integrable absolute value whose Fourier transform has a non-integrable absolute value.** In this sense, the space $L^2(\mathbb{R}^n, d^n x)$ is more appropriate, since it is satisfied that the Fourier transform of any function in $L^2(\mathbb{R}^n, d^n x)$ also belongs to $L^2(\mathbb{R}^n, d^n x)$.

> ### Fourier Transform in $L^2(\mathbb{R}^n, d^n x)$
>
> If $u(x)$ is a square integrable function, $u \in L^2(\mathbb{R}, d^n x)$, meaning that
>
> $$\int_{\mathbb{R}^n} |u(x)|^2 d^n x < \infty,$$
>
> then the Fourier transform $\mathcal{F}(u)$ exists and is also a square integrable function. Additionally, the Fourier transform defines a bijective linear mapping
>
> $$\mathcal{F} : L^2(\mathbb{R}, d^n x) \to L^2(\mathbb{R}, d^n x),$$
>
> that preserves the scalar product:
>
> $$(\mathcal{F}u, \mathcal{F}v) = (u, v),$$
>
> for all $u, v \in L^2(\mathbb{R}, d^n x)$. The inverse of the Fourier transform is what we have defined as the inverse Fourier transformation.
>
> **Theorem**

It is necessary to note that the improper integral accompanying the Fourier transform operation on elements of $L^2(\mathbb{R}^n, d^n x)$ must be understood in a different sense from the usual one. Specifically, it is defined as follows:

$$c(k) = \lim_{R\to\infty} c_R(k),$$

$$c_R(k) := \frac{1}{(2\pi)^{n/2}} \int_{|x|\le R} e^{-ikx} u(x) d^n x,$$

where the limit operation is associated with convergence in the L^2 sense:

$$\lim_{R\to\infty} \int_{\mathbb{R}^n} |c(k) - c_R(k)|^2 d^n k = 0.$$

Another function space in which the Fourier transform has important properties is the **Schwartz space**:

$$S(\mathbb{R}^n) := \left\{ u \in C^\infty(\mathbb{R}^n) : \left|\sup_{x\in\mathbb{R}^n} \left(x^\alpha D^\beta u\right)\right| < \infty, \alpha, \beta \in \mathbb{Z}_+^n \right\},$$

consisting of **test functions** of fast decay. Here, we are using the notation:

$$x^\alpha := x_0^{\alpha_0} x_1^{\alpha_1} \ldots x_{n-1}^{\alpha_{n-1}}.$$

This space is contained in $L^2(\mathbb{R}, d^n x)$.

Fourier Transform in $S(\mathbb{R}^n)$

If $u = u(x)$ belongs to $S(\mathbb{R}^n)$, then **the Fourier transform exists and belongs to $S(\mathbb{R}^n)$**. Moreover, the Fourier transform defines a **bijective linear transformation**

$$\mathcal{F} : S(\mathbb{R}^n) \to S(\mathbb{R}^n).$$

Theorem

Let us now examine some examples of Fourier transforms of single-variable functions.

Examples:

(1) Consider the characteristic function of the interval $[-a, a]$ given by

$$u(x) = \begin{cases} 1, & x \in [-a, a], \\ 0, & x \in \mathbb{R} \setminus [-a, a]. \end{cases}$$

The Fourier transform is easily calculated as follows:

$$c(k) = \frac{1}{\sqrt{2\pi}} \int_{-\infty}^{\infty} e^{-ikx} u(x) dx = \frac{1}{\sqrt{2\pi}} \int_{-a}^{a} e^{-ikx} dx = -\frac{1}{\sqrt{2\pi}} \left[\frac{e^{-ikx}}{ik}\right]_{-a}^{a} dx$$

$$= \sqrt{\frac{2}{\pi}} \frac{\sin ka}{k}.$$

(2) Let us now consider the function

$$u(x) = \begin{cases} e^{-ax}, & x > 0, \\ 0, & \text{elsewhere,} \end{cases}$$

where $a > 0$. The Fourier transform is given by:

$$c(k) = \frac{1}{\sqrt{2\pi}} \int_{\mathbb{R}} e^{-ikx} u(x) = \frac{1}{\sqrt{2\pi}} \int_0^\infty e^{-ikx} e^{-ax} = \frac{1}{\sqrt{2\pi}} \int_0^\infty e^{-(a+ik)x}$$

$$= \left[-\frac{1}{\sqrt{2\pi}} \frac{e^{-(a+ik)x}}{(a+ik)} \right]_{x=0}^\infty = \frac{1}{\sqrt{2\pi}} \frac{a-ik}{a^2+k^2}.$$

(3) For $a > 0$, the Lorentzian function is defined as:

$$u(x) = \frac{1}{x^2 + a^2}.$$

Its Fourier transform is given by:

$$c(k) = \frac{1}{\sqrt{2\pi}} \int_{-\infty}^\infty e^{-ikx} \frac{1}{x^2+a^2} dx,$$

and can be it calculated using complex variable techniques. Let us take

$$f(z) = \frac{e^{-ikz}}{z^2+a^2}$$

and let γ be the positive oriented real line. The singularities of f are two simple poles at

$$z = \pm ia,$$

and the residues of $f(z)$ at these poles are:

$$\text{Res}(f, \pm ia) = \pm e^{\pm ka}/(2ia).$$

When $k > 0$, the integral is the limit as $R \to \infty$ of the integral over the lower semi-circle centered at the origin with radius R. For $k < 0$, the upper semi-circle must be chosen instead. Therefore, in each situation, only one pole contributes to the formula

$$\int_\gamma f(z)dz = 2\pi i \sum_{p \in \text{poles}} \text{Res}(f, p),$$

and the result is:

$$c(k) = \sqrt{\frac{\pi}{2}} \frac{e^{-a|k|}}{a}.$$

§5.4.1. Fourier Transform Properties

Under appropriate conditions (for example, if they belong to the Schwartz space), the Fourier transform exhibits a number of basic properties, see [65], which we will describe now. In this first table, we collect the most immediate properties:

Fourier Transform Properties I

(1) **Linearity**

$$u_i(x) \xrightarrow{\mathscr{F}} c_i(k), \quad i = 1,2$$

$$\Downarrow$$

$$\lambda_1 u_1(x) + \lambda_2 u_2(x) \xrightarrow{\mathscr{F}} \lambda_1 c_1(k) + \lambda_2 c_2(k), \quad \forall \lambda_1, \lambda_2 \in \mathbb{C}$$

(2) **Translations**

$$u(x) \xrightarrow{\mathscr{F}} c(k) \implies u(x+a) \xrightarrow{\mathscr{F}} e^{ik \cdot a} c(k)$$

(3) **Scaling**

$$u(x) \xrightarrow{\mathscr{F}} c(k) \implies e^{i\ell \cdot x} u(x) \xrightarrow{\mathscr{F}} c(k - \ell)$$

(4) If $A \in M_N(\mathbb{R})$ is an invertible matrix, then

$$u(x) \xrightarrow{\mathscr{F}} c(k) \implies u(Ax) \xrightarrow{\mathscr{F}} \frac{1}{|\det A|} c\big((A^\top)^{-1} k\big)$$

(5) **Complex conjugation**

$$u(x) \xrightarrow{\mathscr{F}} c(k) \implies \bar{u}(x) \xrightarrow{\mathscr{F}} \bar{c}(-k)$$

The second set of properties, which requires further analysis, is presented below. We will use the notation D_x^α and D_k^α to denote the multiple derivation operators D^α with respect to the variables x or k, respectively. Additionally, we introduce the operation of **convolution of two functions**, as follows:

Convolution

Given two functions $u(x)$ and $v(x)$ that belong to $\mathcal{S}(\mathbb{R}^n)$, then its convolution is

$$(u * v)(x) := \int_{\mathbb{R}^n} u(x-y)v(y)\mathrm{d}^n y = \int_{\mathbb{R}^n} u(y)v(x-y)\mathrm{d}^n y.$$

Definition

Fourier Transform Properties II

(1) **Derivatives:** $\mathcal{F}(D_x^\alpha(u)) = (\mathrm{i}k)^\alpha \mathcal{F}(u)$.
(2) **Multiplication:** $\mathcal{F}(x^\alpha(u)) = (\mathrm{i}D_k)^\alpha \mathcal{F}(u)$.
(3) **Convolution:** $\mathcal{F}(u * v) = (2\pi)^{n/2} \mathcal{F}(u)\mathcal{F}(v)$.
(4) **Parseval identity:**

$$\int_{\mathbb{R}^n} |u(x)|^2 \, \mathrm{d}^n x = \int_{\mathbb{R}^n} |c(k)|^2 \, \mathrm{d}^n k.$$

Let's prove these properties:

(1) Firstly, we have:

$$\mathcal{F}\left(\frac{\partial u}{\partial x_i}\right) = \frac{1}{(2\pi)^{n/2}} \int_{\mathbb{R}^n} e^{-\mathrm{i}k\cdot x} \frac{\partial u}{\partial x_i} \mathrm{d}^n x$$

$$= \frac{1}{(2\pi)^{n/2}} \int_{\mathbb{R}^{n-1}} \left(\left[e^{-\mathrm{i}k\cdot x} u(x) \right]_{x_i=-\infty}^{\infty} \right.$$
$$\left. - \int_{-\infty}^{\infty} (-\mathrm{i}k_i) e^{-\mathrm{i}k\cdot x} u(x) \mathrm{d}x_i \right) \mathrm{d}_i^{n-1} x,$$

here

$$\mathrm{d}_i^n x := \mathrm{d}x_1 \ldots \mathrm{d}x_{i-1} \mathrm{d}x_{i+1} \ldots \mathrm{d}x_n,$$

and we've integrated by parts. As

$$u \in S(\mathbb{R}^n)$$

satisfies

$$\left[u(x) \right]_{x_i=-\infty}^{\infty} = 0,$$

then we obtain

$$\mathcal{F}\left(\frac{\partial u}{\partial x_i}\right) = \mathrm{i}k_i \mathcal{F}(u).$$

Consequently, iterating this result we get

$$\mathcal{F}\left(\frac{\partial^{\alpha_0}}{\partial x_0^{\alpha_0}} \cdots \frac{\partial^{\alpha_{n-1}}}{\partial x_{n-1}^{\alpha_{n-1}}} u\right) = (\mathrm{i}k_0)^{\alpha_0} \cdots (\mathrm{i}k_{n-1})^{\alpha_{n-1}} \mathcal{F}(u).$$

(2) We begin by observing that

$$\mathcal{F}(x_i u) = \frac{1}{(2\pi)^{n/2}} \int_{\mathbb{R}^n} e^{-\mathrm{i}(k\cdot x)} x_i u(x) \mathrm{d}^n x$$

$$= \frac{1}{(2\pi)^{n/2}} \int_{\mathbb{R}^n} \mathrm{i} \frac{\partial(e^{-\mathrm{i}k\cdot x} u(x))}{\partial k_i} \mathrm{d}^n x.$$

If $u \in \mathcal{S}(\mathbb{R}^n)$, then we can extract the derivative with respect to the parameter k_i out of the integral and write

$$\mathcal{F}(x_i u) = i\frac{\partial}{\partial k_i}\mathcal{F}(u).$$

Thus, iterating this we get

$$\mathcal{F}(x_0^{\alpha_0}\ldots x_{n-1}^{\alpha_{n-1}} u) = \left(i\frac{\partial}{\partial k_0}\right)^{\alpha_0}\cdots\left(i\frac{\partial}{\partial k_{n-1}}\right)^{\alpha_{n-1}}\mathcal{F}(u).$$

(3) Let us write the transformation of a convolution as follows:

$$\mathcal{F}(u * v) = \frac{1}{(2\pi)^{n/2}}\int_{\mathbb{R}^n} e^{-ik\cdot x}\int_{\mathbb{R}^n} u(y)v(x-y)d^n y\,d^n x.$$

Since $u, v \in \mathcal{S}(\mathbb{R}^n)$, the multiple integral is independent of the order in which we make the integrals on each variable. Thus, we can write

$$\mathcal{F}(u * v) = \frac{1}{(2\pi)^{n/2}}\int_{\mathbb{R}^n \times \mathbb{R}^n} e^{-ik\cdot x}u(x-y)v(y)d^n y\,d^n x,$$

which, with the change of variables $x = \xi + \eta$ and $y = \eta$, becomes

$$\mathcal{F}(u * v) = \frac{1}{(2\pi)^{n/2}}\int_{\mathbb{R}^n \times \mathbb{R}^n} e^{-ik\cdot(\xi+\eta)}u(\xi)v(\eta)d^n\xi\,d^n\eta$$

$$= (2\pi)^{n/2}\mathcal{F}(u)\mathcal{F}(v),$$

as we wanted to show.

(4) The scalar product in $L^2(\mathbb{R}, d^n x)$ of $u, v \in \mathcal{S}(\mathbb{R}^n)$ can be written as

$$(u, v) = \int_{\mathbb{R}^n} \bar{u}(y)v(y)d^n y = (\overline{Pu} * v)(0),$$

where $Pu(x) = u(-x)$. Therefore, using the result from (3), we have

$$(u, v) = (2\pi)^{n/2}(\mathcal{F}^{-1}(\mathcal{F}(\overline{Pu})\mathcal{F}(v)))|_{x=0},$$

which means

$$(u, v) = \int_{\mathbb{R}^n} e^{ik\cdot x}\bar{c}(k)d(k)d^n k\Big|_{x=0},$$

where c, d are the Fourier transforms of u, v, respectively. Therefore,

$$\int_{\mathbb{R}^n} \bar{u}(x)v(x)d^n x = \int_{\mathbb{R}^n} \bar{c}(k)d(k)d^n k,$$

and in particular, for $u = v$, we get Parseval's identity.

Although in the previous proofs we have focused on fast decaying functions, these properties are valid in more general situations.

Examples:

(1) First, for $a > 0$, let's calculate the Fourier transform of the one-dimensional Gaussian function

$$u(x) = e^{-a^2 x^2}.$$

Its derivative is

$$D_x u(x) = -2a^2 x u(x).$$

Applying the Fourier transform to both sides of this equation and using properties (i) and (ii) that we have proved earlier, we deduce that the transform $c(k)$ of the function $u(x)$ satisfies

$$D_k c(k) = -\frac{1}{2a^2} k c(k).$$

Integrating this equation, we get

$$c(k) = c(0) \exp\left(-\frac{k^2}{4a^2}\right).$$

On the other hand, from the Fourier transform definition, we know that

$$c(0) = \frac{1}{\sqrt{2\pi}} \int_{-\infty}^{+\infty} u(x)\mathrm{d}x = \frac{1}{\sqrt{2\pi}} \int_{-\infty}^{+\infty} e^{-a^2 x^2}\mathrm{d}x$$

$$= \frac{1}{\sqrt{2\pi}a} \int_{-\infty}^{+\infty} e^{-x^2}\mathrm{d}x = \frac{1}{a\sqrt{2}}.$$

Therefore, we obtain

$$\boxed{\mathscr{F}\left(e^{-a^2 x^2}\right) = \frac{1}{a\sqrt{2}} \exp\left(-\frac{k^2}{4a^2}\right).}$$

In other words, the Fourier transformation of a Gaussian is also a Gaussian.

(2) Let's now consider the Fourier transform of the Gaussian function in n variables:

$$u(x) = \exp\left(-\frac{1}{2} \sum_{i,j=0}^{n-1} A_{ij} x_i x_j\right),$$

where $A = (A_{ij})$ is a positive definite matrix ($A = A^\mathsf{T}$ and $(x, Ax) > 0$ for $x \neq 0$).

Taking partial derivatives with respect to x_i, we get

$$\frac{\partial u}{\partial x_i} = -u(x) \sum_{j=0}^{n-1} A_{ij} x_j.$$

Therefore, using properties (1) and (2), we find that

$$ik_i c(k) = -\sum_{j=0}^{n-1} A_{ij} i \frac{\partial c}{\partial k_j},$$

and so

$$\frac{\partial c}{\partial k_j} = -\Big(\sum_{j=0}^{n-1} (A^{-1})_{ji} k_i\Big) c.$$

This first-order partial differential equation has the following solution

$$c(k) = c(0) \exp\Big(-\frac{1}{2} \sum_{i,j=0}^{n-1} (A^{-1})_{ji} k_i k_j \Big).$$

Let us note that

$$c(0) = \frac{1}{(2\pi)^{n/2}} \int_{\mathbb{R}^n} \exp\Big(-\frac{1}{2} \sum_{i,j=0}^{n-1} A_{ij} x_i x_j \Big) \mathrm{d}^n x.$$

Since A is positive definite, it can be written as $A = O\Lambda O^{\mathrm{T}}$, where O is an orthogonal matrix and $\Lambda = \mathrm{diag}(\lambda_0, \dots, \lambda_{n-1})$ is the diagonal matrix of eigenvalues of A, with $\lambda_i > 0$, $i = 0, \dots, n-1$. Thus, after the change of coordinates $x \to x = O^{-1}x$, with unit Jacobian (since O is an orthogonal matrix, you have $|\det O| = 1$), $c(0)$ is expressed as

$$c(0) = \frac{1}{(2\pi)^{n/2}} \prod_{i=0}^{n-1} \int_{\mathbb{R}} e^{-\lambda_i \tilde{x}_i^2/2} \mathrm{d}\tilde{x}_i$$

$$= \frac{1}{(2\pi)^{n/2}} \prod_{i=0}^{n-1} \sqrt{\frac{2}{\lambda_i}} \int_{\mathbb{R}} e^{-\hat{x}_i^2} \mathrm{d}\hat{x}_i.$$

Hence, as $\det A = \lambda_1 \cdots \lambda_n$, we get

$$c(0) = \frac{1}{\sqrt{\det A}},$$

so that

$$c(k) = \frac{1}{\sqrt{(2\pi)^n \det A}} \exp\Big(-\frac{1}{2} \sum_{i,j=0}^{n-1} (A^{-1})_{ji} k_i k_j \Big).$$

For example, the Fourier transform of

$$u(x_1, x_2) = \exp\big(-(x_1^2 + x_2^2 + x_1 x_2)\big)$$

is

$$c(k_1, k_2) = \frac{1}{\sqrt{3}} \exp\Big(-\frac{1}{3}(k_1^2 + k_2^2 - k_1 k_2) \Big).$$

Here, $A = \begin{bmatrix} 2 & 1 \\ 1 & 2 \end{bmatrix}$, $\det A = 3$, and

$$A^{-1} = \begin{bmatrix} 2/3 & -1/3 \\ -1/3 & 2/3 \end{bmatrix},$$

with eigenvalues $\lambda_1 = 1, \lambda_2 = 3$.

(3) Now let's calculate the Fourier transform of

$$u(x) = (x - 1)^2 e^{-(x+1)^2}.$$

We can expand the factor $(x - 1)^2$ as follows:

$$u(x) = (x^2 - 2x + 1)e^{-(x+1)^2}.$$

Using the known Fourier transform of $e^{-(x+1)^2}$, which is given by

$$e^{ik} \frac{\exp\left(-\frac{k^2}{4}\right)}{\sqrt{2}},$$

we find the transform $c(k)$:

$$c(k) = \left(\left(i\frac{\partial}{\partial k}\right)^2 - 2i\frac{\partial}{\partial k} + 1\right) \frac{\exp\left(-\frac{k^2}{4} + ik\right)}{\sqrt{2}}$$

$$= \frac{1}{\sqrt{2}}\left(-\left(-\frac{1}{2} + \left(-\frac{k}{2} + i\right)^2\right) - 2i\left(-\frac{k}{2} + i\right) + 1\right)\exp\left(-\frac{k^2}{4} + ik\right)$$

$$= \frac{-k^2 + 8ik + 18}{4\sqrt{2}} \exp\left(-\frac{k^2}{4} + ik\right).$$

Further reading: For a deeper understanding and insightful expositions of the Fourier transform, we recommend a variety of resources. Explore the work of Körner [46] for clear and instructive explanations of the Fourier Transform, designed to enhance your comprehension. Consider the writings of Reed and Simon, including [58] and [59], for valuable insights and explanations of the Fourier transform that offer different perspectives on the subject.

For authoritative sources, delve into the books by Simon [67] and Schwartz [65]. These texts provide in-depth treatments of the Fourier transform, making them valuable additions to your study of this topic. By exploring these resources, you can gain a well-rounded and comprehensive understanding of the Fourier Transform from various angles and levels of expertise.

§5.5.   Remarkable Lives and Achievements

Who was Fourier?   Jean-Baptiste Joseph
Fourier was a remarkable French mathematician
and physicist, born on March 21, 1768, in Auxerre,
France. He passed away on May 16, 1830, leaving
behind an enduring legacy of immense contributions
spanning various fields of science and mathematics.
However, he is most renowned for his groundbreaking work in heat conduction and the development of
the Fourier series, which revolutionized the study of
mathematical physics.

His most famous achievement lies in the theory
of heat conduction, where he introduced the concept
of the Fourier series. Astoundingly, he demonstrated that any periodic function can be represented as an infinite sum of sine and cosine functions. This
extraordinary insight laid a solid foundation for modern harmonic analysis
and profoundly influenced numerous areas of mathematics and physics.

Building upon his work with Fourier series, he further advanced his ideas
to create the Fourier transform. This transformative mathematical tool en-
abled the analysis of non-periodic functions and found extensive applications
in signal processing, engineering, image processing, and other scientific domains.

Additionally, Fourier made significant strides in the understanding of the
diffusion equation, now known as the heat equation. By solving this partial
differential equation, he provided a deep understanding of heat flow in solids
and introduced the concept of thermal diffusivity.

Moreover, he formulated the fundamental law of heat conduction, now
known as Fourier's law, which describes the rate at which heat transfers
through a medium. This law remains a cornerstone in the study of heat
transfer.

In 1822, Fourier published his influential work, "Théorie Analytique de
la Chaleur" (The Analytical Theory of Heat), which ingeniously combined
his findings on heat conduction and Fourier series. This seminal book had a
profound impact on the development of mathematical physics and served as
a source of inspiration for numerous mathematicians and physicists.

Beyond his scientific achievements, Fourier also distinguished himself as
a prominent administrator and politician. Serving as the Prefect of the Isère
department in France, he played a vital role in the region's development.

During the time of Napoleon Bonaparte, Fourier's life took an interesting
turn when he became involved in Napoleon's military expedition to Egypt. In
1798, Napoleon led a campaign to Egypt with the aim of disrupting British
trade routes to India and expanding French influence in the region. Fourier,
already known for his mathematical prowess and contributions, was appointed
to join the scientific expedition that accompanied the military campaign.

As part of the Commission des Sciences et des Arts (Commission of Sciences and Arts), Fourier embarked on the voyage to Egypt in 1798. The commission was a group of scientists, scholars, and artists who were tasked with studying and documenting the scientific, historical, and cultural aspects of Egypt.

During the expedition, Fourier conducted various scientific studies and measurements, particularly focusing on issues related to Egypt's geography, climate, and other natural phenomena. He engaged in research on the temperature distribution and heat conduction in Egypt, as well as investigations into the flow of heat in the desert sands and the thermal properties of the soil. These studies were highly relevant for agricultural and engineering purposes, given the desert climate and the need for effective irrigation systems.

Fourier's involvement in the Egyptian expedition allowed him to gain valuable insights and practical knowledge in the fields of physics and applied mathematics. His experiences in Egypt provided him with unique data and observations that contributed to his later work on heat conduction and the Fourier series.

However, Napoleon's campaign in Egypt ultimately faced significant challenges and ended in defeat, with French forces forced to withdraw in 1801. Despite this setback, the scientific legacy of the expedition was substantial, with numerous discoveries and findings published in the "Description de l'Égypte" (Description of Egypt), a monumental work that documented the expedition's scientific achievements.

Fourier's work laid the very bedrock for modern applied mathematics and left an indelible imprint on fields such as engineering, physics, and signal processing. His contributions to mathematical physics remain a subject of deep study and application to this day.

In summary, Fourier's remarkable ideas and methods revolutionized the study of heat and mathematical analysis, establishing him as one of the most influential scientists of the 19th century. His profound impact on the modern understanding of mathematics and its applications across various scientific disciplines endures as a lasting tribute to his genius. For more information, see [33].

Who was Wilbraham? Henry Wilbraham (July 25, 1825 - February 13, 1883) was born into a privileged family in Delamere, Cheshire, to George and Lady Anne Wilbraham. His father was a parliamentarian, and his mother was the daughter of the Earl Fortescue. He received his education at Harrow School and was admitted to Trinity College, Cambridge, at the age of 16, where he earned a Bachelor of Arts (BA) degree in 1846 and a Master of Arts (MA) degree in 1849.

He became a barrister at Lincoln's Inn, which is one of the four Inns of Court in London, providing professional associations for barristers in England and Wales. Barristers at Lincoln's Inn undergo legal training and pass bar exams to practice as advocates in courts and provide legal counsel to clients.

The Inn plays a significant role in training barristers and offers a professional and social community for its members.

Wilbraham's intellectual pursuits extended to various fields, and he wrote papers on elasticity, mechanics, the theory of functions, and probability. His work in the theory of probability brought him into a controversy with prominent mathematicians such as Cayley, Boole, and Richard Dedekind. However, he is best known for his remarkable contribution to the field of mathematics and Fourier analysis at the age of 22 when he published a significant paper on the Gibbs phenomenon. This phenomenon involves the observation and analysis of the overshoot that occurs near points of discontinuity in the partial sums of the Fourier series.

Following his studies, Wilbraham became a Fellow at Trinity College, where he remained until 1856. In 1864, he married Mary Jane Marriott, and they had seven children together. Later in life, he served as the District Registrar of the Chancery Court at Manchester.

Although Wilbraham's life and achievements may not be as widely known as those of other prominent mathematicians and scientists of his time, his work on understanding the Gibbs phenomenon has left a lasting impact on the study of Fourier analysis and mathematical physics. His contributions continue to be appreciated in these fields to this day.

Who was Gibbs? Josiah Willard Gibbs was an American theoretical physicist, mathematician, and chemist, born on February 11, 1839, in New Haven, Connecticut, USA. He passed away on April 28, 1903. His contributions to various fields of science have solidified his status as one of the most influential scientists of the 19th century. Gibbs' work laid the foundation for modern statistical mechanics and thermodynamics, significantly impacting the development of theoretical physics.

Gibbs made groundbreaking contributions to the field of thermodynamics, formulating its fundamental principles and introducing the concept of chemical potential. His work provided the groundwork for studying phase transitions, chemical equilibrium, and the behavior of gases and liquids.

In the field of statistical mechanics, Gibbs was a true pioneer. He developed the statistical interpretation of entropy and established a statistical foundation for the laws of thermodynamics. His work deepened the understanding of the microscopic behavior of matter and bridged the gap between macroscopic and microscopic descriptions of physical systems.

Gibbs also left a lasting impact on mathematics, particularly through his development of vector analysis. He introduced the concept of vector calculus and devised a systematic notation for vector operations, now famously known as Gibbs' notation. His vector calculus notation remains widely used in physics and engineering.

In physical chemistry, Gibbs formulated the phase rule, a fundamental principle that relates the number of phases, components, and degrees of freedom in a chemical system at equilibrium. This rule has been instrumental in the study of phase equilibria in various systems.

One of Gibbs' most significant works is his book "On the Equilibrium of Heterogeneous Substances," published in 1876. This masterpiece is considered a landmark in the field of statistical physics and thermodynamics, offering a comprehensive and unified approach to understanding the thermodynamic properties of different systems.

Gibbs' work continues to have a profound and lasting impact on various scientific disciplines, including physics, chemistry, and mathematics. His concepts and methods remain central in research and education, serving as cornerstones of modern physics and statistical mechanics. Moreover, his ideas have inspired future generations of scientists and played a crucial role in shaping the development of scientific knowledge.

Who was Hölder? Otto Hölder, a German mathematician, was born on December 22, 1859, in Stuttgart, Germany, and he passed away on August 29, 1937, in Leipzig, Germany. He made significant contributions to various areas of mathematics, leaving a lasting impact on several fields.

Hölder introduced the concept of Hölder continuity, a type of mathematical continuity that characterizes the behavior of functions in terms of their rates of change and deviations from smoothness.

Widely used in real analysis, Hölder continuity finds important applications in various mathematical and scientific disciplines. Additionally, he made significant contributions to the field of functional analysis, dealing with vector spaces of functions and their transformations. His work in functional analysis laid the foundation for the modern study of function spaces and operators.

Furthermore, Hölder's contributions extended to number theory, a branch of mathematics focusing on the properties and relationships of numbers. His work in this field enhanced the understanding of various number-theoretic concepts and phenomena.

Today, Hölder's work continues to be studied and applied in various areas of mathematics, and he is remembered as one of the influential mathematicians of his time. His concepts, especially Hölder continuity, remain an essential part of modern mathematical analysis and have implications in numerous scientific fields.

§5.6. Exercises

§5.6.1. Exercises with Solutions

(1) Determine the value at $x = 0$ of the Fourier series in sines and cosines of

$$u(x) = \frac{\sin x}{|x|}, \qquad\qquad -1 < x < 1.$$

Solution: The function $u(x)$ is not continuous at the origin; its lateral limits are $\lim_{x \to 0+} u(x) = 1$ and $\lim_{x \to 0-} u(x) = -1$, and therefore its semi-sum is $\frac{u_+ + u_-}{2} = 0$. By recalling Dirichlet's theorem, we can conclude that the series converges pointwise to 0 at $x = 0$.

(2) Consider the Fourier series expansion in sines and cosines

$$x^{10} + 2x^8 + 4 = \frac{a_0}{2} + \sum_{n=1}^{\infty}(a_n \cos nx + b_n \sin nx), \qquad -\pi < x < \pi.$$

Determine which of the following statements is correct:
(a) $b_n = \frac{1}{n^3}$.

(b) $a_n = \begin{cases} \frac{1}{n^3}, & \text{if } n \text{ is even,} \\ 0, & \text{if } n \text{ is odd.} \end{cases}$

(c) $a_n = 0$.
(d) $b_n = 0$.
(e) $a_n = b_n$.

Solution: As the function $u(x) := x^{10} + 2x^8 + 4$ is even with respect to the midpoint of the interval, $x = 0$, we have that $b_n = 0$. Therefore, (a), (c), and (e) are false and (d) is true. The answer, is (b). Indeed, if we assume this to be true, the Fourier series takes the form $\frac{a_0}{2} + \sum_{n=1}^{\infty} a_{2n} \cos(2nx)$. It is important to note that all the functions $\cos(2nx)$ exhibit symmetry with respect to $x = \frac{\pi}{2}$, meaning that $\cos(2n\pi + \frac{\pi}{2} - x) = \cos(\frac{\pi}{2} + x)$. Consequently, since the Fourier series converges pointwise, we can infer, assuming condition (b) holds, that $u\left(\frac{\pi}{4}\right) = u\left(\frac{3\pi}{4}\right)$. However, this is not true.

(3) Given the function

$$u(x) = |x| \cot x, \qquad\qquad -1 < x < 1,$$

determine the value at $x = 0$ of the Fourier series in sines.
Solution: The function

$$u(x) = |x|\frac{\cos x}{\sin x},$$

is continuous throughout the interval, except for $x = 0$ where the lateral limits of the function exist: $u_- = -1$ and $u_+ = 1$. Therefore, Dirichlet's theorem implies the value of the sum of the corresponding Fourier series at $x = 0$ is the average value $\frac{u_- + u_+}{2} = 0$.

(4) Determine the number of points in the interval $\left[-\frac{\pi}{2}, \frac{\pi}{2}\right]$ where the following Fourier series

$$\sum_{n \in \mathbb{Z}} c_n e^{2nix}$$

of the function

$$u(x) = \frac{\sin x}{|x|}$$

is equal to zero.

Solution: Since the Fourier series consists of exponentials

$$e^{2nix},$$

we deduce that $\omega = \frac{2\pi}{b-a} = 2$, and therefore, the length of the interval is $b - a = \pi$. So, we can take $[a, b] = \left[-\frac{\pi}{2}, \frac{\pi}{2}\right]$. The function belongs to the piecewise C^1 class in this interval with a single discontinuity at $x = 0$. At this point, we have the following lateral limits $\lim_{x \to 0^\pm} u(x) = \pm 1$, and their mean is zero. By Dirichlet's theorem, we can deduce that the Fourier series vanishes at $x = 0$. On the edges of the interval, we have $u\left(\pm\frac{\pi}{2}\right) = \pm\frac{2}{\pi}$, and as the Fourier series in exponentials will recover the periodic extension of the function $u(x) = \frac{\sin x}{|x|}$. Hence, the answer is 3.

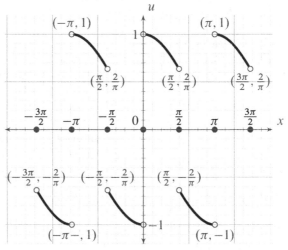

Periodic extension of $u(x) = \frac{\sin x}{|x|}$

(5) For the Fourier series expansion

$$x^2 = a_0 + \sum_{n=1}^{\infty}(a_n \cos nx + b_n \sin nx), \qquad x \in (0, 2\pi),$$

determine the value of the coefficients b_n.

Solution: The coefficients b_n are determined by the formula

$$b_n = \frac{2}{b-a} \int_a^b u(x) \sin n\omega x \, dx$$

$$= \frac{1}{\pi} \int_0^{2\pi} x^2 \sin nx \, dx.$$

This integral, which is straightforward, can be calculated by integrating by parts as follows:

$$x^2 \sin x = \left(-x^2 \frac{\cos nx}{n} \right)' + 2x \frac{\cos nx}{n}$$

$$= \left(-x^2 \frac{\cos nx}{n} + 2x \frac{\sin nx}{n^2} \right)' - 2 \frac{\sin nx}{n^2}$$

$$= \left(-x^2 \frac{\cos nx}{n} + 2x \frac{\sin nx}{n^2} + 2 \frac{\cos nx}{n^3} \right)',$$

so that

$$b_n = \frac{1}{\pi} \left[-x^2 \frac{\cos nx}{n} + 2x \frac{\sin nx}{n^2} + 2 \frac{\cos nx}{n^3} \right]_0^{2\pi},$$

which simplifies to

$$\boxed{ b_n = -\frac{4\pi}{n}. }$$

(6) To determine at how many points in the interval $[0, 3\pi]$ the Fourier series

$$\sum_{n=1}^{\infty} b_n \sin nx$$

of $u(x) = \left(x - \frac{\pi}{2} \right)^2$ is zero, we need to find the values of x for which the coefficients b_n are zero.

Solution: As we are dealing with a sine Fourier series with the functions $\sin nx$, where $n = 1, 2, \ldots$, we have $\omega = 2$ and $a = 0$ (since in the series, the sines appear in the form $\sin n\omega \frac{x-a}{2}$). Thus, the interval where the function $u(x) = \left(x - \frac{\pi}{2} \right)^2$ is defined is $[0, \pi]$.

According to Dirichlet's theorem, the Fourier series of sines will converge pointwise to the periodic extension of the odd extension of u to the interval $[-\pi, \pi]$. From the attached graph of this odd periodic extension, we can observe that the function is discontinuous at the points $\{0, \pi, 2\pi, 3\pi\}$. Applying Dirichlet's theorem, the Fourier series will converge to the average value of the lateral limits, which in this case is 0. Moreover, there are additional roots of the series at $\{\frac{\pi}{2}, \frac{3\pi}{2}, \frac{5\pi}{2}\}$. In summary, the Fourier series is canceled 7 times in the interval $[0, 3\pi]$.

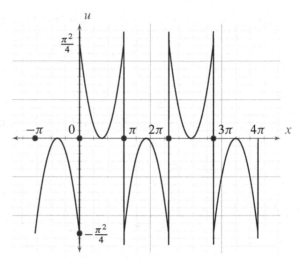

Odd extension of $u(x) = \left(x - \frac{\pi}{2}\right)^2$

(7) The characteristic function of the interval $[-1, 1]$ is given by

$$\chi_{[-1,1]}(x) = \begin{cases} 0, & x \notin [-1, 1], \\ 1, & x \in [-1, 1]. \end{cases}$$

(a) Find the Fourier series $S(x)$ in sines and cosines of $\chi_{[-1,1]}$ over the interval $[-\pi, \pi]$ (be careful not to confuse $[-\pi, \pi]$ with the interval $[-1, 1]!$).

(b) Calculate the value of the series $S(x)$ at $x = 1$.

(c) From the above formula, obtain a series for π in terms of the sines of even numbers.

(d) By evaluating the series at $x = \frac{\pi}{2}$, determine the value of the series

$$\sum_{n=1}^{\infty} (-1)^{n+1} \frac{\sin 2n}{n}.$$

Solution:

(a) In this case, $a = -\pi$ and $b = \pi$, so

$$\omega = \frac{2\pi}{2\pi} = 1,$$

and the requested Fourier series will have the form:

$$S(x) = \frac{a_0}{2} + \sum_{n=1}^{\infty} (a_n \cos nx + b_n \sin nx),$$

where the coefficients are calculated according to the Fourier formulas:

$$a_n = \frac{1}{\pi} \int_{-\pi}^{\pi} \chi_{[-1,1]}(x) \cos nx \, dx$$

$$= \frac{1}{\pi} \int_{-1}^{1} \cos nx \, dx$$

$$= \frac{2}{\pi} \frac{\sin n}{n},$$

$$b_n = \frac{1}{\pi} \int_{-\pi}^{\pi} \chi_{[-1,1]}(x) \sin nx \, dx$$

$$= \frac{1}{\pi} \int_{-1}^{1} \sin nx \, dx$$

$$= 0,$$

for $n \in \{1, 2, \dots\}$, and for $n = 0$, we obtain $a_0 = \frac{2}{\pi}$. Therefore, the requested Fourier series is:

$$S(x) = \frac{1}{\pi} + \frac{2}{\pi} \sum_{n=1}^{\infty} \frac{\sin n}{n} \cos nx.$$

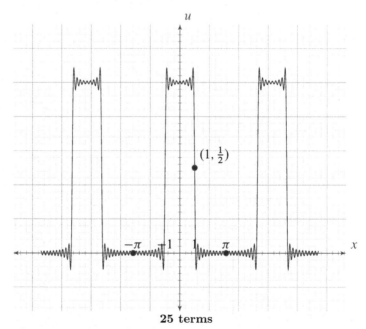

$\left(1, \frac{1}{2}\right)$

25 terms

(b) According to Dirichlet's theorem, the value of the series is the average of the one-sided limits of the function being developed.

In this case, $x = 1$ is a point of discontinuity, and the one-sided limits are distinct: 1 from the left and 0 from the right. Therefore, applying Dirichlet's theorem, we obtain:

$$S(1) = \frac{1}{2}.$$

(c) Evaluating the series at $x = 1$, we obtain

$$\frac{\pi}{2} = 1 + \sum_{n=1}^{\infty} \frac{\sin 2n}{n}.$$

The convergence rate is very slow; in fact, for 5001 terms, the difference is 0.0000306.

(d) To apply the Dirichlet theorem again, we take into account that $x = \frac{\pi}{2}$ is a point of continuity and, therefore, the value of the series is the value of the function. This is $S(\frac{\pi}{2}) = 0$. So, we have

$$0 = \frac{1}{\pi} + \frac{2}{\pi} \sum_{n=1}^{\infty} \frac{\sin n}{n} \cos n\frac{\pi}{2}.$$

As $\cos n\frac{\pi}{2} = 0$ if n is odd and $\cos 2m\frac{\pi}{2} = (-1)^m$, $m \in \mathbb{N}_0$, we conclude that

$$1 = \sum_{n=1}^{\infty} \frac{\sin 2n}{n}(-1)^{n+1}.$$

In this case, convergence is also slow; for 5000 terms, the difference is 0.0001788.

We can write both relationships as follows:

$$\frac{\pi}{2} = 1 + \sin 2 + \frac{\sin 4}{2} + \frac{\sin 6}{3} + \frac{\sin 8}{4} + \frac{\sin 10}{5} + \cdots,$$

$$0 = 1 - \sin 2 + \frac{\sin 4}{2} - \frac{\sin 6}{3} + \frac{\sin 8}{4} - \frac{\sin 10}{5} + \cdots,$$

from which we obtain, using the semi-sum and semi-difference, the following expressions:

$$\frac{\pi}{4} = 1 + \frac{\sin 4}{2} + \frac{\sin 8}{4} + \cdots$$
$$= \sin 2 + \frac{\sin 6}{3} + \frac{\sin 10}{5} + \cdots$$

For 5001 terms, the differences from the exact values are 0.0000477 and 0.0000050.

(8) Let $u(x)$ be a solution of the Poisson equation

$$\Delta u = \rho, \quad x \in \mathbb{R}^3,$$

and

$$\hat{\rho}(k) = \frac{1}{(2\pi)^{3/2}} \int_{\mathbb{R}^3} \rho(x) e^{-ik\cdot x} \, dx,$$

the Fourier transform of $\rho(x)$. Show that the Fourier transform $\hat{u}$ of u obeys the equation: $|k|^2 \hat{u}(k) + \hat{\rho}(k) = 0$

Solution: Performing the Fourier transform of the Poisson equation, we obtain:

$$\mathcal{F}(\Delta u) = \hat{\rho},$$

where we recall that $i k_i \mathcal{F}(u) = \mathcal{F}(D_{x_i} u)$. Substituting this result, we have:

$$\mathcal{F}(\Delta u) = \left((i k_1)^2 + (i k_2)^2 + (i k_3)^2\right)\hat{u} = -|k|^2 \hat{u}.$$

(9) Calculate the Fourier transform of the function

$$u(x) = \begin{cases} xe^x, & -\infty < x < 1, \\ 0, & 1 \le x < \infty. \end{cases}$$

Solution: Let's define

$$v(x) := \begin{cases} e^x, & -\infty < x < 1, \\ 0, & 1 \le x < \infty. \end{cases}$$

We can find the Fourier transform of u by calculating the Fourier transform of xv. Using the property that $\mathcal{F}(xf(x)) = iD_k\mathcal{F}(f(x))$, we have:

$$\mathcal{F}(u) = \mathcal{F}(xv) = iD_k\mathcal{F}(v).$$

Now, let's determine the Fourier transform of v. We have:

$$\mathcal{F}(v) = \frac{1}{\sqrt{2\pi}} \int_{-\infty}^{1} e^{(1-ik)x} dx$$

$$= \frac{e^{1-ik}}{\sqrt{2\pi}(1 - ik)}.$$

Therefore, we can now conclude that

$$\boxed{\mathcal{F}(u) = -\frac{1}{\sqrt{2\pi}} \frac{ik e^{(1-ik)}}{(1 - ik)^2}.}$$

(10) Determine the two-dimensional Fourier transform

$$c(k_1, k_2) = \frac{1}{2\pi} \int_{\mathbb{R}^2} u(x, y) e^{-i(k_1 x + k_2 y)} dx dy,$$

of the function

$$u(x, y) = (x + y)e^{-x^2 - |y|}.$$

Hint: $\int_{-\infty}^{\infty} e^{-x^2} e^{-ikx} dx = \sqrt{\pi} e^{-k^2/4}$.

Solution: In the first place, we must consider that

$$\mathcal{F}(u) = i(D_{k_1} + D_{k_2})\mathcal{F}(e^{-x^2-|y|}).$$

Secondly, we have

$$\mathcal{F}(e^{-x^2-|y|}) = \frac{1}{2\pi} \int_{\mathbb{R}^2} e^{-x^2-|y|} e^{-i(k_1 x + k_2 y)} dx dy$$

$$= \left(\frac{1}{(2\pi)^{1/2}} \int_{\mathbb{R}} e^{-x^2} e^{-ik_1 x} dx \right) \left(\frac{1}{(2\pi)^{1/2}} \int_{\mathbb{R}} e^{-|y|} e^{-ik_2 y} dy \right)$$

$$= \frac{1}{2\sqrt{\pi}} e^{-k_1^2/4} \int_{\mathbb{R}} e^{-|y|} e^{-ik_2 y} dy.$$

Finally,

$$\int_{\mathbb{R}} e^{-|y|} e^{-ik_2 y} dy = \int_{-\infty}^{0} e^{(1-ik_2)y} dy + \int_{0}^{\infty} e^{-(1+ik_2)y} dy$$

$$= \frac{1}{1-ik_2} + \frac{1}{1+ik_2} = \frac{2}{1+k_2^2}.$$

Thus,

$$\mathcal{F}(e^{-x^2-|y|}) = \frac{1}{\sqrt{\pi}} e^{-k_1^2/4} \frac{1}{1+k_2^2}$$

and

$$\boxed{\mathcal{F}(u) = -\frac{i}{\sqrt{\pi}} \frac{e^{-k_1^2/4}}{1+k_2^2} \left(\frac{k_1}{2} + \frac{2k_2}{1+k_2^2} \right).}$$

(11) Determine the Fourier transform $c(k_1, k_2, k_3)$ of the following function

$$u(x, y, z) := \begin{cases} \dfrac{e^{-x-y}}{z^2 + 4}, & \text{if } x > 0 \text{ and } y > 0, \\ 0, & \text{in the remaining of space.} \end{cases}$$

Hint: Whenever $a > 0$ we have that

$$\frac{1}{2\pi} \int_{-\infty}^{\infty} \frac{e^{-ikx}}{x^2 + a^2} dx = \frac{e^{-a|k|}}{2a}.$$

Solution: Since the function factors as

$$u = u_1(x) u_2(y) u_3(z),$$

we have

$$\mathcal{F}^{(3)}(u) = \mathcal{F}^{(1)}(u_1)\mathcal{F}^{(1)}(u_2)\mathcal{F}^{(1)}(u_3).$$

Now, we know that

$$\mathcal{F}^{(1)}(u_1) = \frac{1}{\sqrt{2\pi}} \int_{0}^{\infty} e^{-ik_1 x} e^{-x} dx = \frac{1}{\sqrt{2\pi}} \frac{1}{1+ik_1},$$

$$\mathcal{F}^{(1)}(u_2) = \frac{1}{\sqrt{2\pi}} \int_0^\infty e^{-ik_2 y} e^{-y} dy = \frac{1}{\sqrt{2\pi}} \frac{1}{1 + ik_2},$$

$$\mathcal{F}^{(1)}(u_3) = \frac{1}{\sqrt{2\pi}} \int_0^\infty e^{-ik_3 z} \frac{1}{z^2 + 4} dz = \sqrt{\frac{\pi}{2}} \frac{e^{-2|k_3|}}{4},$$

and, therefore,

$$\boxed{c(k_1, k_2, k_3) = \frac{1}{2\sqrt{2\pi}} \frac{e^{-2|k_3|}}{(1 + ik_1)(1 + ik_2)}.}$$

(12) Find the Fourier transform $c(k_1, k_2)$ of the following function

$$u(x, y) := \begin{cases} xye^{-x-y}, & \text{if } x > 0 \text{ and } y > 0, \\ 0, & \text{in the remaining plane.} \end{cases}$$

Solution: Let us denote

$$G(x, y) := \begin{cases} e^{-x-y}, & \text{if } x > 0 \text{ and } y > 0, \\ 0, & \text{in the remaining plane.} \end{cases}$$

We know that

$$\mathcal{F}^{(2)}(G) = \frac{1}{2\pi} \frac{1}{(1 + ik_1)(1 + ik_2)},$$

as well as

$$\mathcal{F}^{(2)}(xyG) = iD_{k_1} iD_{k_2} \mathcal{F}^{(2)}(G).$$

Thus, as $u(x, y) = xyG(x, y)$, we deduce

$$\boxed{\mathcal{F}^{(2)}(u) = \frac{1}{2\pi} \frac{1}{(1 + ik_1)^2 (1 + ik_2)^2}.}$$

(13) Determine the Fourier transform $c(k)$ of the following function

$$u(x) = x \exp\left(-\frac{|x + 2|}{4}\right).$$

Solution: First, we calculate

$$\mathcal{F}\left(\exp\left(-\frac{|x|}{4}\right)\right) = \frac{1}{\sqrt{2\pi}} \left(\int_{-\infty}^0 \exp\left(-ikx + \frac{x}{4}\right) dx \right.$$

$$\left. + \int_0^\infty \exp\left(-ikx - \frac{x}{4}\right) dx \right)$$

$$= \frac{1}{\sqrt{2\pi}} \left(\frac{1}{\frac{1}{4} - ik} + \frac{1}{\frac{1}{4} + ik} \right)$$

$$= \frac{1}{\sqrt{2\pi}} \frac{8}{1 + 16k^2}.$$

Secondly, we remind that $\mathcal{F}(u(x+a)) = e^{ika}\mathcal{F}(u(x))$ and, consequently,

$$\mathcal{F}\left(\exp\left(-\frac{|x+2|}{4}\right)\right) = \frac{8}{2\sqrt{\pi}}\frac{1}{1+16k^2}e^{i2k}.$$

We also observe that

$$\mathcal{F}\left(x\exp\left(-\frac{|x+2|}{4}\right)\right) = iD_k\mathcal{F}\left(\exp\left(-\frac{|x+2|}{4}\right)\right)$$

$$= iD_k\frac{8}{\sqrt{2\pi}}\frac{1}{1+16k^2}e^{i2k}.$$

Therefore, we arrive at the final result

$$\boxed{\mathcal{F}(u) = -\frac{16}{\sqrt{2\pi}}\frac{16k^2+16ik+1}{(16k^2+1)^2}e^{i2k}.}$$

(14) Consider the function defined by

$$u(x) = \int_{-\infty}^{\infty}\frac{e^{-(x-y)^2}}{1+y^2}dy, \qquad -\infty < x < \infty.$$

(a) Determine functions $f(x)$ and $g(x)$ such that $u(x)$ is expressed as the convolution

$$u(x) = (f * g)(x).$$

(b) Calculate the Fourier transform of $u(x)$.

Solution:

(a) If we choose the functions $f(x)$ and $g(x)$ to be the following Gaussian and Lorentzian functions:

$$f(x) = e^{-x^2}, \quad g(x) = \frac{1}{1+x^2},$$

then their convolution is given by

$$(f * g)(x) = \int_{-\infty}^{\infty} f(x-y)g(y)dy = u(x).$$

(b) Therefore, by using the fact that

$$\mathcal{F}(f * g) = \sqrt{2\pi}\mathcal{F}(f)\mathcal{F}(g)$$

and recalling that

$$\mathcal{F}(f) = \frac{1}{\sqrt{2}}\exp\left(-\frac{k^2}{4}\right), \quad \mathcal{F}(g) = \sqrt{\frac{\pi}{2}}e^{-|k|},$$

we obtain

$$\boxed{\mathcal{F}(u) = \frac{\pi}{\sqrt{2}}\exp\left(-\frac{k^2}{4}-|k|\right).}$$

§5.6.2. Exercises

(1) Expand the function $u(x) = 1$ into a sine Fourier series over the interval $[0, l]$. Analyze whether you can derive the expansion term-by-term. Also, determine the expansion into a cosine Fourier series of this function.

(2) Let $u(x) = x$.
  (a) Find its sine Fourier series expansion over the interval $[0, l]$.
  (b) Find its cosine Fourier series expansion over the interval $[0, l]$.
  (c) Find the Fourier series expansion in sines and cosines over the interval $[-l, l]$.

(3) Determine the Fourier series of sines and cosines of the function

$$u(x) = \begin{cases} 0 & \text{if } -1 \le x \le 0, \\ x & \text{if } 0 \le x \le 1. \end{cases}$$

What value does the series take at $x = 1$?

(4) Determine the Fourier series of sines and cosines of the function $u(x) = e^{ix}$, over the interval $[0, \pi]$.

(5) Determine the Fourier series of sines and cosines of the function

$$u(x) := \begin{cases} 1, & \text{if } -2 \le x \le 0, \\ x, & \text{if } 0 \le x \le 2, \end{cases}$$

and find the corresponding sum of the series at $x = 0$.

(6) Determine the Fourier transforms of the following functions:
  (a) $u(x) = \dfrac{1}{x + \alpha}$, where $\alpha \in \mathbb{C} \setminus \mathbb{R}$.
  (b) $u(x) = \begin{cases} x & \text{if } -a < x < a, \\ 0 & \text{elsewhere.} \end{cases}$
  (c) $u(x) = \dfrac{\sin ax}{x}$, where $a \in \mathbb{R}$.
  (d) $u(x) = (x^2 + 1)e^{-a^2 x^2}$, where $a > 0$.
  (e) $u(x) = \dfrac{x}{x^2 + a^2}$, where $a \in \mathbb{R}$.

(7) Determine the Fourier transform of the function

$$u(x, y) = \begin{cases} 1, & \text{if } -a < x < a \text{ and } -b < y < b, \\ 0, & \text{elsewhere.} \end{cases}$$

(8) Consider the Fourier Transform

$$c(\mathbf{k}) = \frac{1}{2\pi} \int_{\mathbb{R}^2} e^{-i\mathbf{k} \cdot \mathbf{x}} u(\mathbf{x}) d^2 \mathbf{x}$$

of a function $u = u(\mathbf{x})$ defined on the plane $\mathbf{x} \in \mathbb{R}^2$.
  (a) Write the integral that defines $c(\mathbf{k})$ in terms of polar coordinates.

(b) Determine the Fourier transform of the function

$$u(x, y) = \begin{cases} y, & \text{if } x \text{ is in the first quadrant and } |x| < 1, \\ 0, & \text{elsewhere.} \end{cases}$$

Hint: Take the x-axis in the direction and orientation of the vector k.

(9) Determine the Fourier transform

$$c(k) = \frac{1}{(2\pi)^{3/2}} \int_{\mathbb{R}^3} e^{-ik \cdot x} u(x) d^3 x$$

of the function

$$u(x) = \frac{e^{-a|x|}}{|x|}, \quad a > 0.$$

Hint: Use spherical coordinates and take the z-axis in the direction and orientation of the vector k.

6. Eigenfunction Expansion Method

Contents

IGENFUNCTION expansions represent a powerful and versatile approach to tackle problems that extend beyond the confines of homogeneous scenarios. In essence, we elevate our exploration of variable separation techniques to the realm of eigenvalues and eigenfunctions. To embark on this journey, it is imperative that the differential operator at hand be inherently symmetric. The symmetric character ensures that the corresponding eigenfunctions gracefully coalesce into a complete orthogonal set, a pivotal characteristic of this method.

The notion of complete orthogonal set of eigenfunctions propels us into the realm of expansions founded upon these mathematical building blocks. It is within this framework that we unlock the full potential of the separation of variables method. Traditionally reserved for homogeneous problems, SVM now emerges as a powerful tool applicable to the realm of nonhomogeneous problems.

Herein lies the crux of our approach: we harness the power of eigenfunctions to create meaningful expansions that faithfully represent the complexities of the problem at hand. By employing these expansions, rooted in the associated eigenfunctions, we embark on a transformative journey of problem-solving. This methodology empowers us to dissect and comprehend nonhomogeneous scenarios, allowing us to disentangle intricate dynamics and unravel the underlying mathematics governing the system in question.

The objective is to present a diverse range of examples covering various scenarios, all within the context of Cartesian coordinates. In Chapter 8, we will explore boundary value problems that do not fit well in Cartesian coordinates but are naturally described in curvilinear coordinates. These more complex boundary value problems will require the use of special functions, which will be discussed in Chapter 7.

§6.1. Preliminary Discussion: Restricted Inhomogeneities

———◦○◦———

E IGENFUNCTION expansion method (EEM) is a classical tool for solving boundary and/or initial value problems of linear type where the equations are inhomogeneous, and it relies on the use of expansions in complete sets of eigenfunctions of one of the operators appearing in the corresponding partial differential equation.

In this section, we aim to introduce the method and tools applicable to a specific class of problems, allowing for a simplified discussion. Specifically, the boundary conditions imply that the solution lay in the domain of the differential operator B. So we have only restricted inhomogeneities, as the boundary conditions for the spatial variables are homogeneous.

§6.1.1. Restricted Inhomogeneous Problems

The class of inhomogeneous problems we are discussing here belongs to the following type:

$$\begin{cases} Au + Bu = f, & x \in \Omega, \\ \begin{cases} a_i(u) = g_i, & i \in \{1,\ldots,r\}, \\ b_j(u) = 0, & j \in \{1,\ldots,s\}. \end{cases} \end{cases}$$

Here the variables are represented as

$$x = (x_0, \boldsymbol{x}),$$

where

$$\boldsymbol{x} = (x_1, \ldots, x_{n-1}),$$

and they belong to a domain

$$\Omega = I \times \Lambda,$$

for

$$x_0 \in I$$

and

$$\boldsymbol{x} \in \Lambda,$$

being I an open interval of $\mathbb{R}$ and Λ an open set of $\mathbb{R}^{n-1}$. The PDE and the boundary and/or initial conditions must satisfy:

Condition for the PDE

The PDE has the form:
(6.1)
$$A\left(x_0; \frac{\partial}{\partial x_0}\right) u + B\left(x_1, \ldots, x_{n-1}; \frac{\partial}{\partial x_1}, \ldots, \frac{\partial}{\partial x_{n-1}}\right) u = f(x_0, \boldsymbol{x}).$$

The linear conditions at the boundaries and or at initial times must be of a special form. Not all have to be inhomogeneous. Let's see:

Restriction for the BVs

The system of boundary and/or initial conditions consists of two subsystems. One of them is formed by conditions of inhomogeneous type:

(6.2) $a_i(u) = g_i(x), \quad i \in \{1, \ldots, r\},$

which contains **operators a_i that act only on the variable x_0**, while the other is made up of conditions of restricted homogeneous type:

$$b_j(u) = 0, \quad j \in \{1, \ldots, s\},$$

such that **operators b_j act only on the variables $(x_1, \ldots, x_{n-1})$**.

Finally, the nonhomogeneous terms must be of a particular form, and can be expanded in series of eigenfunctions:

Completeness Condition

The inhomogeneous terms f and g_i of the problem admit expansions of the form:

$$f(x) = \sum_m f_m(x_0) w_m(x),$$

$$g_i(x) = \sum_m c_{im} w_m(x), \qquad i \in \{1, \ldots, r\},$$

in a set of eigenfunctions of the operator B: $B w_m = \lambda_m w_m$, that satisfy homogeneous boundary conditions: $b_j(w_m) = 0$, $j = \{1, \ldots, s\}$.

The third condition mentioned above is the most demanding. It is typically fulfilled when the **operator B is symmetric on the domain** given by:

$$\boxed{\mathfrak{D} = \left\{ w \in C^\infty(\overline{\Lambda}) : b_j(w) = 0, \; j \in \{1, \ldots, s\} \right\},}$$

in the space Hilbert space

$$L^2(\overline{\Lambda}, \rho \, \mathrm{d}^n x).$$

The importance of symmetric operators in linear PDE theory lies in this particular property.

Assuming that all three conditions are met, the EEM is implemented through the following process. In the first place, we set the linear superposition principle, **any solution can be expressed as an eigenfunction series expansion**:

Eigenfunction Expansion I

A solution to the problem is sought in the form of a series expansion in eigenfunctions:

$$u(x) = \sum_m v_m(x_0) w_m(x).$$

The coefficients $v_m(x_0)$ of this series are the unknowns to be determined.

Then, for this spectral decomposition, we find the consequences that the differential equation to be satisfied imposes in the unknown coefficients. That is, we obtain the linear ordinary differential equations satisfied by these unknown functions. For that aim, let's project along the eigenfunction w_m.

In the first place, let us project the function Au along the eigenfunction w_m.

Eigenfunction Expansion II

Since the operator A acts solely on the variable x_0 and is linear, we can reasonably expect, under favorable regularity conditions, the following expression:

$$(w_m, Au) = \left(w_m, A\left(\sum_{m'} v_{m'} w_{m'} \right) \right) = (Av_m) \| w_m \|^2.$$

Now, we project Bu along the eigenfunction w_m of the operator B. In this case, it will play an important role the fact that B is symmetric in the domain $\mathfrak{D}$.

Eigenfunction Expansion III

Given that the eigenfunctions w_m as well as the solution u satisfy the homogeneous boundary conditions defining the domain $\mathfrak{D}$ for which B is symmetric, we have that

$$(w_m, Bu) = (Bw_m, u) = \lambda_m (w_m, u) = \lambda_m v_m \| w_m \|^2,$$

where we have used the fact that $Bw_m = \lambda_m w_m$; i.e., w_m are the eigenfunctions of B in the domain $\mathfrak{D}$.

Eigenfunction Expansion IV

If we employ the expansion: $f = \sum_m f_m w_m$, then the PDE is reduced to:

$$(w_m, Au) + (w_m, Bu) = \left(w_m, \sum_{m'} f_{m'} w_{m'} \right).$$

Consequently,

(6.3) $$A v_m + \lambda_m v_m = f_m.$$

Thus, each coefficient v_m will be a solution to this ODE.

Eigenfunction Expansion V

To find a solution $u(x)$ that satisfies the boundary conditions:

$$\begin{cases} a_i(u) = g_i, & i \in \{1, \ldots, r\}, \\ b_j(u) = 0, & j \in \{1, \ldots, s\}, \end{cases}$$

we proceed in a similar manner. Firstly, since the boundary operators a_i only act on the variable x_0:

$$(w_m, a_i(u)) = \left(w_m, a_i \left(\sum_{m'} v_{m'}(x_0) w_{m'}(x) \right) \right) = a_i(v_m) \|w_m\|^2,$$

we can express the functions g_i using their expansions:

$$g_i(x) = \sum_m c_{im} w_m(x),$$

which allows us to rewrite the conditions $a_i(u) = g_i$ as:

$$\left(w_m, \sum_{m'} a_i(v_{m'}) w_{m'} \right) = c_{im} \|w_m\|^2.$$

By identifying the coefficients of the functions $w_m(x)$, we deduce that each coefficient v_m, in addition to satisfying the ODE (6.3), must also meet the following conditions:

(6.4) $$a_i(v_m) = c_{im}, \quad i \in \{1, \ldots, r\}.$$

Observations:

(1) It should be noted that the conditions $b_j(u) = 0$ are automatically fulfilled since:

$$b_j(u) = b_j \left(\sum_m v_m(x_0) w_m(x) \right) = \sum_m v_m(x_0) b_j(w_m)(x) = 0,$$

where we have used the fact that, by the definition of eigenfunctions $w_{\boldsymbol{m}}(\boldsymbol{x})$, we have $b_j(w_{\boldsymbol{m}}) = 0$ for $j \in \{1, \ldots, s\}$.

(2) The method that we have just discussed admits a natural generalization to the case in which the expansions in eigenfunctions of the data f and g_i are generalized linear combinations of eigenfunctions of the operator B, using integration operations such as Fourier transform. In such situations, a solution u is sought in the form of an expansion of the same type, replaced in (6.1) and (6.2), and the coefficients in the expansion obtained are identified to obtain equations analogous to (6.3) and (6.4).

(3) The hypotheses about the existence of expansions of the functions f and g_i in eigenfunctions of B are always fulfilled when the set of eigenfunctions of B is complete, which is true for broad classes of symmetric operators seen in the previous chapter.

Example: Let us consider the following problem with Laplace's equation on a rectangle

$$
\begin{cases}
\dfrac{\partial^2 u}{\partial x^2} + \dfrac{\partial^2 u}{\partial y^2} = 0, & 0 < x < a, \quad 0 < y < b, \\
u(0, y) = 0, \\
u(a, y) = 0, \\
u(x, 0) = \sin \dfrac{\pi x}{a}, \\
u(x, b) = 0.
\end{cases}
$$

This problem concerns the stationary distribution of temperature on a plate with dimensions $[0, a] \times [0, b]$. The plate's boundaries, except for one side, are kept at zero temperature, while the temperature distribution on the remaining side follows a sine pattern. This scenario gives rise to the following spectral problem:

$$
\begin{cases}
-w_{xx} = \lambda^2 w, \\
w|_{x=0} = 0, \\
w|_{x=a} = 0,
\end{cases}
$$

with solutions:

$$
\lambda_n = \frac{n^2 \pi^2}{a^2}, \quad w_n(x) = \sin \frac{n \pi x}{a},
$$

for $n \in \mathbb{N}$. Applying the EEM, we expand the solution as the sine Fourier series:

$$
u(x, y) = \sum_{n=1}^{\infty} v_n(y) \sin \frac{n \pi x}{a}.
$$

For inhomogeneities, we only have the term

$$
\sin \frac{\pi x}{a},
$$

and consequently, we seek our solution in the form:

$$u(x, y) = v(y)\sin\frac{\pi x}{a}.$$

The function v is determined by

$$\begin{cases} v'' = \dfrac{\pi^2}{a^2}v, \\ \begin{cases} v|_{y=0} = 1, \\ v|_{y=b} = 0. \end{cases} \end{cases}$$

Hence,

$$v(y) = Ae^{\frac{\pi}{a}y} + Be^{-\frac{\pi}{a}y},$$

where

$$\begin{cases} A + B = 1, \\ Ae^{\frac{\pi}{a}} + Be^{-\frac{\pi}{a}} = 0. \end{cases}$$

The solution to this system is

$$A = -\frac{e^{-\pi\frac{b}{a}}}{2\sinh\pi\frac{b}{a}}, \qquad\qquad B = \frac{e^{\pi\frac{b}{a}}}{2\sinh\pi\frac{b}{a}}$$

so that the solution reads

$$u(x, y) = \frac{\sinh\frac{\pi}{a}(b - y)}{\sinh\frac{\pi b}{a}}\sin\frac{\pi}{a}x.$$

The graph for this temperature distribution on a plate with $a = \pi, b = 2\pi$ is

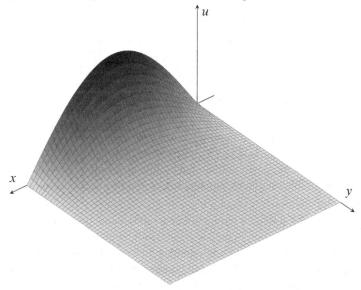

Temperature $u(x, y)$ on the plate $[0, \pi] \times [0, 2\pi]$

§6.1.2. Well- and Ill-Posed Problems

Well-posed in the sense of Hadamard problems in the context of PDEs refer to problems that have certain mathematical properties ensuring their solutions are meaningful, unique, and stable. That is:

(1) The solution exists.
(2) The solution is unique.
(3) The solution depends continuously on the initial and boundary values.

These properties are crucial in the study of PDEs because they guarantee that solutions exist, are unique, and depend continuously on the initial or boundary data. A well-posed problem typically consists of three key components: the governing PDE, initial or boundary conditions, and the domain or spatial region in which the problem is defined. Non well-posed problems are said to be ill-posed.

Here are some examples of well-posed problems in the realm of PDEs:

(1) The one-dimensional heat equation, with initial value and Dirichlet or Neumann BV conditions, is well-posed when the initial condition is sufficiently smooth and the boundary conditions are appropriate.

(2) For the one-dimensional wave equation, well-posedness depends on the smoothness of the initial conditions and the appropriateness of the boundary conditions.

(3) The Dirichlet problem for Laplace's equation in a domain Ω:

$$\begin{cases} \Delta u = 0, & x \in \Omega, \\ u|_{\partial\Omega} = g(x). \end{cases}$$

Well-posedness is ensured when the boundary conditions are consistent and appropriate for the given domain Ω.

(4) The one-dimensional Burgers' equation:

$$\frac{\partial u}{\partial t} + u\frac{\partial u}{\partial x} = \nu\frac{\partial^2 u}{\partial x^2}$$

This equation is well-posed under appropriate conditions on the initial conditions and the viscosity parameter ν.

(5) The Schrödinger equation with appropriate initial value and boundary conditions is well-posed when dealing with quantum systems.

The concept of well-posedness holds a paramount position within the domain of BVPs for PDEs. It serves as a foundational principle and, in turn, imparts physical significance and dependability to solutions, making them applicable across a spectrum of scientific and engineering contexts. Researchers regularly scrutinize the well-posedness of specific PDE problems to ensure the accuracy and trustworthiness of their numerical or analytical solutions, ultimately reflecting the underlying physical processes.

The significance of well-posedness becomes evident when dealing with continuum models, especially when solving BVPs for PDEs through numerical methods. These continuum models describe physical phenomena in a continuous manner, typically relying on differential equations. However, the computational nature of computers necessitates discrete data, mandating the use of numerical methods for approximating solutions. Discretization entails the division of the continuous domain into discrete elements or grid points.

When confronted with finite precision in numerical calculations or data errors, the specter of numerical instability looms. Numerical instability entails the phenomenon where minor errors in initial data or numerical computations lead to exponential error growth within the computed solution. As a result, the reliability of outcomes is called into question.

A pivotal aspect in achieving numerical stability is working with well-posed problems, those characterized by the presence of a unique solution that smoothly varies with input data. This approach lays the foundation for the application of stable algorithms—numerical techniques that mitigate the amplification of errors.

Nevertheless, even in cases where a problem possesses mathematical well-posedness, it may still exhibit ill-conditioning. Ill-conditioning manifests when small errors or perturbations in initial data yield disproportionately large errors in computed solutions. This sensitivity to data perturbations is often quantified by a high condition number. For readers interested in delving further into these numerical issues, we recommend consulting the book by Iserles [40].

Complex systems, particularly nonlinear ones like chaotic systems, epitomize this sensitivity to initial conditions. Chaotic systems are notorious for the phenomenon where tiny changes in initial conditions result in vastly divergent outcomes. This inherent sensitivity can exacerbate numerical instability and ill-conditioning.

In instances where a problem lacks well-posedness or succumbs to ill-posedness due to numerical instability or ill-conditioning, a remedy is required for numerical treatment. This frequently entails the introduction of additional assumptions, such as assuming the smoothness of the solution or imposing constraints on the solution space. This corrective process is referred to as regularization. Among the commonly employed regularization techniques, Tikhonov regularization stands out. It introduces a regularization term into the problem, governing the behavior of the solution and enhancing stability.

Well-posed evolution problems are closely linked to the concept of semigroups comprised of bounded linear operators. This theoretical framework serves as a powerful tool for comprehensively analyzing and addressing such problems. It provides a solid foundation for understanding their behavior and finding solutions.

On the other hand, when dealing with ill-posed problems, the situation becomes more intricate. These problems often demand innovative approaches and specialized extensions to the classical theory. In such cases, one must navigate the challenges presented by instability or underdetermination. Crafting suitable extensions and adaptations of existing methodologies is essential to tackle these complex issues effectively.

As an example of an ill-posed problem, we bring here the classical example by Hadamard [35]. Let's consider the following boundary value problem for the Laplace equation, where k belongs to the set of natural numbers

$$\begin{cases} \dfrac{\partial^2 u}{\partial x^2} + \dfrac{\partial^2 u}{\partial y^2} = 0, \quad 0 < x, y < \pi, \\ u|_{x=0} = 0, \\ u|_{x=\pi} = 0, \\ u|_{y=0} = 0, \\ \dfrac{\partial u}{\partial y}\bigg|_{y=0} = \dfrac{\sin kx}{k}. \end{cases}$$

Notice the unconventional nature of these boundary values, as we are imposing two conditions at the same boundary, $y = 0$. We can proceed by applying the eigenfunction expansion method, taking $B = -\dfrac{\partial^2}{\partial x^2}$ in the Dirichlet domain, and using an orthogonal basis of functions $w_n(x) = \sin nx$ where n is a natural number, with corresponding eigenvalues $\lambda_n = -n^2$. This leads us to express the solution as a sine Fourier series:

$$u = \sum_{k=1} v_n(y) \sin nx.$$

The unknown coefficients $v_n(y)$ are determined by the following BVP:

$$\begin{cases} \dfrac{d^2 v_n}{dy^2} = n^2 v_n, \\ v_n|_{y=0} = 0, \\ \dfrac{\partial v_n}{\partial y}\bigg|_{y=0} = c_n, \end{cases}$$

where c_n represents the n-th Fourier coefficient of the expansion of the inhomogeneous term $\frac{\sin kx}{k}$. Therefore,

$$c_n = \delta_{k,n}\frac{1}{k}.$$

Consequently, among the $v_n(y)$ functions, only $v_k(y)$ is non-null. Therefore, the solutions to the ordinary differential equation are as follows:

$$v_k(y) = A \sinh ky + B \cosh ky.$$

Since we have $v_k(0) = 0$, we must have $B = 0$, and the second boundary condition leads to $A = \frac{1}{k^2}$. Consequently, the sought solution is as follows:

$$\boxed{u = \frac{1}{k^2} \sin kx \sinh ky.}$$

It's noteworthy that as k approaches infinity, the solution becomes unbounded. However, in this limit, the BVP becomes:

$$\begin{cases} \dfrac{\partial^2 u}{\partial x^2} + \dfrac{\partial^2 u}{\partial y^2} = 0, \quad 0 < x, y < \pi, \\ \begin{cases} u|_{x=0} = 0, \\ u|_{x=\pi} = 0, \\ u|_{y=0} = 0, \\ \dfrac{\partial u}{\partial y}\bigg|_{y=0} = 0. \end{cases} \end{cases}$$

The solution to this BVP is $u = 0$. **_Therefore, this is an ill-posed problem, as it lacks continuous dependence on the boundary values._** By the way, as soon as you move one of the two BVs on $y = 0$ to $y = \pi$, the illness disappear, and the problem is well-posed.

§6.2. Application to Evolution Equations

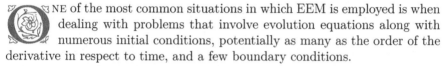

NE of the most common situations in which EEM is employed is when dealing with problems that involve evolution equations along with numerous initial conditions, potentially as many as the order of the derivative in respect to time, and a few boundary conditions.

(1) The given partial differential equation represents an evolution equation in the following form:

$$a \frac{\partial^r u}{\partial t^r} + B\left(x_1, \ldots, x_{n-1}; \frac{\partial}{\partial x_1}, \ldots, \frac{\partial}{\partial x_{n-1}}\right) u = f(t, x),$$

where $a \in \mathbb{C}$.

(2) The solution is subject to r initial conditions:

$$\frac{\partial^i u}{\partial t^i}(t_0, x) = g_i(x), \quad i \in \{0, \ldots, r-1\},$$

along with a set of s boundary conditions:

$$b_j(u) = 0, \quad j \in \{1, \ldots, s\},$$

where the operators b_j act solely on the variables

$$(x_1, \ldots, x_{n-1}).$$

In this case, the operator A is defined as:

$$Au = a \frac{\partial^r u}{\partial t^r},$$

The conditions $a_i(u) = g_i$ are expressed as follows:

$$\frac{\partial^i u}{\partial t^i}(t_0, x) = g_i(x), \quad i \in \{1, \ldots, r\}.$$

Examples:

(1) We now consider an initial/boundary value problem for the Schrödinger equation given by:

$$
\begin{cases}
i\hbar u_t = -\dfrac{\hbar^2}{2m}u_{xx} + V(x)u, \quad x \in (a,b),\ t > 0, \\
\begin{cases}
u\big|_{t=0} = g(x), \\
\begin{cases}
u\big|_{x=a} = 0, \\
u\big|_{x=b} = 0.
\end{cases}
\end{cases}
\end{cases}
$$

This equation describes the evolution of a quantum particle confined within the segment (a,b) and subjected to the potential $V(x)$. We introduce a differential operator, known as the quantum Hamiltonian:

$$
Hu := -\frac{\hbar^2}{2m}u_{xx} + V(x)u,
$$

which simplifies the Schrödinger equation to:

$$
-i\hbar u_t + Hu = 0.
$$

The EEM is based on solving the eigenvalue problem:

$$
\begin{cases}
Hw = \lambda w, \\
\begin{cases}
w\big|_{x=a} = 0, \\
w\big|_{x=b} = 0.
\end{cases}
\end{cases}
$$

Assuming that

$$
\{\lambda_n\}_{n=1}^{\infty}
$$

represents the set of eigenvalues with associated eigenfunctions

$$
\{w_n(x)\}_{n=1}^{\infty},
$$

we seek solutions of the form:

$$
u(x) = \sum_{n=1}^{\infty} v_n(t)w_n(x).
$$

By substituting this into the Schrödinger equation, we obtain:

$$
\sum_{n=1}^{\infty}\left(-i\hbar\frac{dv_n}{dt}(t) + \lambda_n v_n(t)\right)w_n(x) = 0,
$$

which leads to:

$$
-i\hbar\frac{dv_n}{dt}(t) + \lambda_n v_n(t) = 0,
$$

and its solution is:

$$
v_n(t) = a_n \exp\left(-i\frac{\lambda_n}{\hbar}t\right).
$$

If we assume that the initial condition has the expansion:

$$g(x) = \sum_{n=1}^{\infty} c_n w_n(x),$$

then we conclude that

$$a_n = c_n,$$

and, therefore, the solution to our problem is:

$$u(t, x) = \sum_{n=1}^{\infty} c_n \exp\left(-i\frac{\lambda_n}{\hbar}t\right) w_n(x).$$

Let's illustrate this approach with the following examples:
- Let us study now

$$\begin{cases} i\hbar u_t = -\dfrac{\hbar^2}{2m} u_{xx}, & x \in (0, l), \quad t > 0, \\ u\big|_{t=0} = 2\sin\left(\dfrac{\pi}{l}x\right) - \sin\left(3\dfrac{\pi}{l}x\right), \\ \begin{cases} u\big|_{x=0} = 0, \\ u\big|_{x=l} = 0. \end{cases} \end{cases}$$

In other words, we are considering a free quantum particle confined within the segment $(0, l)$. The corresponding eigenvalue problem is given by:

$$\begin{cases} -\dfrac{\hbar^2}{2m} w_{xx} = \lambda w, \\ \begin{cases} w\big|_{x=0} = 0, \\ w\big|_{x=l} = 0. \end{cases} \end{cases}$$

The eigenvalues and corresponding eigenfunctions are found to be:

$$\lambda_n = \frac{\hbar^2}{2m}\frac{n^2\pi^2}{l^2}, \quad w_n(x) = \sin\frac{n\pi}{l}x,$$

for $n \in \mathbb{N}$. Thus, the initial condition can be expanded in these eigenfunctions, allowing us to determine the state of the particle as follows:

$$u(t, x) = 2\sin\left(\frac{\pi}{l}x\right)\exp\left(-i\frac{\hbar}{2m}\frac{\pi^2}{l^2}t\right) - \sin\left(3\frac{\pi}{l}x\right)\exp\left(-i\frac{\hbar}{2m}\frac{9\pi^2}{l^2}t\right).$$

Below, we present the space-time graph for the probability density:

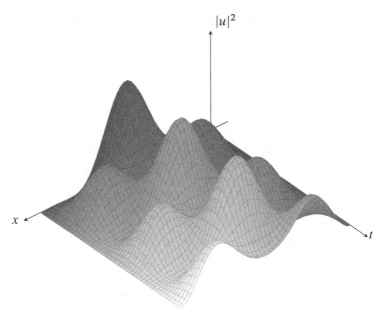

Probability density $|u|^2$ for $l = \pi$

- We will continue with the free case but with a change in the boundary conditions from Dirichlet to periodic conditions. The initial and boundary value problem to solve now becomes:

$$\begin{cases} i\hbar u_t = -\dfrac{\hbar^2}{2m} u_{xx}, \quad x \in (0, l), \quad t > 0, \\ u\big|_{t=0} = 2i - 2\cos\left(\frac{2\pi}{l}x\right) + 7\sin\left(\frac{6\pi}{l}x\right), \\ \begin{cases} u\big|_{x=0} = u\big|_{x=l}, \\ u_x\big|_{x=0} = u_x\big|_{x=l}. \end{cases} \end{cases}$$

The problem of eigenvalues is given by:

$$\begin{cases} -\frac{\hbar^2}{2m} w_{xx} = \lambda w, \\ \begin{cases} w\big|_{x=0} = w\big|_{x=l}, \\ w_x\big|_{x=0} = w_x\big|_{x=l}. \end{cases} \end{cases}$$

The eigenvalues, except for zero which is simple, are all double and are given by:

$$\lambda_n = \frac{\hbar^2}{2m} \frac{4n^2\pi^2}{l^2},$$

for $n \in \mathbb{N}_0$. The corresponding eigenfunctions are:

$$\left\{ w_0(x) = 1, \; w_n^{(+)}(x) = \cos\left(\frac{2\pi}{l} nx\right), \; w_n^{(-)}(x) = \sin\left(\frac{2\pi}{l} nx\right) \right\}_{n=1}^{\infty}.$$

In the initial condition, only the eigenfunctions w_0, $w_1^{(+)}$, and $w_3^{(-)}$ appear. Consequently, the solution is:

$$u(t, x) = 2\mathrm{i} - 2\cos\left(\frac{2\pi}{l}x\right)\exp\left(-\mathrm{i}\frac{\hbar}{2m}\frac{4\pi^2}{l^2}t\right)$$

$$+ 7\sin\left(\frac{6\pi}{l}x\right)\exp\left(-\mathrm{i}\frac{\hbar}{2m}\frac{36\pi^2}{l^2}t\right).$$

(2) We will now find the solution for the heat equation with periodic boundary conditions and initial conditions. The problem can be stated as follows:

$$\begin{cases} u_t = u_{xx}, & x \in (0, l), \quad t > 0, \\ \left. u \right|_{t=0} = \frac{2}{l^2}x^3 - \frac{3}{l}x^2 + x, \\ \begin{cases} \left. u \right|_{x=0} = \left. u \right|_{x=l}, \\ \left. u_x \right|_{x=0} = \left. u_x \right|_{x=l}. \end{cases} \end{cases}$$

Let's take a moment to discuss the consistency between the initial condition and boundary conditions. If we consider

$$g(x) = ax^3 + bx^2 + x$$

and demand that $g(0) = g(l)$ and $g'(0) = g'(l)$, then we arrive at the conditions:

$$al^2 + bl + c = 0,$$
$$3al + 2b = 0.$$

These conditions lead to a uniparametric family of solutions a, b, c, and a possible choice is given by our initial condition.

Next, let's address the problem of eigenvalues given by:

$$\begin{cases} -w_{xx} = \lambda w, \\ \begin{cases} \left. w \right|_{x=0} = \left. w \right|_{x=l}, \\ \left. w_x \right|_{x=0} = \left. w_x \right|_{x=l}. \end{cases} \end{cases}$$

This problem has eigenvalues (all double except for zero):

$$\lambda_n = n^2\omega^2,$$

for $n \in \mathbb{N}_0$ and

$$\omega = \frac{2\pi}{l}.$$

The corresponding eigenfunctions are:

$$w_n(x) = \mathrm{e}^{\mathrm{i}n\omega x}.$$

To apply the EEM, we need to determine the Fourier series:

$$g(x) = \sum_{n\in\mathbb{Z}} c_n \mathrm{e}^{\mathrm{i}n\omega x},$$

$$c_n = \frac{1}{l} \int_0^l g(x) e^{-in\omega x} \, dx.$$

Since our g is a polynomial, we can use the integrals:

$$I_{m,n} := \int_0^l x^m e^{-in\omega x} \, dx = \begin{cases} i\dfrac{1}{n\omega}\left(l^m - m I_{m-1,n}\right), & n \neq 0, \\[2mm] \dfrac{l^{m+1}}{m+1}, & n = 0. \end{cases}$$

Using these integrals, we can find:

$$I_{0,n} = \begin{cases} 0, & n = 0, \\ l, & n \neq 0, \end{cases}$$

$$I_{1,n} = \begin{cases} \dfrac{l^2}{2}, & n = 0, \\[2mm] i\dfrac{l}{n\omega}, & n \neq 0, \end{cases}$$

$$I_{2,n} = \begin{cases} \dfrac{l^3}{3}, & n = 0, \\[2mm] i\dfrac{l^2}{n\omega} + \dfrac{2l}{n^2\omega^2}, & n \neq 0, \end{cases}$$

and

$$I_{3,n} = \begin{cases} \dfrac{l^4}{4}, & n = 0, \\[2mm] i\dfrac{l^3}{n\omega} + \dfrac{3l^2}{n^2\omega^2} - i\dfrac{6l}{n^3\omega^3}, & n \neq 0. \end{cases}$$

Therefore, the Fourier series of the initial condition is:

$$\frac{2}{l^2}x^3 - \frac{3}{l}x^2 + x = \frac{1}{l}\left(\frac{2}{l^2}\frac{l^4}{4} - \frac{3}{l}\frac{l^3}{3} + \frac{l^2}{2}\right.$$

$$+ \sum_{n \neq 0}\left(\frac{2}{l^2}\left(i\frac{l^3}{n\omega} + \frac{3l^2}{n^2\omega^2} - i\frac{6l}{n^3\omega^3}\right)\right.$$

$$\left.\left. - \frac{3}{l}\left(i\frac{l^2}{n\omega} + \frac{2l}{n^2\omega^2}\right) + i\frac{l}{n\omega}\right)e^{in\omega x}\right)$$

$$= \frac{24}{l^2\omega^3}\sum_{n=1}^{\infty}\frac{1}{n^3}\sin n\omega x.$$

Now, let's look for a solution of the form:

$$u(t,x) = \sum_{n=1}^{\infty} v_n(t) \sin n\omega x.$$

This implies:

$$\begin{cases} v_n' = \lambda_n v_n, \\ v_n(0) = \frac{24}{l^2 \omega^3} \frac{1}{n^3}. \end{cases}$$

Hence, the solution for $v_n(t)$ is given by:

$$v_n(t) = \frac{24}{l^2 \omega^3} \frac{1}{n^3} e^{-n^2 \omega^2 t}.$$

Finally, the solution for $u(t, x)$ is:

$$u(t, x) = \frac{24}{l^2 \omega^3} \sum_{n=1}^{\infty} \frac{1}{n^3} e^{-n^2 \omega^2 t} \sin n\omega x.$$

Due to the rapid decay of the coefficients v_n with respect to n and t, a good approximation of the graph can be obtained with just a partial sum of a few terms. Next, we will show the space-time representation of this evolution for $l = 1$ using a partial sum up to 5 terms.

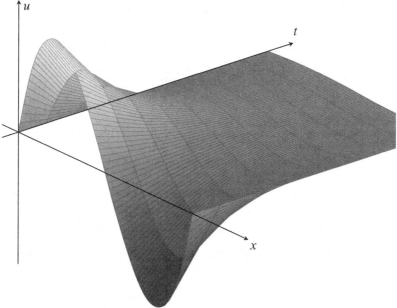

Temperature $u(x, t)$ in a trend $[0, 1]$ with times $0 < t < 0.1$

(3) We will now study the wave equation and the corresponding problem:

$$\begin{cases} u_{tt} = u_{xx}, \quad x \in (-1, 1), \quad t > 0, \\ \begin{cases} \begin{cases} u|_{t=0} = x(x^2 - 1), \\ u_t|_{t=0} = 0, \end{cases} \\ \begin{cases} u|_{x=-1} = u|_{x=1}, \\ u_x|_{x=-1} = u_x|_{x=1}. \end{cases} \end{cases} \end{cases}$$

To find the eigenvalues, we proceed similarly to previous cases. We expand the polynomial

$$g(x) := x(x^2 - 1)$$

in a Fourier series of exponentials:

$$g = \sum_{n=-\infty}^{\infty} c_n e^{in\pi x},$$

where

$$c_n = \frac{1}{2} \int_{-1}^{1} x(x^2 - 1) e^{-in\pi x} dx.$$

After repeated integrations by parts, we get

$$c_n = \begin{cases} 0, & n = 0, \\ (-1)^{n+1} \dfrac{6i}{n^3 \pi^3}, & n \neq 0. \end{cases}$$

Thus, the series becomes

$$g(x) = 12 \sum_{n=1}^{\infty} \frac{(-1)^n}{n^3 \pi^3} \sin n\pi x.$$

Now, considering the eigenfunction expansion of the solution:

$$u(t, x) = \sum_{n=1}^{\infty} v_n(t) \sin n\pi x,$$

we obtain the following differential equation for the coefficients:

$$\begin{cases} v_n'' = -n^2 \pi^2 v_n, \\ v_n(0) = 12 \dfrac{(-1)^n}{n^3 \pi^3}, \\ v_n'(0) = 0. \end{cases}$$

The solution to this equation is

$$v_n(t) = 12 \frac{(-1)^n}{n^3 \pi^3} \cos n\pi t,$$

and, consequently, the final solution becomes

$$u(t, x) = 12 \sum_{n=1}^{\infty} \frac{(-1)^n}{n^3 \pi^3} \cos n\pi t \sin n\pi x.$$

Noting that

$$\cos n\pi t \sin n\pi x = \frac{1}{2}(\sin n\pi(x+t) + \sin n\pi(x-t)),$$

we can rewrite the solution as

$$u(t,x) = \frac{1}{2}\Big(12\sum_{n=1}^{\infty}\frac{(-1)^n}{n^3\pi^3}\sin n\pi(x+t) + 12\sum_{n=1}^{\infty}\frac{(-1)^n}{n^3\pi^3}\sin n\pi(x-t)\Big).$$

Thus, we can express $u(t,x)$ as the semi-sum of two traveling waves with opposite velocities $v_{\pm} = \pm 1$, and each of them has the form of the periodic extension of the initial condition.

The space-time graph of these waves is shown below.

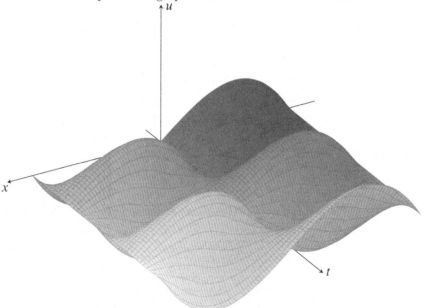

Semi-sum of traveling waves with periodic boundary conditions

(4) The density field $u(t,x)$ of **spinless pions** with mass M in a parallelepipedic box

$$\Lambda = [0, L_1] \times [0, L_2] \times [0, L_3]$$

is described by the following boundary value problem and initial conditions for the Klein–Gordon equation:

$$\begin{cases} u_{tt} - c^2\Delta u + \dfrac{M^2 c^4}{\hbar^2}u = 0, \quad t > 0, \quad x \in \Lambda, \\ \begin{cases} u|_{t=0} = f(x), \\ u|_{\text{walls}} = 0, \end{cases} \end{cases}$$

with
$$f(x) = x(L_1 - x)y(L_2 - y)z(L_3 - z).$$

If we introduce the form of stationary waves
$$u(t, x) = e^{-i\omega t} w(x)$$

into the Klein–Gordon equation, then we obtain
$$-\omega^2 w - c^2 \Delta w + \frac{M^2 c^4}{\hbar^2} w = 0,$$

and since $w(x)$ satisfies the Helmholtz equation
$$\Delta w = -k^2 w,$$

we conclude that
$$\left(-\omega^2 + c^2 k^2 w + \frac{M^2 c^4}{\hbar^2} \right) w = 0,$$

hence the dispersion relation is
$$\boxed{(\hbar\omega)^2 = (\hbar k)^2 c^2 + (Mc^2)^2.}$$

This can be written as the following relativistic energy relation
$$E = \pm\sqrt{p^2 c^2 + (Mc^2)^2},$$

where
$$E = \hbar\omega,$$
$$p = \hbar k.$$

The solutions of the Helmholtz equation with Dirichlet boundary conditions are given by
$$\boxed{w_n(x) = \sin\frac{n_1\pi x}{L_1} \sin\frac{n_2\pi y}{L_2} \sin\frac{n_3\pi z}{L_3},}$$

for $n = (n_1, n_2, n_3) \in \mathbb{N}^3$. The wave vector is
$$k = \pi\left(\frac{n_1}{L_1}, \frac{n_2}{L_2}, \frac{n_3}{L_3}\right).$$

To apply the method of separation of variables, we must first develop the initial data $f(x)$ into a Fourier series of sines. We observe that $f(x)$ is a product of three functions, and each factor depends only on one variable. Furthermore, the factor is essentially the same except for the value of constants. Therefore, we develop the function $x(x - L_1)$ into a Fourier series of sines as follows:
$$x(x - L_1) = \sum_{n=1}^{\infty} b_n \sin\frac{n\pi x}{L_1},$$
$$b_n = \frac{2}{L_1} \int_0^{L_1} x(x - L_1) \sin\frac{n\pi x}{L_1} dx.$$

These integrals are easily calculated as

$$b_n = 4(1 - (-1)^n)\frac{L_1^2}{\pi^3 n^3}$$

$$= \begin{cases} 0, & n \text{ even}, \\ \dfrac{8L_1^2}{\pi^3 n^3}, & n \text{ odd}, \end{cases}$$

and the desired expansion is

$$x(L_1 - x) = \frac{8L_1^2}{\pi^3} \sum_{n \text{ odd}} \frac{1}{n^3} \sin \frac{n\pi x}{L_1}.$$

We then compare the graphs of the function and the truncation to two terms of the sine series.

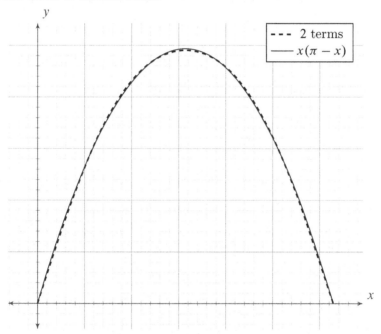

Thus, the coefficients b_n of the Fourier series in sines

$$f(x) = \sum_{n \in \mathbb{N}^3} b_n w_n(x)$$

are

$$b_n = \begin{cases} \dfrac{8^3 L_1^2 L_2^2 L_3^2}{\pi^9} \dfrac{1}{n_1^3 n_2^3 n_3^3}, & n_1, n_2, n_3 \text{ odd}, \\ 0, & \text{in other cases}, \end{cases}$$

and we have the following three-dimensional Fourier series:

$$f(x) = \frac{512 L_1^2 L_2^2 L_3^2}{\pi^9} \sum_{n_1, n_2, n_3 \text{ odd}} \frac{1}{n_1^3 n_2^3 n_3^3} \sin \frac{n_1 \pi x}{L_1} \sin \frac{n_2 \pi y}{L_1} \sin \frac{n_3 \pi z}{L_3}.$$

The series converges uniformly since $f(x)$ satisfies the Dirichlet conditions on the walls of the box.

The method of separation of variables proposes the following expression for the pion field density:

$$u(t, x) = \sum_{n \in \mathbb{N}^3} v_n(t) w_n(x).$$

For this series to be a solution to the boundary value problem and initial conditions of the Klein–Gordon equation, the coefficients $v_n(t)$ must satisfy

$$v_n'' + \omega_n^2 v_n = 0,$$

$$\begin{cases} v_n(0) = b_n, \\ v_n'(0) = 0, \end{cases}$$

where

$$\omega_n = \sqrt{k_n^2 c^2 + \frac{(Mc^2)^2}{\hbar^2}} = \sqrt{\pi^2 c^2 \left(\frac{n_1^2}{L_1^2} + \frac{n_2^2}{L_2^2} + \frac{n_3^2}{L_3^2} \right) + \frac{(Mc^2)^2}{\hbar^2}}$$

The solution to the ODE is

$$v_n(t) = A_n \cos \omega_n t + B_n \sin \omega_n t,$$

where A_n, B_n are constants to be determined by the initial conditions. Since $v_n'(0) = 0$, we must have $B_n = 0$, and to satisfy

$$v_n(0) = b_n,$$

we need to impose $A_n = b_n$. In conclusion,

$$v_n(t) = b_n \cos \omega_n t$$

and the pion field density will have the following three-dimensional Fourier expansion:

$$u(t, x) = \frac{512 L_1^2 L_2^2 L_3^2}{\pi^9} \sum_{\substack{n_1, n_2, n_3 \\ \text{odd}}} \cos \sqrt{\pi^2 c^2 \left(\frac{n_1^2}{L_1^2} + \frac{n_2^2}{L_2^2} + \frac{n_3^2}{L_3^2} \right) + \frac{(Mc^2)^2}{\hbar^2}} t$$

$$\times \frac{1}{n_1^3 n_2^3 n_3^3} \sin \frac{n_1 \pi x}{L_1} \sin \frac{n_2 \pi y}{L_1} \sin \frac{n_3 \pi z}{L_3}.$$

§6.3. General Discussion: Full Inhomogeneities

OW, we confront the more general problem where the system of boundary and/or initial conditions consists of two subsystems. As before, one of them is built up by conditions of inhomogeneous type:

$$a_i(u) = g_i(x), \quad i \in \{1, \ldots, r\}.$$

This subsystem contains **operators a_i that act only on the variable** $x_0 \equiv t$, while the other subsystem, which sets this case apart from previously discussed ones, is also made up of inhomogeneous conditions:

(6.5)
$$b_j(u) = h_j(x), \quad x \in I \times \partial \Lambda \quad j \in \{1, \ldots, s\}.$$

These conditions involve **operators b_j that act only on the variables** $x = (x_1, \ldots, x_{n-1})$.

In order to apply the eigenfunction expansion method, we need to assume that:

(1) The operator B on the given domain $\mathfrak{D}$ is assumed to be symmetric.
(2) We assume that the corresponding set of eigenfunctions, denoted as $\{w_m\}_m \subset \mathfrak{D}$, forms a complete set in the corresponding Hilbert space, denoted as $L^2(\overline{\Lambda}, \rho \, d^n x)$.
(3) We assume the existence of a solution, u, in $L^2(\overline{\Lambda}, \rho \, d^n x)$. As a result, we can express u as:

$$u(t, x) = \sum_m v_m(t) w_m(x).$$

(4) Additionally, we require that these boundary conditions (6.5) are equivalent to the conditions expressed as:

(6.6)
$$(w_m, Bu) = (Bw_m, u) - I_m(h),$$

where the terms $I_m(h)$ must take the following form:

$$I_m(h) = \int_{\partial \Lambda} \sum_j C_{m,j} h_j \, dS.$$

Here, $C_{m,j}$ represents linear operators acting on the boundary data h_j, and it must be determined for each specific case.

Therefore, the partial differential equation $(A + B)u = f$ can be projected onto the eigenfunctions as follows:

$$(w_m, (A + B)u) = (w_m, Au) + (w_m, Bu)$$
$$= (w_m, Au) + (Bw_m, u) + I_m = \|w_m\|^2 f_m.$$

Now, by introducing the representation $u = \sum_{m'} v_{m'}(t) w_{m'}(x)$, and considering the L^2 convergence of this series, we obtain:

$$\sum_{m'} ((w_m, A(v_{m'})w_{m'}) + (\lambda_m w_m, v_{m'} w_{m'})) + I_m = \|w_m\|^2 f_m.$$

As a result, since $(w_m, w_{m'}) = \|w_m\|^2 \delta_{m,m'}$, (6.3) the unknown functions $v_m(t)$ solve:

$$\begin{cases} A v_m + \lambda_m v_m = f_m + \dfrac{I_m(h)}{\|w_m\|^2}, \\ a_i(v_m) = c_{im}, \quad i \in \{1, \dots, r\}. \end{cases}$$

☞   **Observation:** It's important to observe that the expansion of the solution u is expressed in terms of eigenfunctions that belong to the domain $\mathfrak{D}$, i.e., satisfy homogeneous boundary conditions on $\partial \Lambda$. However, the solution u has been constructed with the assumption that it satisfies inhomogeneous boundary conditions on $\partial \Lambda$. Each finite truncation of the expansion, $u = \sum_m v_m w_m$, following the linear superposition principle, will indeed satisfy homogeneous boundary conditions. However, it's in the limit and in integral mean in the spatial domain Λ, specifically through strong convergence in the L^2 Hilbert space, where the solution u adheres to the given inhomogeneous boundary value problem. Pointwise convergence at $\partial \Lambda$ will only result in homogeneous boundary conditions and may potentially exhibit Gibbs oscillations at those boundary points.

This phenomenon is not specific to fully inhomogeneous problems; it also occurs in restricted cases when the function g_i may not belong to the domain $\mathfrak{D}$. The corresponding expansion converges strongly to a solution, but the condition $a_j(u) = g_j$ may not hold pointwise on the boundary $\partial \Lambda$.

In this context, we would like to direct the reader's attention to [23, example 5.0.1]. In this example, a BVP for the heat equation is discussed, which describes a one-dimensional bar insulated at the right edge and subjected to a constant heat flow at the left edge. Green's function method is employed, leading to a solution obtained through a Fourier series expansion.

It's important to note that during this discussion, the author emphasizes the precautions that need to be taken into account, especially regarding the poor convergence of the Fourier series near the edges.

To illustrate this procedure, let's assume that $B = L$ with

$$Lu = \frac{1}{\rho}(-\operatorname{div}(p\nabla u) + qu),$$

is a $(n-1)$-dimensional Sturm–Liouville operator, where ρ, p, q are real smooth functions in Λ and $\rho, p > 0$ in Λ. A Sturm–Liouville-type operator of this kind satisfies the multidimensional Lagrange's identity (remember that we proved it for the three-dimensional case)

$$(w, Lu) = (Lw, u) - \int_{\partial \Lambda} p\left(\frac{\partial u}{\partial n}\bar{w} - u\frac{\partial \bar{w}}{\partial n}\right) dS.$$

This operator will be symmetric in domains where the boundary terms in Lagrange's identity cancel. For example, mixed conditions of Robin type

$$(6.7a) \qquad b(u) := \left(p\frac{\partial u}{\partial n} + \sigma u \right)\Big|_{\partial \Lambda} = 0,$$

$$(6.7b) \qquad b(u) := \left(\sigma p\frac{\partial u}{\partial n} + u \right)\Big|_{\partial \Lambda} = 0,$$

with σ a smooth real function on the boundary. Define a domain

$$\mathfrak{D} = \left\{ w \in C^\infty(\overline{\Lambda}) : b(w) = 0 \right\},$$

such that the Sturm–Liouville operator is symmetric in $\mathfrak{D}$ (prove it).

The first Equation (6.7a), for $\sigma = 0$, delivers the Neumann condition, while the second condition (6.7b) for $\sigma = 0$ reduces to the Dirichlet case.

Lagrange's identity is not only useful for the identification of domains of symmetry of this Sturm–Liouville operator but also to treat with the more general inhomogeneities. Indeed, if Equation (6.5) is determined by $b(u) = h$ with h being a given function defined over $I \times \partial \Lambda$, then we get

$$(w_m, Lu) = (Lw_m, u) - I_m(h)$$

with I_m, in the first case, given by

$$I_m(h) = \int_{\partial \Lambda} p\left(\frac{\partial u}{\partial n}\bar{w}_m - u\frac{\partial \bar{w}_m}{\partial n} \right) dS = \int_{\partial \Lambda} \left(p\frac{\partial u}{\partial n} + \sigma u \right)\bar{w}_m\, dS$$

$$= \int_{\partial \Lambda} h\bar{w}_m\, dS$$

and by

$$I_m(h) = \int_{\partial \Lambda} p\left(\frac{\partial u}{\partial n}\bar{w}_m - u\frac{\partial \bar{w}_m}{\partial n} \right) dS = -\int_{\partial \Lambda} \sigma p\left(\frac{\partial u}{\partial n} + u \right)p\frac{\partial \bar{w}_m}{\partial n}\, dS$$

$$= -\int_{\partial \Lambda} hp\frac{\partial \bar{w}_m}{\partial n}\, dS,$$

for the second case.

§6.3.1. Illustrative Examples

(1) Let's consider the heat distribution on a finite one-dimensional bar modeled as the interval $\Lambda = [0, \pi]$ for $t > 0$. The boundary value

problem we need to handle is as follows:

$$\begin{cases} \dfrac{\partial u}{\partial t} = \dfrac{\partial^2 u}{\partial x^2} + f(t,x), \quad x \in [0,\pi], \quad t > 0, \\ \begin{cases} u|_{t=0} = g(x), \\ \begin{cases} \dfrac{\partial u}{\partial x}\bigg|_{x=0} = h_1(t), \\ \dfrac{\partial u}{\partial x}\bigg|_{x=\pi} = h_2(t). \end{cases} \end{cases} \end{cases}$$

This problem describes a heat-conducting bar with an initial temperature distribution given by the function $g(x)$ and two boundaries where the heat flow is determined by the functions $h_1(t)$ at $x = 0$ and $h_2(t)$ at $x = \pi$. The form of the driven force term f depends on the specific physical situation and the nature of the external source or sink of heat. For example, it could represent a heat source, like a heating element, that releases heat into the bar, or it could represent a heat sink, like a cooling system, that absorbs heat from the bar.

The differential operator $B = -\dfrac{\partial^2}{\partial x^2}$ on the domain of smooth functions that satisfy homogeneous Neumann conditions,

$$\begin{cases} w_x|_{x=0} = 0, \\ w_x|_{x=\pi} = 0, \end{cases}$$

leads to eigenvalues $\lambda_n = n^2$ for $n \in \{0, 1, 2, \dots\}$, and eigenfunctions $w_0 = \frac{1}{2}$ and $w_n = \cos nx$, for $n \in \mathbb{N}$.

Here we have $\|w_0\|^2 = \frac{\pi}{4}$ and $\|w_n\|^2 = \frac{\pi}{2}$, for $n \in \mathbb{N}$.

These eigenfunctions form a complete orthogonal set in the Hilbert space $L^2([0,\pi], dx)$, allowing us to represent the driven force $f(x)$ and initial temperature distribution $g(x)$ as a Fourier series:

(6.8)
$$\begin{cases} f(t,x) = \dfrac{f_0(t)}{2} + \displaystyle\sum_{n=1}^{\infty} f_n(t) \cos nx, \quad f_n(t) = \dfrac{2}{\pi} \displaystyle\int_0^{\pi} f(t,x) \cos nx \, dx, \\ g(x) = \dfrac{c_0}{2} + \displaystyle\sum_{n=1}^{\infty} c_n \cos nx, \qquad\quad c_n = \dfrac{2}{\pi} \displaystyle\int_0^{\pi} g(x) \cos nx \, dx. \end{cases}$$

The Lagrange identity (4.6), obtained by integrating twice by parts, leads to:

$$(w_n, Bu) = (Bw_n, u) - I_n,$$

which is equivalent to Equation (6.6). The term I_n takes the form:

$$I_n = \left[w_n \frac{\partial u}{\partial x} - \frac{\mathrm{d}w_n}{\mathrm{d}x} u \right]_{x=0}^{\pi} = \begin{cases} \dfrac{1}{2}\left(\dfrac{\partial u}{\partial x}\Big|_{x=\pi} - \dfrac{\partial u}{\partial x}\Big|_{x=0} \right), & n = 0, \\[3mm] (-1)^n \dfrac{\partial u}{\partial x}\Big|_{x=\pi} - \dfrac{\partial u}{\partial x}\Big|_{x=0}, & n \in \mathbb{N}, \end{cases}$$

$$= \begin{cases} \dfrac{1}{2}(h_2 - h_1), & n = 0, \\[3mm] (-1)^n h_2 - h_1, & n \in \mathbb{N}. \end{cases}$$

Hence, we find

$$\frac{I_n}{\|w_n\|^2} = \begin{cases} \dfrac{4}{\pi}\dfrac{1}{2}(h_2 - h_1), & n = 0, \\[3mm] \dfrac{2}{\pi}((-1)^n h_2 - h_1), & n \in \mathbb{N}, \end{cases}$$

$$= \frac{2}{\pi}((-1)^n h_2 - h_1).$$

Now, for the unknown function $v_n(t)$ in the eigenfunction expansion of the temperature

$$u = \sum_{n=0}^{\infty} v_n(t)w_n(x),$$

the following ordinary differential equations of the first order are obtained:

$$\begin{cases} \dfrac{\mathrm{d}v_n}{\mathrm{d}t} + n^2 v_n = h_n, \\[3mm] v_n|_{t=0} = c_n, \end{cases}$$

for $n \in \mathbb{N}_0$, and

$$h_n := f_n + \frac{2}{\pi}\left((-1)^n h_2 - h_1\right).$$

From ODE theory, the solution is:

$$v_n(t) = c_n e^{-n^2 t} + \int_0^t e^{-n^2(t-\tau)} h_n(\tau)\mathrm{d}\tau.$$

Finally, in terms of the function

$$J_n(t) := \int_0^t e^{n^2 \tau} h_n(\tau)\mathrm{d}\tau$$

the solution is given by:

(6.9)
$$\boxed{ u = \frac{c_0 + J_0(t)}{2} + \sum_{n=1}^{\infty} (c_n + J_n(t)) e^{-n^2 t} \cos nx. }$$

We observe that when these types of general inhomogeneities come into play, the rule that only the modes involved in the inhomogeneities are excited in the final solution no longer holds. Now, the general inhomogeneous terms h_1 and h_2 excite the mode n as soon as $J_n \neq 0$.

To be more specific, let's consider the functions:

$$f(t,x) = e^{-at} \cos 5x, \qquad\qquad g(x) = 1 + \cos x,$$
$$h_1(t) = -\frac{\pi}{4} e^{-at}, \qquad\qquad h_2(t) = \frac{\pi}{4} e^{-at},$$

where $a > 0$. Then, we find that $h_{2n} = e^{-at}$ and $h_5 = e^{-at}$, while all remaining odd terms cancel out, i.e., $h_{2n+1} = 0$ for $n \neq 2$. Hence,

$$J_{2n} = \frac{e^{(4n^2 - a)t} - 1}{4n^2 - a},$$

$$J_{2n+1} = \begin{cases} \dfrac{e^{(25-a)t} - 1}{25 - a}, & n = 2, \\ 0, & n \neq 2, \end{cases}$$

for $n \in \mathbb{N}$. Here we are assuming for simplicity that $a \neq 25, 4n^2$, for $n \in \mathbb{N}$.

Consequently, we get

$$u = 1 + e^{-t} \cos x + \frac{1 - e^{-at}}{2a} + \frac{e^{-at} - e^{-25t}}{25 - a} \cos 5x$$
$$+ \sum_{n=1}^{\infty} \frac{e^{-at} - e^{-4n^2 t}}{4n^2 - a} \cos 2nx.$$

The subsequent Fourier series:

$$-\frac{\pi}{4\sqrt{a} \sin \frac{\pi \sqrt{a}}{2}} \cos\left(\sqrt{a}\left(x - \frac{\pi}{2}\right)\right) = -\frac{1}{2a} + \sum_{n=1}^{\infty} \frac{1}{4n^2 - a} \cos 2nx,$$

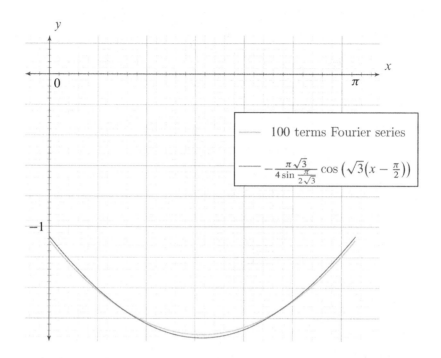

brings us to the following solution:

$$u = 1 + \frac{1}{2a} + e^{-t}\cos x - \frac{\pi}{4\sqrt{a}\sin\frac{\pi\sqrt{a}}{2}}\cos\left(\sqrt{a}\left(x - \frac{\pi}{2}\right)\right)e^{-at}$$

$$+ \frac{e^{-at} - e^{-25t}}{25 - a}\cos 5x - \sum_{n=1}^{\infty}\frac{e^{-4n^2 t}}{4n^2 - a}\cos 2nx.$$

We observe that the modes $n = 0, 1, 5$ are excited, as expected for the restricted inhomogeneous BVP. Additionally, as this problem goes beyond restricted inhomogeneities, all the even modes also come into play. Moreover, the equilibrium state temperature that the system approaches as $t \to \infty$ is

$$u_{\text{equi}} = 1 + \frac{1}{2a}.$$

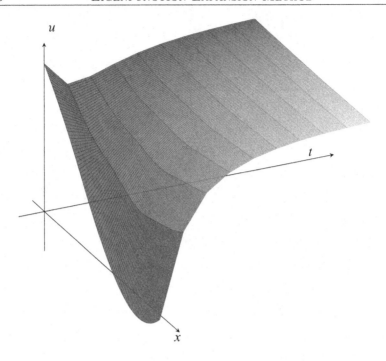

$u(x,t)$ **in a bar** $[0, \pi]$ **with** $0 < t < 12$ **and** $a = \frac{1}{3}$

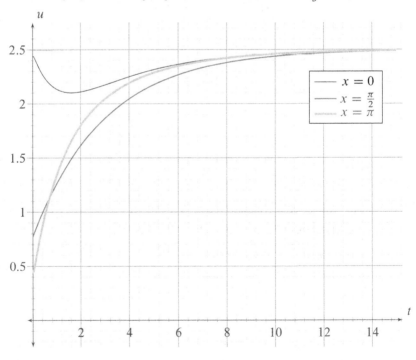

When $a \to 0$, so that $h_1 = -\frac{\pi}{4}$ and $h_2 = \frac{\pi}{4}$, we have that

$$\frac{1}{2a} - \frac{\pi}{4\sqrt{a}\sin\frac{\pi\sqrt{a}}{2}}\cos\left(\sqrt{a}\left(x - \frac{\pi}{2}\right)\right)e^{-at}$$

$$= \frac{1}{2a} - \left(1 - \frac{a}{2}\left(x - \frac{\pi}{2}\right)^2 + O(a^2)\right)\frac{1}{2a}\left(1 + \frac{\pi^2 a}{24} + O(a^2)\right)(1 - at + O(a^2))$$

$$= \frac{1}{4}\left(x - \frac{\pi}{2}\right)^2 + \frac{t}{2} - \frac{\pi^2}{24} + O(a).$$

Hence, the solution reduces to:

$$u = 1 + e^{-t^2}\cos x + \frac{1}{4}\left(x - \frac{\pi}{2}\right)^2 + \frac{t}{2} - \frac{\pi^2}{48} + \frac{1 - e^{-25t}}{25}\cos 5x$$

$$- \sum_{n=1}^{\infty}\frac{e^{-4n^2 t}}{4n^2}\cos 2nx.$$

We observe that the temperature is unbounded in time, growing linearly. This has an easy physical interpretation: the inhomogeneous boundary conditions continuously supply heat into the bar from both the left and right edges. It is important to recall that at the right edge, a positive gradient indicates incoming heat, while a negative gradient indicates outgoing heat. Conversely, at the left edge, the situation is reversed.

To verify our procedure, let's prove that (6.9) is indeed the solution obtained using Green's function method. To achieve this, we will present Green's function expression for the solution and proceed to derive the corresponding Green's function for this case. We will then incorporate it back into the solution formula, resulting in (6.9).

For this purpose, let's follow Duffy's approach in [23, chapter 5, problem 1]. Green's function (fundamental solution) $G(x, t | \chi, \tau)$ for the heat equation, which satisfies the equation:

$$\begin{cases} \dfrac{\partial G}{\partial t} - \dfrac{\partial^2 G}{\partial x^2} = \delta(x - \chi)\delta(t - \tau), & x, \xi \in (0, \pi), \quad t, \tau > 0, \\[2mm] \dfrac{\partial G}{\partial x}\Big|_{x=0} = 0, \quad t > 0, \\[2mm] \dfrac{\partial G}{\partial x}\Big|_{x=\pi} = 0, \quad t > 0, \end{cases}$$

leads to the solution of the problem through convolution:

$$u(x,t) = \int_0^{t^+} \int_0^{\pi} f(\xi,\tau)G(x,t\mid\xi,\tau)d\xi d\tau$$

$$+ \int_0^{t^+} \left[G(x,t\mid\xi,\tau)\frac{\partial u(\xi,\tau)}{\partial\xi} - u(\xi,\tau)\frac{\partial G(x,t\mid\xi,\tau)}{\partial\xi} \right]_{\xi=0}^{\xi=\pi} d\tau$$

$$+ \int_0^{\pi} u(\xi,0)G(x,t\mid\xi,0)d\xi.$$

After substitution, this becomes:

(6.10)
$$u(x,t) = \int_0^{t^+} \int_0^{\pi} f(\xi,\tau)G(x,t\mid\xi,\tau)d\xi d\tau$$

$$+ \int_0^{t^+} \left(G(x,t\mid\pi,\tau)h_2(\tau) - G(x,t\mid 0,\tau)h_1(\tau)\right)d\tau$$

$$+ \int_0^{\pi} g(\xi)G(x,t\mid\xi,0)d\xi.$$

Moreover, this Green's function, expressed in terms of eigenfunctions, reads [23, example 5.2.1]:

$$G(x,t|\xi,\tau) = H(t-\tau) \sum_{n=1}^{\infty} \frac{w_n(\xi)w_n(x)}{\|w_n\|^2} e^{-\lambda_n(t-\tau)}.$$

In our case, it is given by:

(6.11)
$$G(x,t|\xi,\tau) = H(t-\tau)\left(\frac{1}{\pi} + \frac{2}{\pi}\sum_{n=1}^{\infty} \cos n\xi \, \cos nx \, e^{-n^2(t-\tau)}\right),$$

where H is Heaviside's step function. Now, let's introduce (6.11) into (6.10) and recall the Fourier expansions (6.8) to obtain:

$$u(x,t) = \int_0^{t^+} \left(\frac{f_0(\tau)}{2} + \sum_{n=1}^{\infty} f_n(\tau)e^{-n^2(\tau-t)} \cos nx\right)d\tau$$

$$+ \int_0^{t^+} \left(\frac{1}{\pi} + \frac{2}{\pi}\sum_{n=1}^{\infty}(-1)^n \cos nx \, e^{-n^2(t-\tau)}\right) h_2(\tau)d\tau$$

$$- \int_0^{t^+} \left(\frac{1}{\pi} + \frac{2}{\pi}\sum_{n=1}^{\infty} \cos nx \, e^{-n^2(t-\tau)}\right) h_1(\tau)d\tau$$

$$+ \frac{c_0}{2} + \sum_{n=1}^{\infty} c_n \cos nx e^{-n^2 t},$$

for $t > 0$, which is precisely (6.9) as claimed.

(2) Let us consider the electric potential, denoted as $u(x, y)$, on a square plate with dimensions $[0, \pi] \times [0, \pi]$. The potential satisfies:

$$\begin{cases} u_{xx} + u_{yy} = f(x, y), & (x, y) \in [0, \pi] \times [0, \pi], \\ \begin{cases} u\big|_{x=0} = h_1(y), \\ u\big|_{x=\pi} = h_2(y), \\ u\big|_{y=0} = g_1(x), \\ u\big|_{y=\pi} = g_2(x). \end{cases} \end{cases}$$

In this scenario, the charge distribution is represented by $f(x, y)$, while the boundaries are maintained at given potentials h_1, h_2, g_1, and g_2, correspondingly. To tackle this problem, we employ the method of separation of variables for the Laplacian

$$-\Delta = A + B,$$

where we define the following operators:

$$A = -\frac{\partial^2}{\partial y^2}, \qquad B = -\frac{\partial^2}{\partial x^2}.$$

Given the Dirichlet-type conditions that define the domain of B, we immediately observe that the eigenvalues are $\lambda_n = n^2$, and the corresponding eigenfunctions are $\sin nx$ for $n \in \mathbb{N}$. Note that $\|w_n\|^2 = \frac{\pi}{2}$. The inhomogeneous terms can be expanded as follows:

$$f(x, y) = \sum_{n=1}^{\infty} f_n(y) \sin nx, \qquad f_n(y) = \frac{2}{\pi} \int_0^\pi f(x, y) \sin nx \, dx,$$

$$g_1(x) = \sum_{n=1}^{\infty} c_{1,n} \sin nx, \qquad c_{1,n} = \frac{2}{\pi} \int_0^\pi g_1(x) \sin nx \, dx,$$

$$g_2(x) = \sum_{n=1}^{\infty} c_{2,n} \sin nx, \qquad c_{2,n} = \frac{2}{\pi} \int_0^\pi g_2(x) \sin nx \, dx.$$

Again, we have $(w_n, Bu) = (Bw_n, u) - I_n$, where, for $n \in \mathbb{N}$:

$$I_n = \left[w_n \frac{\partial u}{\partial x} - \frac{dw_n}{dx} u \right]_{x=0}^{\pi} = n\big((-1)^{n+1} h_2 + h_1\big),$$

$$\frac{I_n}{\|w_n\|^2} = \frac{2}{\pi} n\big((-1)^{n+1} h_2 + h_1\big).$$

The unknown function $v_n(y)$ in the eigenfunction expansion of the electrostatic potential

$$u = \sum_{n=0}^{\infty} v_n(y) w_n(x)$$

satisfies the following ODEs:

$$(6.12) \qquad \begin{cases} \dfrac{d^2 v_n}{dy^2} - n^2 v_n = H_n, \\ \begin{cases} v_n|_{y=0} = c_{1,n}, \\ v_n|_{y=\pi} = c_{2,n}, \end{cases} \end{cases}$$

for

$$H_n := f_n + \frac{2}{\pi} n((-1)^n h_2 - h_1).$$

A fundamental set of solutions $\{v_{n,1}, v_{n,2}\}$ to the homogeneous version of (6.12) is given by: $v_{n,1} = \sinh ny$ and $v_{n,2} = \sinh n(\pi - y)$. The Wronskian of this fundamental set is

$$\begin{aligned} W_n &:= W(v_{n,1}, v_{n,2}) \\ &= v_{n,1} v'_{n,2} - v'_{n,1} v_{n,2} \\ &= -n \sinh n\pi. \end{aligned}$$

To find a solution to (6.12), we apply the method of variation of parameters, and a particular solution is given by:

$$\begin{aligned} v_{\mathrm{p},n}(y) = \frac{\sinh ny}{n \sinh n\pi} \int_0^y H_n(s) \sinh n(\pi - s) ds \\ - \frac{\sinh n(\pi - y)}{n \sinh n\pi} \int_0^y H_n(s) \sinh ns \, ds. \end{aligned}$$

Using the identity:

$$\sinh ny \sinh n(\pi - s) - \sinh ns \sinh n(\pi - y) = \sinh n\pi \sinh n(y - s),$$

we get:

$$v_{\mathrm{p},n}(y) = \frac{1}{n} \int_0^y H_n(s) \sinh n(y - s) ds.$$

This particular solution satisfies:

$$v_{\mathrm{p},n}(0) = 0, \quad v_{\mathrm{p},n}(\pi) = \frac{1}{n} \int_0^\pi H_n(s) \sinh n(\pi - s) ds.$$

Then, a general solution of the ODE in (6.12) will have the form:

$$v_n(y) = A \sinh ny + B \sinh n(\pi - y) + v_{\mathrm{p},n}(y).$$

Imposing the boundary conditions that determine (6.12), we find:

$$c_{1,n} = B \sinh n\pi, \quad c_{2,n} = A \sinh n\pi + v_{\mathrm{p},n}(\pi).$$

Thus, the solution is:

$$v_n(y) = \frac{c_{2,n} - v_{\mathrm{p},n}(\pi)}{\sinh n\pi} \sinh ny + \frac{c_{1,n}}{\sinh n\pi} \sinh n(\pi - y) + v_{\mathrm{p},n}(y).$$

Therefore, the electrical potential is given by:

$$u = \sum_{n=1}^{\infty} \sin nx \left(\frac{c_{2,n} - \frac{1}{n}\int_0^\pi H_n(s)\sinh n(\pi - s)ds}{\sinh n\pi} \sinh ny \right.$$
$$+ \frac{c_{1,n}}{\sinh n\pi}\sinh n(\pi - y)$$
$$\left. + \frac{1}{n}\int_0^y \left(f_n(s) + \frac{2}{\pi}n\big((-1)^n h_2(s) - h_1(s)\big)\right)\sinh n(y - s)ds \right).$$

As a specific case, we take

$$f = \sin x \sin y, \quad g_1 = g_2 = \sin x, \quad -h_1 = h_2 = \sin y.$$

Hence,

$$f_1 = \sin y, \quad c_{1,1} = c_{2,1} = 1,$$

and

$$f_n = 0, \quad c_{1,n} = c_{2,n} = 0$$

for $n \in \{2, 3, \dots\}$. Therefore, for $n \in \mathbb{N}$, we get:

$$H_1 = \sin y, \quad H_{2n+1} = 0, \quad H_{2n} = \frac{8}{\pi}n\sin y.$$

We find:

$$\int_0^y \sin(s)\sinh n(y - s)ds = \frac{\sinh ny - n\sin y}{n^2 + 1}.$$

Consequently, for $n \in \mathbb{N}$, the particular solutions are:

$$v_{p,1}(y) = \frac{\sinh y - \sin y}{2},$$
$$v_{p,2n+1}(y) = 0,$$
$$v_{p,2n}(y) = \frac{4}{\pi}\frac{\sinh 2ny - 2n\sin y}{4n^2 + 1}.$$

Then, we obtain the following solutions:

$$v_1(y) = \left(\frac{1}{\sinh \pi} - \frac{1}{2}\right)\sinh y + \frac{1}{\sinh \pi}\sinh(\pi - y) + \frac{\sinh y - \sin y}{2}$$
$$= \frac{\sinh y + \sinh(\pi - y)}{\sinh \pi} - \frac{\sin y}{2},$$
$$v_{2n}(y) = -\frac{4}{\pi}\frac{1}{4n^2 + 1}\sinh 2ny + \frac{4}{\pi}\frac{\sinh 2ny - 2n\sin y}{4n^2 + 1}$$
$$= -\frac{8}{\pi}\frac{n}{4n^2 + 1}\sin y,$$
$$v_{2n+1}(y) = 0.$$

Thus, the solution is given by:

$$u = \left(\frac{\sinh y + \sinh(\pi - y)}{\sinh \pi} - \frac{\sin y}{2} \right) \sin x - \frac{8 \sin y}{\pi} \sum_{n=1}^{\infty} \frac{n}{4n^2 + 1} \sin 2nx.$$

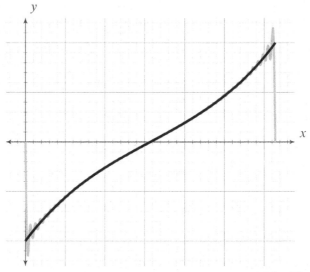

5th truncation of the sine series of $\dfrac{\sinh\left(x - \frac{\pi}{2}\right)}{\sinh\left(\frac{\pi}{2}\right)}$

Recalling our example e^x as a case study from the beginning of Chapter 4 on Fourier Analysis, we can observe that for the Fourier series mentioned earlier, the following relationship holds:

$$\frac{\sinh\left(x - \frac{\pi}{2}\right)}{\sinh\left(\frac{\pi}{2}\right)} = -\frac{8 \sin y}{\pi} \sum_{n=1}^{\infty} \frac{n}{4n^2 + 1} \sin 2nx$$

Finally, the solution is

$$u = \left(\frac{\sinh y + \sinh(\pi - y)}{\sinh \pi} - \frac{\sin y}{2} \right) \sin x + \sin y \frac{\sinh\left(x - \frac{\pi}{2}\right)}{\sinh\left(\frac{\pi}{2}\right)}.$$

Electric potential density plot

(3) Finally, let's consider the study of the temperature distribution $u(x, y, t)$ on a square plate with dimensions $[0, 1] \times [0, 1]$. The temperature distribution satisfies the following conditions:

$$
\left\{
\begin{array}{l}
u_t = u_{xx} + u_{yy}, \quad (x, y) \in [0, 1] \times [0, 1], \quad t > 0, \\
\left\{
\begin{array}{l}
u\big|_{t=0} = xy, \\
\left\{
\begin{array}{l}
u\big|_{x=0} = \alpha y t, \\
u\big|_{x=1} = 0, \\
u\big|_{y=0} = \beta x t, \\
u\big|_{y=1} = 0,
\end{array}
\right.
\end{array}
\right.
\end{array}
\right.
$$

with $\alpha, \beta \in \mathbb{R}$, the operator $L = -\Delta$, in terms of the Laplacian Δ, and the associated homogeneous boundary value problem being of Dirichlet type, are provided with a complete set of eigenfunctions:

$$
w_{\boldsymbol{m}} = \sin(m_1 \pi x) \sin(m_2 \pi y), \quad \boldsymbol{m} = (m_1, m_2) \in \mathbb{N}^2,
$$

corresponding to the following eigenvalues:

$$
\lambda_{\boldsymbol{m}} = \pi^2 (m_1^2 + m_2^2), \quad m_1, m_2 \in \mathbb{N}.
$$

We observe that $\|w_m\|^2 = \frac{1}{4}$. The initial condition can be expressed as a two-dimensional sine Fourier series:

$$xy = 4 \sum_{m_1,m_2=1}^{\infty} \frac{(-1)^{m_1+m_2}}{\pi^2 m_1 m_2} \sin(m_1 \pi x)\sin(m_2 \pi y),$$

where Euler's formula for the Fourier coefficients is employed:

$$xy = \sum_{m_1,m_2=1}^{\infty} c_m w_m, \quad c_m = \int_{[0,1]^2} xy \frac{w_m(x,y)}{\|w_m\|^2} dxdy.$$

Let us now address the new inhomogeneities that arise in the two boundary conditions of Dirichlet type when the inhomogeneity is taken into account. Thus, our task is to compute:

$$I_m(h) = -\int_{\Gamma} h \frac{\partial w_m}{\partial n} dl,$$

where Γ is the boundary of the unit square, and we traverse it anticlockwise. The inhomogeneities are only nonzero on two segments of this boundary: Γ_1, which corresponds to $y = 0$, and Γ_4, which corresponds to $x = 0$. The unit normal vectors are $n = -j$ on Γ_1 and $n = -i$ on Γ_4. Consequently, since

$$\nabla w_m = (m_1 \pi \cos(m_1 \pi x)\sin(m_2 \pi y), m_2 \pi \sin(m_1 \pi x)\cos(m_2 \pi y)),$$

we have:

$$\frac{\partial w_m}{\partial n} = \begin{cases} -m_2 \pi \sin(m_1 \pi x), & \text{on } \Gamma_1, \\ -m_1 \pi \sin(m_2 \pi y), & \text{on } \Gamma_4. \end{cases}$$

Hence,

$$I_m = \beta \int_0^1 m_2 \pi \sin(m_1 \pi x) xt dx + \alpha \int_0^1 m_1 \pi \sin(m_2 \pi y) yt dy$$
$$= -\left((-1)^{m_1} \frac{m_2}{m_1} \beta + (-1)^{m_2} \frac{m_1}{m_2} \alpha\right)t.$$

The ODE satisfied by v_m is

$$\begin{cases} \frac{dv_m}{dt} + \pi^2(m_1^2+m_2^2)v_m = -4\left((-1)^{m_1}\frac{m_2}{m_1}\beta + (-1)^{m_2}\frac{m_1}{m_2}\alpha\right)t, \\ v_m(0) = 4\frac{(-1)^{m_1+m_2}}{\pi^2 m_1 m_2}, \end{cases}$$

whose solution is

$$v_m(t) = 4\left((-1)^{m_1}\frac{m_2}{m_1}\beta + (-1)^{m_2}\frac{m_1}{m_2}\alpha\right)\frac{1-\pi^2(m_1^2+m_2^2)t}{\pi^4(m_1^2+m_2^2)^2}$$
$$+ 4\left(\frac{(-1)^{m_1+m_2}}{\pi^2 m_1 m_2} - \left((-1)^{m_1}\frac{m_2}{m_1}\beta + (-1)^{m_2}\frac{m_1}{m_2}\alpha\right)\frac{1}{\pi^4(m_1^2+m_2^2)^2}\right)$$

$$\times \exp\left(-\pi^2(m_1^2 + m_2^2)t\right).$$

Finally, the temperature $u(t, x, y)$ of the plate will be

$$u = 4 \sum_{m_1,m_2=1}^{\infty} \left(\left((-1)^{m_1} \frac{m_2}{m_1} \beta + (-1)^{m_2} \frac{m_1}{m_2} \alpha \right) \frac{1 - \pi^2(m_1^2 + m_2^2)t}{\pi^4(m_1^2 + m_2^2)^2} \right.$$
$$+ \left(\frac{(-1)^{m_1+m_2}}{\pi^2 m_1 m_2} - \left((-1)^{m_1} \frac{m_2}{m_1} \beta + (-1)^{m_2} \frac{m_1}{m_2} \alpha \right) \frac{1}{\pi^4(m_1^2 + m_2^2)^2} \right)$$
$$\left. \times \exp\left(-\pi^2(m_1^2 + m_2^2)t\right) \right) \sin(m_1 \pi x) \sin(m_2 \pi y).$$

§6.4.   Unbounded Domains and Fourier Transform

⟨—◦◦—⟩

UNBOUNDED domain problems can be addressed in the EEM by utilizing the Fourier transform. Specifically, when dealing with two independent variables (x_0, x_1), the Fourier transform is applied concerning the variable x_1 under the following circumstances:

(1) The PDE involves constant coefficients.
(2) The variable x_1 varies over the entire real line

$$-\infty < x_1 < \infty.$$

(3) There are no boundary conditions with respect to the variable x_1 or the conditions are of the form

$$\lim_{x_1 \to \pm\infty} u = 0.$$

In such situations, the inhomogeneous terms of the problem are expanded using their Fourier transform, and a similar approach is employed to seek an expansion for the unknown solution. This scenario commonly arises in problems with initial conditions that depend on both a time variable and a spatial variable $(x_0, x_1) = (t, x)$, which belong to unbounded domains.

We now present three boxes detailing the strategy to follow for a two-dimensional case where the partial differential equation is linear with constant coefficients, having order N in time variable t and an order of M in space variable x. The objective is to solve this equation subject to N initial conditions.

First, we elucidate how the Fourier transform, interpreted as an eigenfunction expansion, leads to a solution expressed in terms of certain functions determined through inverse Fourier transformations.

To solve it we use the Fourier transformation in the way we now explain. Let us proceed according to the strategy detailed in the following box:

Fourier Transform Strategy

Let's consider the problem:

(6.13)
$$
\begin{cases}
\dfrac{\partial^N u}{\partial t^N} + \displaystyle\sum_{n=0}^{M} a_n \dfrac{\partial^n u}{\partial x^n} = f, \quad x \in \mathbb{R}, \quad t > 0, \\[2mm]
\left. u \right|_{t=0} = g_1(x), \\[2mm]
\left. \dfrac{\partial u}{\partial t} \right|_{t=0} = g_2(x), \\
\qquad \vdots \\
\left. \dfrac{\partial^{N-1} u}{\partial t^{N-1}} \right|_{t=0} = g_N(x).
\end{cases}
$$

We *expand* the inhomogeneous terms in the following manner:

$$
g_n(x) = \frac{1}{\sqrt{2\pi}} \int_{-\infty}^{\infty} c_n(k) e^{ikx} dk, \quad f(t,x) = \frac{1}{\sqrt{2\pi}} \int_{-\infty}^{\infty} c(t,k) e^{ikx} dk,
$$

where c_n and c are the Fourier transforms of g_n and f, respectively:

$$
c_n(k) = \frac{1}{\sqrt{2\pi}} \int_{-\infty}^{\infty} g_n(x) e^{-ikx} dx = \mathcal{F}(g_n),
$$

$$
c(t,k) = \frac{1}{\sqrt{2\pi}} \int_{-\infty}^{\infty} f(t,x) e^{-ikx} dx = \mathcal{F}(f(t,\cdot)).
$$

Next, we look for a solution with the same form:

$$
u(t,x) = \frac{1}{\sqrt{2\pi}} \int_{-\infty}^{\infty} v(t,k) e^{ikx} dk,
$$

where v is the Fourier transform of u with respect to the variable x:

$$
v(t,k) = \frac{1}{\sqrt{2\pi}} \int_{-\infty}^{\infty} u(t,x) e^{-ikx} dx = \mathcal{F}(u(t,\cdot)).
$$

Substituting this expression in the equations of the problem (6.13) and identifying coefficients in the exponentials e^{ikx}, we find the following N-th order ODE for $v(t,k)$ with respect to t, with N initial conditions

(6.14)
$$
\begin{cases}
\dfrac{\partial^N v(t,k)}{\partial t^N} + \displaystyle\sum_{n=0}^{M} a_n (ik)^n v(t,k) = c(t,k), \quad t > 0, \\[2mm]
\left. v \right|_{t=0} = c_1(k), \\[2mm]
\left. \dfrac{\partial v}{\partial t} \right|_{t=0} = c_2(k), \\
\qquad \vdots \\
\left. \dfrac{\partial^{N-1} v}{\partial t^{N-1}} \right|_{t=0} = c_N(k).
\end{cases}
$$

We are presented with the challenge of solving the initial value problem for an N-th order ordinary differential equation, as detailed in Equation (6.14). While the homogeneous counterpart of this problem offers a straightforward solution, the presence of multiple nonhomogeneous terms introduces complexity that renders the solution more difficult to get.

Fourier Transform Solution

The solution to this problem takes the form:

$$v(t,k) = v_0(t,k) + \sum_{n=1}^{N} e_n(t,k)\, c_n(k),$$

where $v_0(t,k)$ is the solution of the PDE (6.14) with homogeneous initial conditions, and $e_n(t,k)$ is a solution of the PDE obtained by eliminating the inhomogeneous terms in (6.14), with the additional condition that its n-th time derivative, evaluated at $t = 0$, is equal to 1, and all other derivatives are taken as 0. With this solution, we have:

$$u(t,x) = \frac{1}{\sqrt{2\pi}} \int_{-\infty}^{\infty} v(t,k) e^{ikx}\, dk$$

$$= u_0(t,x) + \sum_{n=1}^{N} \frac{1}{\sqrt{2\pi}} \int_{-\infty}^{\infty} e_n(t,k)\, c_n(k) e^{ikx}\, dk,$$

where

$$u_0(t,x) = \frac{1}{\sqrt{2\pi}} \int_{-\infty}^{\infty} v_0(t,k) e^{ikx}\, dk = \mathcal{F}^{-1}(v_0(t,\cdot)),$$

is the inverse Fourier transform of $v_0(t,k)$. If we know the inverse Fourier transform of the functions $e_n(t,k)$ given by:

$$G_n(t,x) = \frac{1}{\sqrt{2\pi}} \int_{-\infty}^{\infty} e_n(t,k) e^{ikx}\, dk = \mathcal{F}^{-1}(e_n(t,\cdot)),$$

then

$$u(t,x) = u_0(t,x) + \sum_{n=1}^{N} \mathcal{F}^{-1}\Big(\mathcal{F}(G_n) \cdot \mathcal{F}(g_n)\Big)$$

$$= u_0(t,x) + \frac{1}{\sqrt{2\pi}} \sum_{n=1}^{N} G_n(t,x) * g_n(x),$$

and we can express the solution (6.13) in the following form:

$$(6.15) \quad u(t,x) = u_0(t,x) + \sum_{n=1}^{N} \frac{1}{\sqrt{2\pi}} \int_{-\infty}^{\infty} G_n(t,x-y)\, g_n(y)\, dy.$$

Observations:

(1) The method that we have just described really follows the EEM strategy because what we do basically is to *expand* with respect to a set of eigenfunctions

$$\{w_k(x) = e^{ikx}\}_{k \in \mathbb{R}},$$

now a continuous set, of the operator with constant coefficients

$$B = \sum_{n=1}^{M} a_n \frac{\partial^n u}{\partial x^n}.$$

(2) Equation (6.15) is a Green's function representation of the solution.

Examples:

(1) We apply these ideas to the **heat equation**

$$\begin{cases} \dfrac{\partial u}{\partial t} = a^2 \dfrac{\partial^2 u}{\partial x^2}, \\ u\big|_{t=0} = f(x), \end{cases}$$

with $t > 0$ and $-\infty < x < \infty$. We begin by expanding the initial condition $f(x)$ as follows:

$$f(x) = \frac{1}{\sqrt{2\pi}} \int_{-\infty}^{\infty} c(k) e^{ikx} dk, \quad c(k) = \mathcal{F}(f),$$

and then seek a solution in the form:

$$u(t,x) = \frac{1}{\sqrt{2\pi}} \int_{-\infty}^{\infty} v(t,k) e^{ikx} dk, \quad v(t,k) = \mathcal{F}(u(t,\cdot)).$$

Substituting this into the equation and identifying coefficients with respect to the exponentials e^{ikx}, we arrive at the following problem:

$$\begin{cases} \dfrac{\partial v(t,k)}{\partial t} = -a^2 k^2 v(t,k), \\ v\big|_{t=0} = c(k). \end{cases}$$

The solution to this problem is:

$$v(t,k) = e(t,k)c(k),$$

with $e(t,k) := e^{-a^2 t k^2}$. And, since we know the inverse Fourier transform of $e_n(t,k)$; i.e.,

$$\mathcal{F}^{-1}\left(e^{-a^2 t k^2}\right) = \sqrt{\frac{1}{2ta^2}} \exp\left(-\frac{x^2}{4a^2 t}\right),$$

we can express the solution as follows:

$$u(t,x) = \frac{1}{2a\sqrt{\pi t}} \int_{-\infty}^{\infty} \exp\left(-\frac{(x-y)^2}{4a^2 t}\right) f(y)\,\mathrm{d}y.$$

This result can be generalized to the multidimensional case:

$$\begin{cases} u_t = a^2 \Delta u, & t > 0, \quad x = (x_1, \ldots, x_n) \in \mathbb{R}^n, \\ u\big|_{t=0} = f(x). \end{cases}$$

In this case, the EEM leads to the solution:

$$u(t,x) = \frac{1}{(2a\sqrt{\pi t})^n} \int_{\mathbb{R}^n} \exp\left(-\frac{\|x-y\|^2}{4a^2 t}\right) f(y)\,\mathrm{d}^n y.$$

This a Green function representation of the solution. Recall that

$$\|x\| := +\sqrt{x_1^2 + \cdots + x_n^2}.$$

(2) We apply the Fourier transform to the free Schrödinger equation. The previously derived results for the heat equation can be extended to obtain analogous formulas for solutions of the free Schrödinger equation in any dimension:

$$\begin{cases} i\hbar u_t = -\frac{\hbar^2}{2m}\Delta u, & t > 0, \quad x \in \mathbb{R}^n, \\ u|_{t=0} = f(x). \end{cases}$$

Note that this initial condition problem for the Schrödinger equation is obtained from the heat equation by setting:

$$a^2 = i\frac{\hbar}{2m}.$$

Therefore, making that identification of parameter a, we deduce the following expression of the solution

$$u(t,x) = \left(\frac{m}{2\pi i\hbar t}\right)^{\frac{n}{2}} \int_{\mathbb{R}^n} \exp\left(i\frac{m\|x-y\|^2}{2\hbar t}\right) f(y)\,\mathrm{d}^n y.$$

This a Green's function representation of the solution.

§6.5.   Exercises

§6.5.1.   Exercises with Solutions

(1) For the one-dimensional quantum harmonic oscillator, we are tasked with solving the following initial and boundary value problem:

$$\begin{cases} i\hbar u_t = -\dfrac{\hbar^2}{2m}u_{xx} + \dfrac{1}{2}kx^2u, & -\infty < x < \infty, t > 0, \\ \left\{ \begin{array}{l} u|_{t=0} = g(x), \\ \lim\limits_{x\to\pm\infty} u = 0. \end{array} \right. \end{cases}$$

Solution: The eigenvalue equation for this problem is given by:

$$-\frac{\hbar^2}{2m}w_{xx} + \frac{1}{2}kx^2w = \lambda w, -\infty < x < \infty.$$

This equation is known as the Weber equation, who was the first person to study it in 1869. By introducing the variables $K := \frac{mk}{\hbar}$, $X := \sqrt[4]{K}x$, and $\Lambda := \frac{2m}{\hbar\sqrt{K}}\lambda$, we can rewrite the eigenvalue problem as the well-known Hermite equation:

$$-w_{XX} + X^2w = \Lambda w,$$

which we will explore in detail in the next chapter. The boundary conditions at infinity lead to the eigenvalues:

$$\Lambda_n = 2n + 1, \quad n = 0, 1, 2, \ldots,$$

with corresponding eigenfunctions:

$$w_n(X) = H_n(X)e^{-\frac{X^2}{2}},$$

where $H_n(X), n = 0, 1, \ldots$, are the Hermite polynomials forming an orthogonal basis $\{H_n(x)\}_{n=0}^{\infty}$ in $L^2(\mathbb{R}, e^{-x^2}dx)$. The eigenvalues are given by:

$$\boxed{\lambda_n = \hbar\sqrt{\frac{k}{2m}}(2n + 1)}$$

and the corresponding eigenfunctions, known as Hermite or Hermite-Weber functions, are:

$$\boxed{w_n(x) = H_n\left(\sqrt{\frac{\sqrt{mk}}{\hbar}}x\right)\exp\left(-\frac{\sqrt{mk}}{2\hbar}x^2\right),}$$

for $n \in \mathbb{N}_0$. Therefore, if $g(x) \in L^2(\mathbb{R}, dx)$, we can express it as an expansion:

$$\boxed{g(x) = \sum_{n=0}^{\infty} c_n w_n(x)}$$

and the solution to the initial and boundary value problem will be:

$$u(t,x) = \sum_{n=0}^{\infty} c_n \exp\left(-i\frac{\lambda_n}{\hbar}t\right) w_n(x).$$

If we set $\sqrt{mk}\hbar = 1$, then we obtain the functions known as Weber or cylindrical-parabolic functions:

$$\psi_n(x) = \left(2^n n! \sqrt{\pi}\right)^{-\frac{1}{2}} \exp\left(-\frac{x^2}{2}\right) H_n(x),$$

which form an orthonormal basis for the Hilbert space $L^2(\mathbb{R}, dx)$.

(2) Using the Fourier transform, solve the following initial value problem for the wave equation on the real line:

$$\begin{cases} u_{tt} = u_{xx}, & x \in \mathbb{R}, \quad t > 0, \\ u|_{t=0} = f(x), \\ u_t|_{t=0} = g(x). \end{cases}$$

Explain the connection between this method and the d'Alembert formula.

Solution: We can utilize the eigenfunction $\{e^{ikx}\}_{k \in \mathbb{R}}$ to solve the spectral problem:

$$w_{xx} = -k^2 w, \quad x \in \mathbb{R}.$$

This leads us to the following expression for the solution:

$$u(t,x) = \frac{1}{\sqrt{2\pi}} \int_{\mathbb{R}} v(t,k) e^{ikx} dk.$$

Substituting this expression into the wave equation, we get:

$$\int_{\mathbb{R}} (v_{tt}(k,t) + k^2 v(k,t)) e^{ikx} dk = 0.$$

For this reason, we impose the following equation:

$$v_{tt}(k,t) + k^2 v(k,t) = 0,$$

whose general solution is:

$$v(k,t) = A(k) e^{ikt} + B(k) e^{-ikt}.$$

Consequently, the EEM leads to a solution of the form:

$$u(t,x) = \frac{1}{\sqrt{2\pi}} \int_{\mathbb{R}} \left(A(k) e^{ik(x+t)} + B(k) e^{ik(x-t)}\right) dk,$$

$$u_t(t,x) = \frac{1}{\sqrt{2\pi}} \int_{\mathbb{R}} ik\left(A(k) e^{ik(x+t)} - B(k) e^{ik(x-t)}\right) dk,$$

which, at $t = 0$, reduces to:

$$f(x) = \frac{1}{\sqrt{2\pi}} \int_{\mathbb{R}} (A(k) + B(k)) e^{ikx} dk,$$

$$g(x) = \frac{1}{\sqrt{2\pi}} \int_{\mathbb{R}} ik\big(A(k) - B(k)\big)e^{ikx}dk.$$

To find the Fourier transforms of the initial data, $\hat{f}(k)$ and $\hat{g}(k)$, we obtain:

$$\hat{f}(k) = A(k) + B(k), \quad \hat{g}(k) = ik\big(A(k) - B(k)\big),$$

which can be solved immediately:

$$A(k) = \frac{1}{2}\left(\hat{f}(k) + \frac{\hat{g}(k)}{ik}\right), \quad B(k) = \frac{1}{2}\left(\hat{f}(k) - \frac{\hat{g}(k)}{ik}\right).$$

Finally, we can express the solution in terms of the Fourier transforms of the initial data as follows:

$$u(t,x) = \frac{1}{2}\left(\frac{1}{\sqrt{2\pi}} \int_{\mathbb{R}} \hat{f}(k)e^{ik(x+t)}dk + \frac{1}{\sqrt{2\pi}} \int_{\mathbb{R}} \hat{f}(k)e^{ik(x-t)}dk\right)$$
$$+ \frac{1}{2}\left(\frac{1}{\sqrt{2\pi}} \int_{\mathbb{R}} \hat{g}(k)\frac{e^{ik(x+t)}}{ik}dk - \frac{1}{\sqrt{2\pi}} \int_{\mathbb{R}} \hat{g}(k)\frac{e^{ik(x-t)}}{ik}dk\right).$$

Using the formula $\frac{e^{ik(x\pm t)}}{ik} = \frac{1}{ik} + \int_0^{x\pm t} e^{iks}ds$, we can further simplify this expression to:

$$u(t,x) = \frac{1}{2}\left(\frac{1}{\sqrt{2\pi}} \int_{\mathbb{R}} \hat{f}(k)e^{ik(x+t)}dk + \int_{\mathbb{R}} \hat{f}(k)e^{ik(x-t)}dk\right)$$
$$+ \frac{1}{2}\int_{x-t}^{x+t} \frac{1}{\sqrt{2\pi}} \int_{\mathbb{R}} \hat{g}(k)e^{iks}dk.$$

Finally, using the inverse Fourier transform, this expression leads to the well-known d'Alembert formula:

$$u(t,x) = \frac{f(x+t) + f(x-t)}{2} + \frac{1}{2}\int_{x-t}^{x+t} g(s)ds.$$

(3) Using the EEM, solve the problem of initial and boundary conditions for the heat equation in an infinite plate of width L as follows:

$$\begin{cases} u_t = u_{xx} + u_{yy}, & (x,y) \in \mathbb{R} \times (0,L), \quad t > 0, \\ u|_{t=0} = g, \\ \quad \begin{cases} u|_{y=0} = 0, \\ u|_{y=L} = 0, \end{cases} \end{cases}$$

with

$$g = \begin{cases} \sin\frac{\pi y}{L}, & x \in [-1,1], \\ 0, & x \notin [-1,1]. \end{cases}$$

Solution: The associated spectral problem is

$$\begin{cases} -(w_{xx} + w_{yy}) = \lambda w, \quad \lambda = k_1^2 + k_2^2, \\ \begin{cases} w|_{y=0} = 0, \\ w|_{y=L} = 0. \end{cases} \end{cases}$$

Which, in turn, is a homogeneous boundary value problem, for the two-dimensional Helmholtz equation, in which we can separate variables into Cartesian coordinates. Thus, the solutions are written as

$$w(x, y) = \left(a_1 e^{ik_1 x} + b_1 e^{-ik_1 x}\right)\left(a_2 e^{ik_2 y} + b_2 e^{-ik_2 y}\right).$$

The boundary conditions give

$$\begin{bmatrix} 1 & 1 \\ e^{ik_2 L} & e^{-ik_2 L} \end{bmatrix} \begin{bmatrix} a_2 \\ b_2 \end{bmatrix} = \begin{bmatrix} 0 \\ 0 \end{bmatrix}.$$

Nonzero solutions of this system exist as long as $k_2 = n\dfrac{\pi}{L}$, $n \in \mathbb{N}$. Therefore, taking $k = k_1$, the eigenvalues are

$$\left\{ k^2 + n^2 \frac{\pi^2}{L^2} \right\}_{k \in \mathbb{R}, n \in \mathbb{N}}$$

with corresponding eigenfunction given by

$$\left\{ \frac{1}{\sqrt{2\pi}} e^{ikx} \sin n\pi y \right\}_{k \in \mathbb{R}, n \in \mathbb{N}}.$$

Note that there is a continuous set of eigenvalues, since we have no boundary conditions in the variable x. The solution will be expressed according to the EEM as follows

$$u(t, x, y) = \sum_{n=1}^{\infty} \frac{1}{\sqrt{2\pi}} \int_{\mathbb{R}} v_n(k, t) e^{ikx} \sin \frac{n\pi y}{L} dk.$$

Inserting this expression in the PDE delivers the following ODE for $v_n(k, t)$

$$\frac{\partial v_n}{\partial t}(t, k) + \left(k^2 + \frac{n^2 \pi^2}{L^2} \right) v_n(t, k) = 0.$$

whose solution is

$$v_n(k, t) = a_n(k) \exp\left(-\left(k^2 + \frac{n^2 \pi^2}{L^2} \right) t \right).$$

To find out which eigenfunctions are involved in this expansion, we analyze the corresponding expansions of the inhomogeneities. In this case, it is only the function $g(x, y)$, whose form indicates that can be expressed in terms of $\left\{ \frac{1}{\sqrt{2\pi}} e^{ikx} \sin \pi y \right\}_{k \in \mathbb{R}}$ only. Hence

$$g(x, y) = \left(\frac{1}{\sqrt{2\pi}} \int_{\mathbb{R}} c(k) e^{ikx} dk \right) \sin \frac{\pi y}{L},$$

$$c(k) = \frac{1}{\sqrt{2\pi}} \int_{-1}^{1} e^{-ikx} dx = \sqrt{\frac{2}{\pi}} \frac{\sin k}{k}.$$

This is,

$$g(x, y) = \left(\int_{\mathbb{R}} \frac{\sin k}{\pi k} e^{ikx} dk \right) \sin \frac{\pi y}{L},$$

that leads to

$$\boxed{\begin{aligned} u(t, x, y) &= e^{-\frac{\pi^2}{L^2} t} I(t, x) \sin \frac{\pi}{L} y, \\ I(t, x) &:= \int_{\mathbb{R}} e^{-k^2 t + ikx} \frac{\sin k}{\pi k} dk. \end{aligned}}$$

We now evaluate $I(t, x)$ that we write in the form

$$I(t, x) = \frac{1}{2\pi} \int_{\mathbb{R}} e^{-k^2 t} \left(\frac{e^{ik(x+1)} - e^{ik(x-1)}}{ik} \right) dk,$$

and recalling that

$$\frac{e^{ik(x\pm 1)}}{ik} = \frac{1}{ik} + \int_{0}^{x\pm 1} e^{iks} ds$$

it is expressed as follows

$$I(t, x) = \frac{1}{\sqrt{2\pi}} \int_{x-1}^{x+1} \left(\frac{1}{\sqrt{2\pi}} \int_{\mathbb{R}} e^{-k^2 t} e^{iks} dk \right) ds.$$

Now, as we have already calculated, the inverse FT of a Gaussian is another Gaussian; i.e.,

$$\frac{1}{\sqrt{2\pi}} \int_{\mathbb{R}} e^{-k^2 t} e^{iks} dk = \frac{1}{\sqrt{2t}} e^{-\frac{s^2}{4t}}$$

which implies

$$I(t, x) = \frac{1}{\sqrt{\pi}} \int_{x-1}^{x+1} \frac{1}{2\sqrt{t}} \exp\left(-\frac{s^2}{4t} \right) ds \quad \text{(change of variable } \frac{s}{2\sqrt{t}} = \zeta)$$

$$= \frac{1}{\sqrt{\pi}} \int_{\frac{x-1}{2\sqrt{t}}}^{\frac{x+1}{2\sqrt{t}}} e^{-\zeta^2} d\zeta.$$

In terms of the error function

$$\boxed{\operatorname{erf}(x) := \frac{2}{\sqrt{\pi}} \int_{0}^{x} e^{-\zeta^2} d\zeta,}$$

we can express it as follows

$$I(t, x) = \frac{1}{2}\left(\operatorname{erf}\left(\frac{x+1}{2\sqrt{t}} \right) - \operatorname{erf}\left(\frac{x-1}{2\sqrt{t}} \right) \right).$$

In short, we have proven that the solution to the problem is

$$u(t,x,y) = \frac{1}{2}\exp\left(-\frac{\pi^2}{L^2}t\right)\left(\operatorname{erf}\left(\frac{x+1}{2\sqrt{t}}\right) - \operatorname{erf}\left(\frac{x-1}{2\sqrt{t}}\right)\right)\sin\frac{\pi}{L}y.$$

Notice that for $x > 0$ the error function has the following asymptotic behavior

$$\operatorname{erf} x = 1 + \frac{1}{\sqrt{\pi}x}e^{-x^2}\left(1 - O\left(\frac{1}{x^2}\right)\right), \qquad x \to \infty,$$

and also that it is an odd function

$$\operatorname{erf} x = -\operatorname{erf}(-x).$$

Therefore,

$$\lim_{t\to 0+}\frac{1}{2}\left(\operatorname{erf}\left(\frac{x+1}{2\sqrt{t}}\right) - \operatorname{erf}\left(\frac{x-1}{2\sqrt{t}}\right)\right) = \begin{cases} 1, & x \in [-1,1], \\ 0, & x \notin [-1,1], \end{cases}$$

and the initial condition is recovered $\lim_{t\to 0+} u(t,x,y) = g(x,y)$.

(4) Using the Fourier transform, let's find the solution to the following initial value problem for the heat equation on the straight line:

$$\begin{cases} u_t = u_{xx}, & x \in \mathbb{R}, \quad t > 0, \\ u|_{t=0} = g(x) := xe^{-|x|}. \end{cases}$$

Solution: Using the Fourier transform, we obtain the following expression for the solution:

$$u(t,x) = \frac{1}{\sqrt{2\pi}}\int_{\mathbb{R}} c(k)e^{-k^2 t + ikx}dk,$$

where $c(k)$ is the Fourier transform of the initial data $g(x)$ given by:

$$c(k) = \mathcal{F}(g) = iD_k\mathcal{F}\left(e^{-|x|}\right)$$

$$= \frac{1}{\sqrt{2\pi}}iD_k\left(\int_{-\infty}^{0} e^{(1-ik)x}dx + \int_{0}^{\infty} e^{-(1+ik)x}dx\right)$$

$$= \frac{1}{\sqrt{2\pi}}iD_k\left(\frac{1}{1-ik} + \frac{1}{1+ik}\right) = -\frac{1}{\sqrt{2\pi}}\frac{4ik}{(1+k^2)^2}$$

Thus, the solution can be written as:

$$u(t,x) = -\frac{2i}{\pi}\int_{\mathbb{R}} \frac{k}{(1+k^2)^2}e^{-k^2 t + ikx}dk.$$

However, this integral is difficult to evaluate directly. So, we adopt an alternative strategy, based on Green's function, to express the solution in a different form:

$$u(t, x) = \frac{1}{2\sqrt{\pi t}} \int_{\mathbb{R}} s e^{-|s|} \exp\left(-\frac{(x-s)^2}{4t}\right) ds.$$

Now, we analyze this integral by splitting it into two parts:

$$u(t, x) = F(t, x) - F(t, -x),$$

$$F(t, x) := \frac{1}{2\sqrt{\pi t}} \int_0^\infty s \exp\left(-s - \frac{(x-s)^2}{4t}\right) ds.$$

Using some relationships, we can express $F(t, x)$ as:

$$F(t, x) = \frac{x - 2t}{2} e^{t-x} \operatorname{erfc}\left(-\frac{x - 2t}{2\sqrt{t}}\right),$$

where $\operatorname{erfc}(x) := 1 - \operatorname{erf}(x)$ is the complementary error function. The final expression for the solution is then:

$$u(t, x) = \frac{x - 2t}{2} e^{t-x} \operatorname{erfc}\left(-\frac{x - 2t}{2\sqrt{t}}\right) + \frac{x + 2t}{2} e^{t+x} \operatorname{erfc}\left(\frac{x + 2t}{2\sqrt{t}}\right).$$

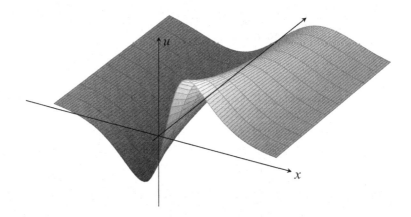

Temperature $u(x, t)$ in the infinite trend with times $0 < t < 1.8$

As t approaches zero from the positive side, we find that converges to the initial temperature data

$$\lim_{t \to 0^+} u(t, x) = g(x) = x e^{-|x|}.$$

(5) Solve, by EEM, for the vibrating string the following boundary and initial value problem

$$\begin{cases} u_{tt} = u_{xx}, \quad x \in \mathbb{R}, \quad t > 0, \\ \begin{cases} \begin{cases} u|_{t=0} = \sin \pi x, \\ u_t|_{t=0} = |x|, \\ u|_{x=-1} = u|_{x=1}, \\ u_x|_{x=-1} = u_x|_{x=1}. \end{cases} \end{cases} \end{cases}$$

Solution: In order to apply the EEM, we must solve the following eigenvalue problem

$$\begin{cases} -w_{xx} = \lambda w, \quad x \in [-1, 1], \\ \begin{cases} w|_{x=-1} = w|_{x=1}, \\ w_x|_{x=-1} = w_x|_{x=1}. \end{cases} \end{cases}$$

The corresponding eigenvalues are $\lambda_n = n^2 \pi^2$ with associated eigenfunctions given by

$$\{1, \cos n\pi x, \sin n\pi x\}_{n=1}^{\infty}.$$

The solution is

$$u(t, x) = v_0(t) + \sum_{n=1}^{\infty} \left(v_n(t) \cos n\pi x + \tilde{v}_n(t) \sin n\pi x \right).$$

By introducing this expression into the wave equation and decoupling the different modes, we get

$$v_{0,tt} = 0,$$
$$v_{n,tt} + n^2 \pi^2 v_n = 0,$$
$$\tilde{v}_{n,tt} + n^2 \pi^2 \tilde{v}_n = 0,$$

that leads to

$$\begin{cases} v_0(t) = A_0 + B_0 t, \\ v_n(t) = A_n \cos n\pi t + B_n \sin n\pi t, \quad n \in \mathbb{N}, \\ \tilde{v}_n(t) = \tilde{A}_n \cos n\pi t + \tilde{B}_n \sin n\pi t, \quad n \in \mathbb{N}. \end{cases}$$

We must now expand the initial conditions. As $u|_{t=0} = f(x) := \sin \pi x$ is already expanded, it is enough to calculate the Fourier series of $u_t|_{t=0} = g(x) := |x|$. Being the function $g(x)$ even in $[-1, 1]$ its expansion is reduced to a Fourier series in cosines. The corresponding coefficients will be

$$a_0 = 2 \int_0^1 x \, dx = 1,$$

$$a_n = 2 \int_0^1 x \cos n\pi x \, dx = \begin{cases} 0, & \text{if } n \text{ is even,} \\ -\dfrac{4}{\pi^2 n^2}, & \text{if } n \text{ is odd.} \end{cases}$$

Then, in the eigenfunction expansion of $u(t, x)$ should only appear

$$1, \sin \pi x, \cos \pi x, \cos 3\pi x, \cos 5\pi x, \dots.$$

Therefore, the only nonzero coefficients are $\tilde{v}_1, v_0, v_1, v_3, v_5, \dots$. Consequently, for $n \in \mathbb{N}_0$, we have

$$
\begin{cases}
A_0 = v_0(0) = 0, \\
A_{2n+1} = v_{2n+1} = 0, \\
\tilde{A}_1 = \tilde{v}_1(0) = 1, \\
\end{cases}
$$

$$
\begin{cases}
B_0 = v_0'(0) = \frac{1}{2}, \\
(2n+1)\pi B_{2n+1} = v_{2n+1}'(0) = -\frac{4}{\pi(2n+1)^2}, \\
\pi \tilde{B}_1 = \tilde{v}_1' = 0.
\end{cases}
$$

We conclude that the Fourier series for the sought solution is

$$
u(t, x) = \frac{t}{2} + \cos \pi t \sin \pi x
$$
$$
- \frac{4}{\pi^3} \sum_{n=0}^{\infty} \frac{1}{(2n+1)^3} \sin\big((2n+1)\pi t\big) \cos\big((2n+1)\pi x\big).
$$

This solution can be interpreted as follows. First of all, using the addition formulas of the trigonometric functions, we write it as follows:

$$
u(t, x) = \frac{1}{2}\Big(\sin \pi(x - t) + \sin \pi(x + t) + t
$$
$$
- \frac{4}{\pi^3} \sum_{n=0}^{\infty} \frac{1}{(2n+1)^3} \sin\big((2n+1)\pi(x+t)\big)
$$
$$
+ \frac{4}{\pi^3} \sum_{n=0}^{\infty} \frac{1}{(2n+1)^3} \sin\big((2n+1)\pi(x-t)\big)\Big).
$$

According to Dirichlet's theorem, the Fourier series

$$
\frac{1}{2} - \frac{4}{\pi^2} \sum_{n=0}^{\infty} \frac{1}{(2n+1)^2} \cos(2n+1)\pi x
$$

converges pointwise, for all $x \in \mathbb{R}$ to the periodic extension $g_{\text{per}}(x)$ of

$$
g(x) = |x|, \quad x \in [-1, 1].
$$

The $|x|$ function is continuous and of bounded variation. Indeed, it is the difference of two nondecreasing functions

$$|x| = u_2(x) - u_1(x),$$

where

$$u_1(x) = \begin{cases} x, & 0 < x < 1, \\ 0, & -1 < x \leq 0, \end{cases}$$

$$u_2(x) = \begin{cases} 0, & 0 \leq x < 1, \\ x, & -1 < x < 0. \end{cases}$$

Hence, this Fourier series on sines and cosines converges uniformly. Therefore, it is possible to derive and integrate term-by-term

$$\int_0^x g_{\text{per}}(s)ds = \frac{x}{2} - \frac{4}{\pi^3} \sum_{n=0}^{\infty} \frac{1}{(2n+1)^3} \sin(2n+1)\pi x.$$

Next, we show the graphs of the 12 term truncations of the series of Fourier of $g_{\text{per}}(x)$ and of $\int_0^x g_{\text{per}}(s)ds$

$[|x|]_{\text{per}}$

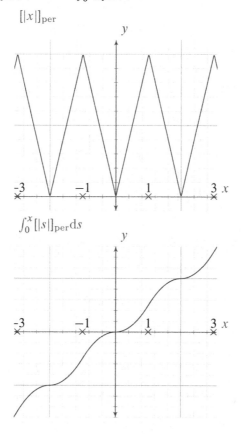

$\int_0^x [|s|]_{\text{per}} ds$

We also show a space-time graph of the solution of the d'Alembert solution and another of the truncation to ten terms of the corresponding Fourier series.

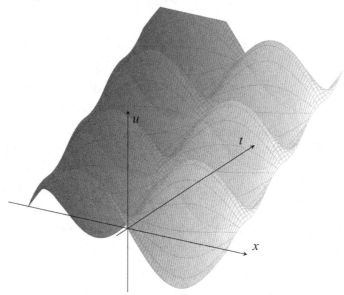

Waves with periodic boundary conditions

(6) Applying the eigenfunction expansion method, we aim to solve the following initial and boundary value problem for the vibrating string:

$$\left\{ \begin{array}{l} u_{tt} = u_{xx}, \quad x \in [0, 2\pi], \quad t > 0, \\ \left\{ \begin{array}{l} \left\{ \begin{array}{l} u|_{t=0} = x(x - 2\pi), \\ u_t|_{t=0} = 0, \\ u|_{x=0} = 0, \\ u|_{x=2\pi} = 0. \end{array} \right. \end{array} \right. \end{array} \right.$$

Solution: To apply the EEM, we first need to find the solution of the eigenvalue problem:

$$\left\{ \begin{array}{l} -w_{xx} = \lambda w, \\ \left\{ \begin{array}{l} w|_{x=0} = 0, \\ w|_{x=2\pi} = 0, \end{array} \right. \end{array} \right.$$

which has eigenvalues $\lambda_n = \frac{n^2}{4}$ with corresponding eigenfunctions $w_n(x) = \sin \frac{nx}{2}$, where $n \in \mathbb{N}$. Next, we expand the initial conditions $u|_{t=0} = f(x) := x(x - 2\pi)$ and $u_t|_{t=0} = g(x) = 0$ in terms of these eigenfunctions.

For the Fourier expansion of $f(x)$, we have:

$$f(x) = \sum_{n=1}^{\infty} b_n \sin \frac{nx}{2},$$

$$b_n := \frac{1}{\pi} \int_0^{2\pi} x(x - 2\pi) \sin \frac{nx}{2} dx = \begin{cases} 0, & \text{if } n \text{ is even,} \\ -\dfrac{32}{\pi n^3}, & \text{if } n \text{ is odd.} \end{cases}$$

The eigenfunction expansion of the solution is:

$$u(t, x) := \sum_{n=1}^{\infty} v_n(t) \sin \frac{nx}{2},$$

where $v_n(t)$ solves:

$$v_n''(t) + \frac{n^2}{4} v_n(t) = 0,$$

$$\begin{cases} v_n(0) = b_n, \\ v_n'(0) = 0. \end{cases}$$

Solving the above equation, we find that:

$$v_n(t) = b_n \cos \frac{nt}{2},$$

and, therefore, the solution is given by:

$$u(t, x) = -\frac{32}{\pi} \sum_{n \text{ is odd}} \frac{1}{n^3} \cos \frac{nt}{2} \sin \frac{nx}{2}.$$

Using trigonometric addition formulas, we can rewrite the solution as:

$$u(t, x) = -\frac{1}{2}\left(\frac{32}{\pi} \sum_{n \text{ is odd}} \frac{1}{n^3} \sin \frac{n(x - t)}{2} + \frac{32}{\pi} \sum_{n \text{ is odd}} \frac{1}{n^3} \sin \frac{n(x + t)}{2} \right).$$

Therefore, by employing the Fourier series expansion of $f_{\text{odd}}(x)$, we obtain the d'Alembert formula:

$$u(t, x) = \frac{f_{\text{odd}}(x - t) + f_{\text{odd}}(x + t)}{2}.$$

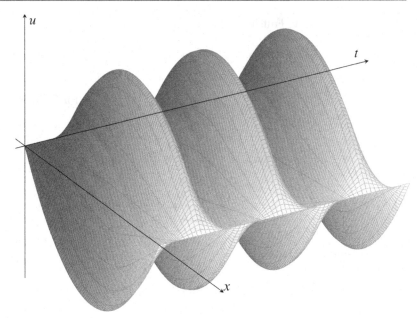

(7) Solve by Fourier transform the initial value problem:

$$\begin{cases} u_t + u_x - u_{xx} = 0, & x \in \mathbb{R}, \quad t > 0, \\ u(x,0) = e^{-x^2}. \end{cases}$$

Solution: The functions e^{ikx} are eigenfunctions of the operator

$$B - \frac{\partial}{\partial x} - \frac{\partial^2}{\partial x^2},$$

with eigenvalues

$$\lambda = ik + k^2.$$

Hence, we seek a solution in the form:

$$u = \int_{\mathbb{R}} c(k) e^{-(ik+k^2)t} e^{ikx} dk,$$

where

$$c(k) = \frac{1}{2\pi} \int_{\mathbb{R}} e^{-x^2} e^{-ikx} dx = \frac{1}{2\sqrt{\pi}} e^{-\frac{k^2}{4}}.$$

Thus, we have:

$$u = \frac{1}{2\sqrt{\pi}} \int_{\mathbb{R}} e^{-k^2(t+\frac{1}{4})} e^{-ik(t-x)} dk,$$

which can be further simplified with the change of variables:

$$X = k, \quad K = t - x, \quad a = \sqrt{t + \frac{1}{4}},$$

to obtain:

$$u = \frac{1}{2\sqrt{\pi}} \int_{\mathbb{R}} e^{-a^2 X^2} e^{-iKX} dX,$$

and this leads to the final solution:

$$\boxed{u(t,x) = \frac{1}{\sqrt{1+4t}} \exp\left(-\frac{(t-x)^2}{1+4t}\right).}$$

This equation is the heat equation to which a convectional or drag term has been added, which accounts for a forced drag, for some reason, of the heat flow. The space-time graph is shown below, where the drag-out together with the typical dissipation of calorific phenomena can be appreciated.

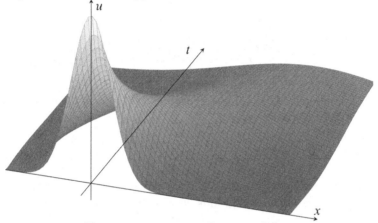

Temperature space-time graph

(8) Using the EEM, determine the solution of the following boundary and initial value problem for the heat equation

$$\left\{ \begin{array}{l} u_t = u_{xx}, \quad x \in \mathbb{R}, \quad t > 0, \\ \left\{ \begin{array}{l} u|_{t=0} = x^2 \sin \pi x, \\ \left\{ \begin{array}{l} u|_{x=-1} = 0, \\ u|_{x=1} = 0. \end{array} \right. \end{array} \right. \end{array} \right.$$

Hint:

$$\int \sin ax \sin bx dx = \left\{ \begin{array}{ll} \dfrac{\sin(a-b)x}{2(a-b)} + \dfrac{\sin(a+b)x}{2(a+b)}, & a \neq b \\ \dfrac{x}{4} - \dfrac{\sin 2ax}{2a}, & a = b. \end{array} \right.$$

Solution: The application of the EEM in this case requires the resolution of the following eigenvalue problem: $-\dfrac{d^2 u}{dx^2} = \lambda u$ with $u(-1) = u(1) = 0$. This Dirichlet problem has eigenvalues

$\lambda_n = \frac{n^2\pi^2}{4}$ with corresponding eigenfunctions $w_n(x) = \sin\frac{n\pi(x+1)}{2}$.
Therefore, we look for the solution $u(x,t)$ in the form of a sine
Fourier series:

$$u(x,t) = \sum_{n=1}^{\infty} v_n(t) \sin\frac{n\pi(x+1)}{2}.$$

Hence, $v_n(t)$ must fulfill

$$v' = -\lambda v,$$

so that

$$v(t) = c\exp\left(-\frac{n^2\pi^2 t}{4}\right)$$

and the expansion is:

$$u(x,t) = \sum_{n=1}^{\infty} c_n \exp\left(-\frac{n^2\pi^2 t}{4}\right) \sin\frac{n\pi(x+1)}{2}.$$

The coefficients c_n are the Fourier coefficients of the initial condition.
Thus, we have:

$$c_n = \int_{-1}^{1} x^2 \sin\pi x \sin\frac{n\pi(x+1)}{2} dx.$$

Expanding the last factor of the integrand and analyzing the parity
of the resulting integrands, we conclude that:

$$c_n = \begin{cases} 0, & n \text{ is odd,} \\ (-1)^m \int_{-1}^{1} x^2 \sin\pi x \sin m\pi x dx, & n = 2m. \end{cases}$$

Integrating twice by parts, with the help of the hint provided, we
conclude that:

$$c_n = \begin{cases} 0, & n \text{ is odd,} \\ -\dfrac{2\pi^2-3}{6\pi^2}, & n = 2, \\ -\dfrac{8m}{(m^2-1)^2\pi^2}, & n = 2m, m > 1. \end{cases}$$

This leads us to the final result:

$$u(t,x) = -\frac{2\pi^2-3}{6\pi^2}e^{-\pi^2 t}\sin\pi(x+1)$$

$$-\sum_{m=2}^{\infty} \frac{8m}{(m^2-1)^2\pi^2}e^{-m^2\pi^2 t}\sin m\pi(x+1).$$

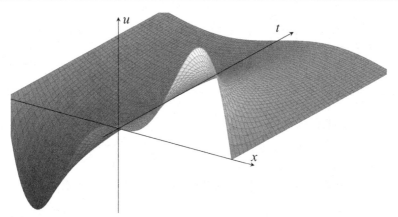

(9) Solve, using Fourier transform, the following initial value problem for the heat equation in the plane:

$$\begin{cases} u_t = u_{xx} + u_{yy}, & (x, y) \in \mathbb{R}^2, \quad t > 0, \\ u|_{t=0} = x^2 e^{-x^2 - y^2}. \end{cases}$$

Hint: $\displaystyle\int_{-\infty}^{\infty} e^{-x^2} e^{-ikx} dx = \sqrt{\pi} e^{-\frac{k^2}{4}}.$

Solution: The solution is expressed in terms of the Fourier transform as follows:

$$u(t, x, y) = \frac{1}{2\pi} \int_{\mathbb{R}^2} v(t, k, q) e^{i(kx+qy)} dk\,dq,$$

where

$$\frac{\partial v}{\partial t} = -(k^2 + q^2)v$$

and, consequently,

$$v(t, k, q) = c(k, q) e^{-(k^2+q^2)t}.$$

Hence,

$$u(t, x, y) = \frac{1}{2\pi} \int_{\mathbb{R}^2} c(k, q) e^{-(k^2+q^2)t} e^{i(kx+qy)} dk\,dq,$$

where $c(k, q)$ satisfies

$$c(k, q) = -\frac{d^2}{dk^2} \mathcal{F}(e^{-x^2-y^2})$$

$$= -\frac{k^2 - 2}{8} e^{-\frac{k^2+q^2}{4}}.$$

Therefore, we conclude that

$$u(t, x, y) = -\int_{\mathbb{R}^2} \frac{k^2 - 2}{16\pi} e^{-\frac{k^2+q^2}{4}} e^{-(k^2+q^2)t} e^{i(kx+qy)} dk\,dq,$$

this is

$$u(t,x,y) = -\frac{1}{8}\frac{1}{\sqrt{2\pi}}\int_{\mathbb{R}} (k^2 - 2)e^{-k^2(t+\frac{1}{4})}e^{ikx}dk\,\frac{1}{\sqrt{2\pi}}\int_{\mathbb{R}} e^{-q^2(t+\frac{1}{4})}e^{iqy}dq,$$

from where, using the basic properties of the Fourier transform, we obtain

$$\boxed{u(t,x,y) = \frac{2t(1+4t)+x^2}{(1+4t)^3}\exp\left(-\frac{x^2+y^2}{1+4t}\right).}$$

(10) Solve the following initial and boundary value problem for the heat equation in a one-dimensional bar:

$$\begin{cases} u_t = u_{xx}, & x \in \mathbb{R}, \quad t > 0, \\ \begin{cases} u|_{t=0} = \sin \pi x, \\ \begin{cases} u_x|_{x=0} = 0, \\ u|_{x=1} = 0. \end{cases} \end{cases} \end{cases}$$

Note that the Neumann-type boundary condition at $x = 0$ models, according to Fourier's law, the thermal insulation of that boundary. In fact, the Neumann-type conditions for the heat equation represent conditions on the heat flow over the boundary, in this case, indicating thermal insulation at the boundary. Therefore, this problem involves studying the thermal evolution of a bar with one end thermally insulated and the other end at zero temperature.

Solution: To apply the eigenfunction expansion method, we first consider the eigenvalue problem for the differential operator

$$B = -\frac{d^2}{dx^2},$$

that is, we seek for solutions for $Bw = \lambda w$, with $\lambda = k^2$ in the space

$$\{w \in C^\infty([0,1]), w_x|_{x=0} = 0, w|_{x=1} = 0\}.$$

It is easy to deduce that the eigenvalues are

$$\lambda_n = k_n^2,$$

$$k_n := \frac{(2n+1)\pi}{2},$$

and the corresponding eigenfunctions

$$w_n(x) := \cos k_n x.$$

Since the boundary conditions are separated for a regular Sturm–Liouville operator, the spectrum is not degenerated, and the eigenfunction set forms an orthogonal basis. Therefore, we can expand the inhomogeneity $\sin \pi x$ on this basis:

$$\sin \pi x = \sum_{n=0}^{\infty} c_n \cos \frac{(2n+1)\pi}{2}x,$$

with

$$c_n = \frac{(w_n, \sin \pi x)}{\|w_n\|^2}$$

$$= 2 \int_0^1 \sin \pi x \, \cos \frac{(2n+1)\pi}{2} x \, \mathrm{d}x$$

$$= -\frac{8}{\pi} \frac{1}{4n^2 + 4n - 3}.$$

The expansion in eigenfunctions will be

$$u(t, x) = \sum_{n=0}^{\infty} v_n(t) \cos \frac{(2n+1)\pi}{2} x,$$

where $v_n(t)$ solves

$$\frac{\mathrm{d}v_n}{\mathrm{d}t} + \frac{(2n+1)^2 \pi^2}{4} v_n = 0,$$

$$v_n(0) = c_n.$$

In short, the expansion for the solution is given by

$$u(t, x) = -\frac{8}{\pi} \sum_{n=0}^{\infty} \frac{1}{4n^2 + 4n - 3} e^{-\frac{(2n+1)^2 \pi^2 t}{4}} \cos \frac{(2n+1)\pi}{2} x.$$

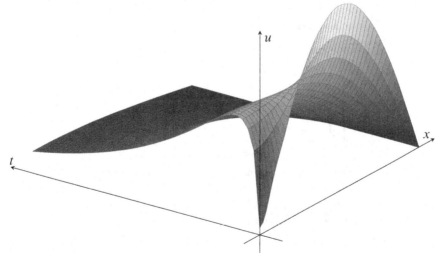

Temperature $u(x, t)$: **16th truncation**

The following graph shows the edge temperature with thermal insulation as a function of time.

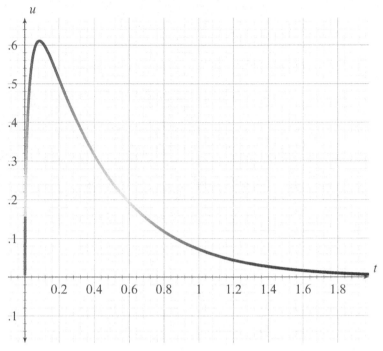

Fourier Series, truncated to 150 terms

(11) The steady-state temperature $u(x, y)$ of a two-dimensional heat-conducting rectangular plate, with two of its opposite edges subject to a constant heat flux over time and the other two edges at zero temperature, is described by the following boundary value problem for the Laplace equation:

$$\begin{cases} u_{xx} + u_{yy} = 0, \quad 0 < x < a, \quad 0 < y < b, \\ u_x|_{x=0} = \sin\dfrac{\pi y}{b}, \\ u_x|_{x=a} = \sin\dfrac{\pi y}{b}, \\ u|_{y=0} = 0, \\ u|_{y=b} = 0. \end{cases}$$

(a) Determine the steady-state temperature $u(x, y)$ that solves this problem.

(b) Find the solution in the case where the condition $u_x|_{x=a} = \sin\frac{\pi y}{b}$ is replaced by $u_x|_{x=a} = \sin\frac{3\pi y}{b}$.

Solution:

(a) In this case, we separate the two-dimensional Laplacian as $\Delta = A + B$, where $A = \frac{\partial^2}{\partial x^2}$ and $B = \frac{\partial^2}{\partial y^2}$, where the inhomogeneous conditions only appear in the variable x. The eigenvalue problem

$$Bw(y) = \lambda w(y),$$

$$\begin{cases} w|_{y=0} = 0, \\ w|_{y=b} = 0. \end{cases}$$

This leads to the following eigenfunctions and eigenvalues:

$$w_n(y) = \sin\frac{n\pi y}{b}, \quad \lambda_n = -\frac{n^2\pi^2}{b^2},$$

for $n \in \mathbb{N}$. The method of eigenfunction expansion involves seeking a solution in the form

$$u(x, y) = \sum_{n=1}^{\infty} v_n(x)\sin\frac{n\pi y}{b},$$

where $v_n'' = \frac{n^2\pi^2}{b^2}v_n$. We observe that the inhomogeneous terms only excite the fundamental mode, and therefore, the solution will take the form

$$u(x, y) = v(x)\sin\frac{\pi y}{b}, \quad v''(x) = \frac{\pi^2}{b^2}v(x).$$

This implies,

$$v(x) = A\sinh\frac{\pi x}{b} + B\cosh\frac{\pi x}{b},$$

where A and B are constants to be determined from the inhomogeneous boundary conditions, i.e.,

$$\begin{cases} v'(0) = 1, \\ v'(a) = 1. \end{cases}$$

To implement them, we compute the derivative

$$v'(x) = \frac{\pi}{b}\left(A\cosh\frac{\pi x}{b} + B\sinh\frac{\pi x}{b}\right),$$

and evaluate it at $x = 0$ to obtain $A = \frac{b}{\pi}$ and at $x = a$ to determine B from the equation

$$\cosh\frac{\pi a}{b} + B\frac{\pi}{b}\sinh\frac{\pi a}{b} = 1$$

yielding

$$B = \frac{b}{\pi}\frac{1 - \cosh\frac{\pi a}{b}}{\sinh\frac{\pi a}{b}}.$$

Thus, for the function $v(x)$, we found the following expression:

$$v(x) = \frac{b}{\pi}\left(\sinh\frac{\pi x}{b} + \frac{1 - \cosh\frac{\pi a}{b}}{\sinh\frac{\pi a}{b}}\cosh\frac{\pi x}{b}\right)$$

$$= \frac{b}{\pi}\frac{1}{\sinh\frac{\pi a}{b}}\left(\cosh\frac{\pi x}{b} - \cosh\frac{\pi(x-a)}{b}\right).$$

In conclusion, the requested solution is:

$$u(x,y) = \frac{b}{\pi}\frac{1}{\sinh\frac{\pi a}{b}}\left(\cosh\frac{\pi x}{b} - \cosh\frac{\pi(x-a)}{b}\right)\sin\frac{\pi y}{b}.$$

Temperature $u(x,y)$ on the plate

The interpretation of the diagram is as follows. Fourier's law tells us that the heat flux is $-\kappa u'$, where κ is the material's conductivity and u' is the temperature gradient. In our case, we have the same gradient value, which is $\sin\frac{\pi y}{b}$, on both the right and left edges. This means that we have the same heat flux on both sides, but on the left, it is outgoing from the plate, while on the right, it is entering the plate. Therefore, the right edge is warmer, and the left edge is colder.

(b) We apply the method of eigenfunction expansion, as in the previous case. Now, the inhomogeneous terms excite two different modes, and our solution will take the form

$$u = v_1(x) \sin \frac{\pi y}{b} + v_3(x) \sin \frac{3\pi y}{b},$$

where

$$v_1(x) = A_1 \sinh \frac{\pi x}{b} + B_1 \cosh \frac{\pi x}{b}, \quad v_3(x) = A_3 \sinh \frac{3\pi x}{b} + B_3 \cosh \frac{3\pi x}{b},$$

to which we have to impose the following boundary conditions

$$\begin{cases} v_1'(0) = 1, \\ v_1'(a) = 0, \end{cases}$$

$$\begin{cases} v_3'(0) = 0, \\ v_3'(a) = 1. \end{cases}$$

Temperature $u(x, y)$ on the plate

As the derivatives are

$$v_1'(x) = \frac{\pi}{b}\left(A_1 \cosh \frac{\pi x}{b} + B_1 \sinh \frac{\pi x}{b} \right),$$

$$v_3'(x) = \frac{3\pi}{b}\left(A_3 \cosh \frac{3\pi x}{b} + B_3 \sinh \frac{3\pi x}{b} \right),$$

the following coefficients are determined:

$$A_1 = \frac{b}{\pi}, \qquad B_1 = -\frac{b}{\pi} \tanh \frac{\pi a}{b}, \qquad A_3 = 0, \qquad B_3 = \frac{b}{3\pi} \operatorname{csch} \frac{3\pi a}{b}.$$

Thus, the functions take the following form:

$$v_1(x) = -\frac{b}{\pi} \frac{1}{\sinh \frac{\pi a}{b}} \cosh \frac{\pi(x-a)}{b}, \qquad v_3(x) = \frac{b}{3\pi} \frac{1}{\sinh \frac{3\pi a}{b}} \cosh \frac{3\pi x}{b},$$

and finally, we obtain the solution:

$$u(x,y) = \frac{b}{\pi} \left(-\frac{1}{\sinh \frac{\pi a}{b}} \cosh \frac{\pi(x-a)}{b} \sin \frac{\pi y}{b} + \frac{1}{3} \frac{1}{\sinh \frac{3\pi a}{b}} \cosh \frac{3\pi x}{b} \sin \frac{3\pi y}{b} \right).$$

§6.5.2. Exercises

(1) Apply the eigenfunction expansion method to solve the initial and boundary value problem for the vibrating string:

$$\left\{ \begin{array}{l} u_{tt} = u_{xx}, \quad x \in \mathbb{R}, \quad t > 0, \\ \left\{ \begin{array}{l} u|_{x=0} = 0, \\ u|_{x=1} = 0, \end{array} \right. \\ \left\{ \begin{array}{l} u|_{t=0} = g_1(x), \\ u_t|_{t=0} = g_2(x). \end{array} \right. \end{array} \right.$$

Calculate the solution expansion for the following initial data:

$$g_1(x) = x(x-1), \quad g_2(x) = 0.$$

(2) The Fourier transform method applied to the initial value problem for the heat equation:

(6.16)
$$\left\{ \begin{array}{l} u_t = a^2 u_{xx}, \quad x \in \mathbb{R}, \quad t > 0 \\ u|_{t=0} = f(x), \end{array} \right.$$

provides the following expression for the solution:

$$u(t,x) = \frac{1}{2a\sqrt{\pi t}} \int_{-\infty}^{\infty} \exp\left(-\frac{(x-y)^2}{4a^2 t} \right) f(y) dy.$$

(a) Prove the following alternative expression:

$$u(t,x) = \frac{1}{\sqrt{\pi}} \int_{-\infty}^{\infty} e^{-s^2} f(x + 2a\sqrt{t}s) ds.$$

(b) Determine the solution of (6.16) in the following two cases:

$$f(x) = \left\{ \begin{array}{ll} 1, & 0 < x < 1, \\ 0, & \text{otherwise}, \end{array} \right. \qquad f(x) = \left\{ \begin{array}{ll} x, & 0 < x < 1, \\ 0, & \text{otherwise}. \end{array} \right.$$

Hint: Use the error function $\operatorname{erf}(x) = \frac{2}{\sqrt{\pi}} \int_0^x e^{-s^2} ds$.

7. Special Functions

Contents

OURIER series and the Fourier transform play pivotal roles in solving linear partial differential equations using the eigenfunction expansion approach, particularly when accompanied by appropriate boundary conditions. It is important to highlight that this method relies on the boundaries possessing a natural representation within Cartesian (or rectangular) coordinates. The fundamental justification for this approach stems from the correlation between the associated eigenvalue problems and differential operators with constant coefficients.

However, complications arise when the boundaries of the regions are not easily expressible in Cartesian coordinates. In such scenarios, a transition to curvilinear coordinate systems becomes necessary, allowing for a more natural representation of the boundaries. This shift necessitates the utilization of eigenvalue problems associated with Sturm-Liouville-type differential operators featuring variable coefficients. This situation arises notably in cases where the Laplacian is involved, and the boundaries are naturally defined in cylindrical or spherical coordinates.

This chapter is devoted to the exploration of functions required to address these specific cases. In particular, the focus will be on their expansion in terms of series centered around specific points.

§7.1. Frobenius Series

OLVING eigenvalue problems for these boundary value scenarios, which naturally find their formulation within cylindrical or spherical coordinates, needs the solution of a specific type of second-order linear ordinary differential equation (ODE) featuring nonconstant coefficients.

(7.1)
$$u'' + a(x)u' + b(x)u = 0,$$

where $a = a(x)$ and $b = b(x)$ are functions of the independent variable x. Drawing upon the principles of linear ODEs, we can ascertain that the general solution of Equation (7.1) takes the form of an arbitrary linear combination:

$$u(x) = C_1 u_1(x) + C_2 u_2(x),$$

where $u_1(x)$ and $u_2(x)$ are two linearly independent solutions.

In this chapter, our objective revolves around identifying specific solutions for Equation (7.1) by initiating expansions centered around a designated fixed point x_0. This expansion takes the following form:

$$u(x) = (x - x_0)^\alpha \sum_{n=0}^{\infty} c_n (x - x_0)^n$$
$$= c_0 (x - x_0)^\alpha + c_1 (x - x_0)^{\alpha+1} + \cdots.$$

Notably, the exponent α featured in the factor

$$(x - x_0)^\alpha$$

is not restricted to being an integer number. The coefficients c_n need to be determined, and series of this nature are referred to as **Frobenius series**. Analyzing such series expansions necessitates extending our considerations of Equation (7.1) into the complex plane $\mathbb{C}$. Consequently, the sought-after solutions $u = u(z)$, as well as the coefficients $a = a(z)$ and $b = b(z)$, all become functions of the complex variable $z \in \mathbb{C}$. Here, the derivative with respect to the complex variable z is denoted as

$$u' = \frac{du}{dz}.$$

When seeking solutions expressed as Frobenius series, denoted by $u = u(z)$, it becomes essential to ascertain the type of point around which the series expansion occurs. This point, which we shall designate as z_0, plays a crucial role in this determination.

Ordinary Point

A point $z_0 \in \mathbb{C}$, where $z_0 \neq \infty$, is classified as an **ordinary point** of Equation (7.1) if the functions $a(z)$ and $b(z)$ exhibit analytic behavior at z_0. In more concrete terms, this means that there exist expansions in the form of power series:

$$a(z) = a_0 + a_1(z - z_0) + a_2(z - z_0)^2 + \cdots,$$
$$b(z) = b_0 + b_1(z - z_0) + b_2(z - z_0)^2 + \cdots,$$

These series converge within a certain disk centered around z_0.

Definition

Regular Singular Point

A point z_0 in the complex plane $\mathbb{C}$, excluding infinity, is termed a **regular singular point** of Equation (7.1) if either $a(z)$ or $b(z)$ is singular at $z = z_0$, yet the modified functions $(z - z_0)a(z)$ and $(z - z_0)^2 b(z)$ are analytic at z_0. This entails expansions in power series

$$(z - z_0)a(z) = a_0 + a_1(z - z_0) + a_2(z - z_0)^2 + \cdots,$$
$$(z - z_0)^2 b(z) = b_0 + b_1(z - z_0) + b_2(z - z_0)^2 + \cdots.$$

These series converge within a disk centered around z_0. It's important to note that some coefficients, namely a_0, b_0, and b_1, must be nonzero; otherwise, the functions $a(z)$ and $b(z)$ would both be analytic at $z = z_0$.

Definition

We will focus solely on ODEs that feature these two specific types of points, excluding from our study the irregular singular points which do not fall within the aforementioned categories.

To **categorize** the types of points arising in a given equation, we need to consider the potential singularities of the functions $a(z)$ (coefficient of u') and $b(z)$ (coefficient of u).

Ordinary points encompass instances where both functions $a(z)$ and $b(z)$ are analytic.

Regular singular points are points at which either function $a(z)$ or $b(z)$ exhibits a pole, with the maximum order being one for $a(z)$ and two for $b(z)$.

Let's showcase a selection of notable second-order ODEs that align with our prior analysis and find extensive applications across various scientific and engineering domains.

Hermite Equation: $u'' - 2zu' + 2\nu u = 0$

Considering ν as a complex number, in this particular scenario, where

$$a(z) = -2z,$$
$$b(z) = 2\nu,$$

it is worth noting that these functions are analytical across the entire complex plane $\mathbb{C}$. Consequently, all points within this context are classified as regular.

The Hermite equation holds significant importance in various mathematical and scientific disciplines, primarily within the realms of differential equations, quantum mechanics, and mathematical physics.

Named after Charles Hermite, a French mathematician, the Hermite equation and its associated Hermite polynomials are essential tools with broad applications in mathematics and across multiple scientific disciplines.

Airy Equation: $u'' - zu = 0$

In this case, where
$$a(z) \equiv 0,$$
$$b(z) = -z,$$
both of which are analytical functions across the entire complex plane $\mathbb{C}$, all points are regular.

The Airy equation appears in various scientific and mathematical contexts, particularly in the study of wave phenomena and oscillatory behavior.

It is named after the British mathematician and astronomer Sir George Biddell Airy, who contributed to the understanding of the equation and its solutions. The Airy equation is a versatile mathematical tool that plays a crucial role in understanding wave phenomena, diffraction, and oscillatory behavior in various scientific fields.

Its solutions, the Airy functions, have applications in optics, quantum mechanics, mathematical analysis, asymptotic approximations, and more. The equation's significance lies in its ability to capture complex behaviors that involve oscillations and diffraction patterns.

Legendre Equation: $((1 - z^2)u')' + \lambda u = 0$

Considering λ as a complex number, upon solving for the second derivative, the equation takes the form:
$$u'' - \frac{2z}{1 - z^2}u' + \frac{\lambda}{1 - z^2}u = 0.$$
As a result, we have
$$a(z) = -\frac{2z}{1 - z^2},$$
$$b(z) = \frac{\lambda}{1 - z^2}.$$
The points $z_0 = 1$ and $z_0 = -1$ are characterized as simple poles for $a(z)$ and $b(z)$, consequently qualifying as regular singular points. The remaining points are regular.

The Legendre equation is named after the French mathematician Adrien-Marie Legendre and arises in various areas of mathematics and physics. This equation and its solutions, the Legendre polynomials among them, have diverse applications in mathematics and physics. They are especially valuable in describing systems with spherical symmetry, including problems in electrostatics, quantum mechanics, potential theory, and more.

The equation's significance lies in its ability to provide solutions that capture the behavior of physical systems with spherical symmetry and its role in expanding functions using orthogonal polynomials.

Laguerre Equation: $zu'' + (1 - z)u' + ku = 0$

The equation simplifies to:

$$u'' + \left(\frac{1}{z} - 1\right)u' + \frac{k}{z}u = 0,$$

which yields $a(z) = \frac{1}{z} - 1$ and $b(z) = \frac{k}{z}$. Here, the point $z_0 = 0$ is a simple pole for both $a(z)$ and $b(z)$. This point represents the only regular singular point, while the remaining points in the complex plane are regular.

The Laguerre equation is a differential equation that is named after the French mathematician Edmond Laguerre. It arises in various mathematical and scientific contexts, particularly in problems involving exponential growth or decay and in the study of quantum mechanics.

Bessel Equation: $z^2u'' + zu' + (z^2 - v^2)u = 0$

We have $a(z) = \frac{1}{z}$ and $b(z) = 1 - \frac{v^2}{z^2}$, with v being a complex number. As a result, $z_0 = 0$ is a simple pole for $a(z)$ and a double pole for $b(z)$. It stands as the sole regular singular point, while all other points in the complex plane are regular.

Bessel functions are a family of special functions that arise in solving problems involving circular and cylindrical symmetry. They are named after Friedrich Bessel, a German mathematician who made significant contributions to their study. Bessel functions are a versatile mathematical tool with wide-ranging applications in physics, engineering, and mathematics, and they play a crucial role in understanding wave phenomena, diffraction, heat conduction, and quantum mechanics in such systems.

§7.2. Ordinary Points

 RDINARY points, in the context of ordinary differential equations, are specific values of the independent variable for which the coefficients of the ODE remain finite and wellbehaved; i.e., they are analytic functions. The concept of ordinary points is particularly important when analyzing the behavior and solutions of linear differential equations.

The behavior of solutions near an ordinary point can be analyzed using power series methods. If z_0 is an ordinary point of the ODE, then it is possible to find a power series solution centered at z_0 that converges within a certain disk around z_0. This power series solution can provide insights into the behavior of the solution near the ordinary point and may be used for its approximation.

On the other hand, if an ODE has a singular point (either regular singular points or irregular points) at a certain value of z, then the coefficients of the ODE become unbounded at that point, is a point of nonanalytical character. Analyzing solutions around singular points can be more complex and often involves special series like the Frobenius method to find solutions.

By definition, if z_0 is an ordinary point for the ordinary differential equation

$$(7.2) \qquad u'' + a(z)u' + b(z)u = 0,$$

then the functions $a(z)$ and $b(z)$ can be expressed through power series expansions:

$$(7.3) \qquad \begin{aligned} a(z) &= a_0 + a_1(z - z_0) + a_2(z - z_0)^2 + \cdots, \\ b(z) &= b_0 + b_1(z - z_0) + b_2(z - z_0)^2 + \cdots. \end{aligned}$$

The subsequent theorem succinctly summarizes the essential properties concerning the solutions of Equation (7.2) in the vicinity of z_0.

Cauchy Theorem for Ordinary Points

If z_0 is an ordinary point of Equation (7.2), then the following assertions hold true:

(1) Any solution u of (7.2) is analytic at $z = z_0$.

(2) The radius of convergence for the associated power series is greater than or equal to the distance between z_0 and the nearest singularity of the functions $a(z)$ and $b(z)$.

(3) The coefficients within the power series expansion of a solution to (7.2), as depicted in

$$(7.4) \quad u = \sum_{n=0}^{\infty} c_n(z - z_0)^n = c_0 + c_1(z - z_0) + c_2(z - z_0)^2 + \cdots,$$

can be determined by substituting the series (7.3) and (7.4) into the ODE (7.2), subsequently identifying the coefficients corresponding to the powers of $(z - z_0)^n$.

Theorem

Observations: It's worth noting that the first aspect of the theorem suggests that the singularities of a solution u of (7.2) might possess are exclusively tied to the singularities of the functions $a(z)$ and $b(z)$. Consequently, if neither $a(z)$ nor $b(z)$ exhibits any singular points, the solutions of (7.2) will also be devoid of singularities.

Moreover, the theorem equips us with a methodology to compute solutions of (7.1) in the form of power series centered around ordinary points. The ensuing example serves to illustrate this approach.

The fundamental concept underlying Cauchy's theorem is that starting from (7.2), one can derive higher-order derivatives as analytic functions. Consequently, we can readily calculate all the Taylor power series coefficients at the specified ordinary point. The challenging aspect lies in demonstrating the region where this series converges.

§7.2.1. Hermite Equation

The quantum version of the harmonic oscillator, upon simplifying the corresponding Schrödinger equation, yields the following ordinary differential equation:

$$v'' - (z^2 - 2v - 1)v = 0.$$

Upon performing the transformation $u = \exp(z^2/2)v$, this equation transforms into Hermite's ODE:

$$(7.5) \qquad\qquad u'' - 2zu' + 2vu = 0.$$

It can be formulated as an eigenvalue problem for a Sturm–Liouville operator (nonregular due to an open-ended domain) in the following manner:

$$-\frac{1}{\exp(-z^2)}\big(\exp(-z^2)u'\big)' = 2vu,$$

where $\rho(x) = p(x) = \exp(-x^2)$ and $q(x) = 0$.

This equation aligns with our discussion with the identification $a(z) = -2z$ and $b(z) = 2v$. Let's consider $z_0 = 0$. According to the preceding theorem, we can conclude that the power series solutions converge over the entire complex plane, thanks to the absence of singularities in the functions $a(z)$ and $b(z)$. Our goal is to discover a solution in the form of a power series expansion centered at $z_0 = 0$

$$u = \sum_{n=0}^{\infty} c_n z^n = c_0 + c_1 z + c_2 z^2 + \cdots.$$

After substituting this into (7.5) and considering that $a(z)$ and $b(z)$ are already expanded as power series in terms of z, we arrive at:

$$\sum_{n=0}^{\infty} \big((n+2)(n+1)c_{n+2} - 2(n-v)c_n\big)z^n = 0.$$

Identifying coefficients for the powers of z^n, we deduce:

$$(n + 2)(n + 1)c_{n+2} - 2(n - v)c_n = 0, \quad n \in \mathbb{N}_0.$$

From this, we establish the recurrence relation for the unknown coefficients of u:

(7.6)
$$c_{n+2} = -\frac{2(v - n)}{(n + 2)(n + 1)}c_n.$$

Hence, each coefficient c_n determines c_{n+2}, and through iteration, we derive that:

$$c_{2n} = \frac{(-2)^n}{(2n)!}(v - 2n + 2)_{2n-1}c_0, \quad c_{2n+1} = \frac{(-2)^n}{(2n + 1)!}(v - 2n + 1)_{2n-1}c_1.$$

Where we have used the Pochhammer symbol $(z)_n := \frac{\Gamma(z+n)}{\Gamma(z)} = z(z + 1)\cdots(z + n - 1)$. In conclusion, setting $(c_0, c_1) = (1, 0)$ and $(c_0, c_1) = (0, 1)$, we obtain two linearly independent solutions of the Hermite equation:

$$u_0(z) = \sum_{n=0}^{\infty} \frac{(-2)^n}{(2n)!}(v - 2n + 2)_{2n-1}z^{2n},$$

$$u_1(z) = \sum_{n=0}^{\infty} \frac{(-2)^n}{(2n + 1)!}(v - 2n + 1)_{2n-1}z^{2n+1}.$$

Hermite Polynomials From (7.6), we can infer that when a coefficient c_n within the solution series is zero, all coefficients c_{n+2k} with $k \in \mathbb{N}_0$ will also be zero. When n is even, this implies that $u_0(x)$ forms a polynomial. Conversely, when n is odd, it signifies that $u_1(x)$ takes the form of a polynomial. Considering this observation alongside (7.6), it's evident that for $v = 2k$ and $v = 2k + 1$, the solutions $u_0(z)$ and $u_1(z)$ become polynomial functions. These polynomials are proportional to the Hermite polynomials, commonly referred to as such.

Hermite Polynomials

The Hermite polynomials can be determined using the ***Rodrigues formula***:

$$H_n(z) = (-1)^n \exp(z^2)\frac{d^n}{dz^n} \exp(-z^2).$$

The initial Hermite polynomials are as follows:

$$H_0(z) = 1, \qquad\qquad H_1(z) = 2z,$$
$$H_2(z) = 4z^2 - 2, \qquad\qquad H_3(z) = 8z^3 - 12z,$$
$$H_4(z) = 16z^4 - 48z^2 + 12, \qquad H_5(z) = 32z^5 - 160z^3 + 120z.$$

Hermite Polynomials Properties

(1) **Recurrence Formula:**
$$H_n(z) = 2z\,H_{n-1}(z) - 2(n-1)H_{n-2}(z).$$

(2) **Generating Function:**
$$e^{2tx-t^2} = \sum_{n=0}^{\infty} \frac{t^n}{n!} H_n(x).$$

(3) **ODE:**
$$H_n'' - 2z\,H_n' + 2nH_n = 0.$$

(4) **Orthogonality:**
$$\int_{\mathbb{R}} H_n(x)H_m(x)e^{-x^2}\,\mathrm{d}x = \sqrt{\pi}2^n n!\delta_{n,m},$$
for $n, m \in \{0, 1, 2, \dots\}$.

(5) **Completeness:** The Hermite polynomial set $\{H_n(x)\}_{n=0}^{\infty}$ is a **complete orthogonal set** in $L^2(\mathbb{R}, e^{-x^2}\mathrm{d}x)$.

Hermite Polynomials

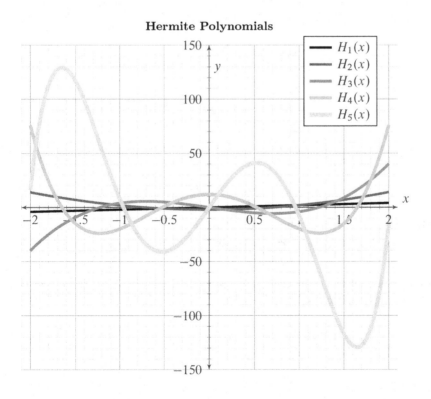

The Hermite polynomials hold significance not only in quantum physics, as demonstrated by their role in solving the quantum harmonic oscillator—a vital component in the foundation of quantum field theory—but also in various other mathematical disciplines. These encompass probability theory, combinatorics, numerical analysis, and system theory.

☞   **Observations:** What happens to $u_0(z)$ and $u_1(z)$ when $v \notin \{0, 1, 2, \dots\}$? In accordance with Cauchy's theorem, we understand that both solutions are analytic across $\mathbb{C}$, making them inherently valid. However, considering Liouville's theorem, a singularity will manifest at $z = \infty$. In specific applications, especially in the context of quantum physics, we require solutions $u_0(x)$ and $u_1(x)$ that fall within

$$L^2\left(\mathbb{R}, e^{-x^2} dx\right) = \left\{ f(x) : \mathbb{R} \to \mathbb{C} : \int_{-\infty}^{\infty} |f(x)|^2 \exp(-x^2) dx < \infty \right\}.$$

This is fulfilled when series are truncated, leading to simple polynomials. Contrastingly, when series extend infinitely,

$$u_0(x), u_1(x) \notin L^2\left(\mathbb{R}, e^{-x^2} dx\right).$$

While we won't prove this, a plausible argument can be made, even in rigorous mathematical terms.

Let's compare coefficients c_n of $u_0(z)$ with those of $\exp(z^2)$ as $n \to \infty$. For $a > 0$, the ensuing power series emerges:

$$\exp(a^2 z^2) = \sum_{n=0}^{\infty} C_n z^n,$$

$$C_n = \begin{cases} \dfrac{a^2}{\frac{n}{2}!}, & \text{for even } n, \\ 0, & \text{for odd } n. \end{cases}$$

The even coefficients C_n exhibit a d'Alembert-like quotient (inapplicable for odd n):

$$\frac{C_{n+2}}{C_n} = \frac{a^2}{\frac{n}{2} + 1}.$$

For large, even n, the behavior unfolds as:

$$\frac{C_{n+2}}{C_n} \sim \frac{2a^2}{n}, \quad n \to \infty.$$

Conversely, for the exclusively nonzero even coefficients c_n of the power series $u_0(x)$, the scenario reads:

$$\frac{c_{n+2}}{c_n} = \frac{2(v - n)}{(n + 2)(n + 1)}$$

$$\sim \frac{2}{n},$$

for $n \to \infty$. Hence, it's reasonable to deduce that the power series of $u_0(x)$ mimics the power series of $\exp(x^2)$—a distinction arises due to the nature of ν being an even number. For noneven ν and x on the real line, we observe

$$|u_0(x)|^2 \sim \exp(2x^2),$$

consequently yielding

$$u_0(x) \notin L^2\left(\mathbb{R}, e^{-x^2} dx\right).$$

Analogously, for nonodd ν, a similar argument leads to

$$|u_1(x)|^2 \sim x^2 \exp(2x^2)$$

and

$$u_1(x) \notin L^2\left(\mathbb{R}, e^{-x^2} dx\right).$$

§7.2.2. Legendre Equation

The Legendre equation we encountered earlier takes the form:

$$u'' - \frac{2z}{1 - z^2}u' + \frac{\lambda}{1 - z^2}u = 0,$$

with regular singular points at $z_0 = \pm 1$, while all other points remain regular. When expressed as an eigenvalue problem for an operator of Sturm–Liouville type, it appears as:

$$-((1 - x^2)u')' = \lambda u,$$

with

$$\rho = 1, \quad p(x) = 1 - x^2, \quad q(x) \equiv 0.$$

This is not a regular Sturm–Liouville problem due to its behavior at $x = \pm 1$.

Let's choose the origin $z_0 = 0$ as the center for the expansions. Each solution can be represented by a power series expansion:

$$u = \sum_{n=0}^{\infty} c_n z^n = c_0 + c_1 z + c_2 z^2 + \cdots,$$

with a convergence radius greater than or equal to 1. To determine the coefficients, we can rewrite the Legendre equation as:

$$(1 - z^2)u'' - 2zu' + \lambda u = 0.$$

Substituting the series of u into the Legendre equation and grouping coefficients in powers of z^n, we obtain:

$$\sum_{n=0}^{\infty} \left((n + 2)(n + 1)c_{n+2} - n(n - 1)c_n - 2nc_n + \lambda c_n\right)z^n = 0.$$

This gives rise to the following recurrence relation:

$$(7.7) \qquad c_{n+2} = \frac{n(n + 1) - \lambda}{(n + 2)(n + 1)}c_n,$$

for $n \in \mathbb{N}_0$. By considering the two choices

$$(c_0, c_1) = (1, 0),$$
$$(c_0, c_1) = (0, 1),$$

we derive two linearly independent solutions for the Legendre equation:

$$u_0(z) = \sum_{n=0}^{\infty} c_{2n} z^{2n},$$

$$u_1(z) = \sum_{n=0}^{\infty} c_{2n+1} z^{2n+1}.$$

Legendre Polynomials Building upon the recurrence relation (7.7), we can employ a similar approach to our treatment of the Hermite functions. By selecting

$$\lambda = n(n + 1),$$

for a fixed value of n, we can deduce the following:

(1) When n is even ($n = 2k$), all coefficients c_{2m} with $m > k$ are eliminated, resulting in $u_0(z)$ being a polynomial.
(2) When n is odd ($n = 2k + 1$), all coefficients c_{2m+1} with $m > k$ vanish, leading to $u_1(z)$ being a polynomial.

The polynomials generated through this process are referred to as the **Legendre polynomials**.

Rodrigues' Formula

Legendre polynomials can be computed using the Rodrigues formula:

$$P_n(z) = \frac{1}{2^n n!} \frac{\mathrm{d}^n}{\mathrm{d}z^n} (z^2 - 1)^n,$$

for $n \in \mathbb{N}_0$.

Initial Legendre Polynomials

The initial Legendre polynomials can be computed using the Rodrigues formula:

$$P_0(z) = 1, \qquad\qquad P_1(z) = z,$$

$$P_2(z) = \frac{1}{2}(3z^2 - 1), \qquad P_3(z) = \frac{1}{2}(5z^3 - 3z),$$

$$P_4(z) = \frac{1}{8}(35z^4 - 30z^2 + 3), \qquad P_5(z) = \frac{1}{8}(63z^5 - 70z^3 + 15z).$$

Properties of Legendre Polynomials

(1) **Bonnet Recurrence Formula:**

$$(n+1)P_{n+1}(z) = (2n+1)zP_n(z) - nP_{n-1}(z).$$

(2) **Generating function:**

$$\frac{1}{\sqrt{1-2tx+t^2}} = \sum_{n=0}^{\infty} P_n(x)t^n$$

for $|t| < 1$ and $|x| \leq 1$.

(3) **Second-order ODE:**

$$((1-z^2)P_n')' + n(n+1)P_n = 0.$$

(4) **Orthogonality:**

$$\int_{-1}^{1} P_n(x)P_m(x)\,\mathrm{d}x = \frac{1}{n+\frac{1}{2}}\delta_{n,m},$$

for $n, m \in \{0, 1, 2 \dots\}$.

(5) **Completeness:** The set

$$\{P_n(x)\}_{n=0}^{\infty} \subset L^2([-1,1], \mathrm{d}x)$$

is a complete orthogonal set.

Observations: What about the solutions $u_0(z)$ y $u_1(z)$ when $\lambda \neq n(n+1)$, $n \in \{0, 1, 2, \dots\}$? In this scenario, the discussion is simpler compared to the Hermite case. We are aware that our solutions, $u_0(z)$ and $u_1(z)$, uniformly converge within the unit disk.

However, when n is even, the series for $u_0(z)$ terminates, resulting in $u_0(z)$ being a polynomial and thereby defined across the entirety of $\mathbb{C}$. Similarly, for odd n, the series for $u_1(z)$ concludes, also yielding a polynomial and, consequently, encompassing all of $\mathbb{C}$.

For cases where the series do not conclude, it's evident that the convergence radius for both $u_0(z)$ and $u_1(z)$ remains at 1.

Applications: Legendre polynomials emerge when Newtonian or Coulombian potentials are expanded in the form:

(7.8)
$$\frac{1}{|\mathbf{r} - \mathbf{r_0}|} = \frac{1}{\sqrt{r^2 + r_0^2 - 2rr_0\cos\theta}}$$

$$= \begin{cases} \frac{1}{r}\sum_{\ell=0}^{\infty} \left(\frac{r_0}{r}\right)^{\ell} P_\ell(\cos\theta), & r > r_0, \\ \frac{1}{r_0}\sum_{\ell=0}^{\infty} \left(\frac{r}{r_0}\right)^{\ell} P_\ell(\cos\theta), & r < r_0. \end{cases}$$

Here, $\theta = \widehat{r r_0}$ represents the latitude of point r concerning the axis established by the vector $r - r_0$. This series is applicable when considering the multipolar expansion of a point load outside the origin.

Excellent books to study orthogonal polynomials are [18], and the encyclopedic [41]. Other good references are [6, 53], and on the historical side refer to the Szegő authoritative book [70]. With a more historical flavor, see [15] and [42].

Legendre Polynomials

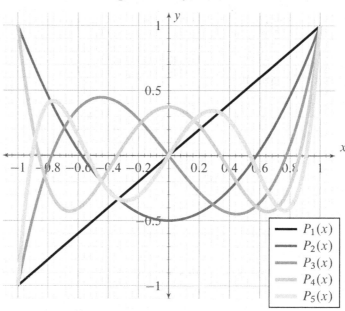

§7.3. Regular Singular Points

REGULAR singular point, in the context of ordinary differential equations, is a specific type of singular point that behaves in a somewhat controlled manner when solving linear second-order ODEs. Understanding regular singular points is crucial for solving ODEs in various scientific and engineering applications.

To find solutions near a regular singular point, mathematicians and scientists often use the Frobenius method. This method involves assuming that the solution can be expressed as a Frobenius series, plugging this series into the ODE, and solving for the coefficients of the series. The solutions obtained using the Frobenius method are known as Frobenius series solutions.

Regular singular points often appear in physics and engineering problems, particularly when dealing with differential equations that describe physical phenomena with singular behavior at specific points. Examples include problems involving quantum mechanics, fluid dynamics, and heat conduction in materials with specific properties.

We begin by studying a straightforward case, the Euler equation.

§7.3.1. Euler Equation

The Euler equation serves as a simplified model for understanding the behavior of solutions around regular singular points. The equation is given by:

$$u'' + \frac{q_1}{z} u' + \frac{q_0}{z^2} u = 0,$$

where q_1 and q_0 are nonzero complex numbers. It's clear that $z_0 = 0$ is a regular singular point. By multiplying the equation by z^2, we obtain:

$$z^2 u'' + q_1 z u' + q_0 u = 0,$$

or equivalently:

$$Lu = 0,$$

where L is the differential operator:

$$L = z^2 \frac{d^2}{dz^2} + q_1 z \frac{d}{dz} + q_0.$$

It's noteworthy that:

$$Lz^\alpha = P(\alpha) z^\alpha, \quad P(\alpha) := \alpha(\alpha - 1) + q_1 \alpha + q_0.$$

This equation tells us that z^α is always an eigenfunction of L with eigenvalue $\lambda = P(\alpha)$, and that $u = z^\alpha$ is a solution to Euler's equation when $P(\alpha) = 0$. The function $P(\alpha)$ is a second-degree polynomial, known as **indicial polynomial**, and will thus have two roots, counting multiplicities. If the two roots α_1 and α_2 are distinct, then z^{α_1} and z^{α_2} are linearly independent solutions, and a general solution of the Euler equation is:

$$u = c_1 z^{\alpha_1} + c_2 z^{\alpha_2}.$$

The situation becomes more involved if the indicial polynomial has a double root α_1, and one solution is $u = z^{\alpha_1}$. In this case, we need to find another independent solution. A method to find this solution is as follows. Since the root is double, we must have $P(\alpha) = (\alpha - \alpha_1)^2$, and consequently,

$$\left. \frac{dP}{d\alpha} \right|_{\alpha = \alpha_1} = 0.$$

Now, if we differentiate the equation with respect to α, we get:

(7.9)
$$\frac{\partial L(z^\alpha)}{\partial \alpha} = L\left(\frac{dz^\alpha}{d\alpha} \right) = L(z^\alpha \log z)$$
$$= P'(\alpha) z^\alpha + P(\alpha) z^\alpha \log z.$$

Since α_1 is a double root, we have: $P(\alpha_1) = P'(\alpha_1) = 0$. Then, (7.9) for $\alpha = \alpha_1$ simplifies to:

$$L(z^{\alpha_1} \log z) = 0,$$

meaning that $u = z^{\alpha_1} \log z$ is another solution to the equation. In summary, $z^{\alpha_1}, z^{\alpha_1} \log z$ forms a system of two linearly independent solutions, and a general solution is:

$$u = c_1 z^{\alpha_1} + c_2 z^{\alpha_1} \log z.$$

It's important to note the appearance of potential noninteger exponent powers and the logarithm function in the solutions. This illustrates the behavior we can anticipate in such situations.

§7.3.2. Frobenius Series

When dealing with a regular singular point $z_0 \in \mathbb{C}$, a solution u of (7.1) isn't always analytically expressible within a disk centered at z_0, as it could exhibit singularities like poles or logarithmic branches. In such cases, the solution doesn't conform to a power series expansion centered around z_0; instead, it takes on a more complex structure. Fuchs discovered that there's invariably at least one solution that can be represented in the form of a factor $(z - z_0)^\alpha$, where α is generally a complex number that may not necessarily be an integer. This factor is then multiplied by a power series centered around z_0, giving rise to an expression like:

$$
\begin{aligned}
u &= (z - z_0)^\alpha \sum_{n=0}^{\infty} c_n (z - z_0)^n \\
&= c_0 (z - z_0)^\alpha + c_1 (z - z_0)^{\alpha+1} + \cdots, \quad c_0 \neq 0.
\end{aligned}
$$

(7.10)

This kind of expansion is referred to as a **Frobenius series**.

Moreover, Fuchs' findings imply the existence of another solution, which can either follow the Frobenius pattern (7.10), or assume the form:

$$u = (z - z_0)^{\alpha_1} \log(z - z_0) \sum_{n=0}^{\infty} b_n (z - z_0)^n + (z - z_0)^{\alpha_2} \sum_{n=0}^{\infty} c_n (z - z_0)^n,$$

where the power series converge within a disk whose radius is greater than or equal to the distance between z_0 and the nearest singularity of $a(z)$ or $b(z)$.

Example: Consider Euler's equation

$$u'' + \frac{1}{4z^2} u = 0,$$

which possesses a regular singular point at $z_0 = 0$. If we seek a solution in the form of a power series around $z_0 = 0$

$$u = c_0 + c_1 z + \cdots,$$

substituting the series into the equation yields

$$\left(n - \frac{1}{2}\right)^2 c_n = 0,$$

for $n \in \mathbb{N}_0$. This implies that $c_n = 0$, leading to the trivial solution $u \equiv 0$. Considering this as an Euler equation, we recognize that a general solution is

$$u = c_1 \sqrt{z} + c_2 \sqrt{z} \log z.$$

Let's consider that z_0 is a regular singular point of the differential equation

(7.11) $$u'' + a(z)u' + b(z)u = 0.$$

According to the definition of a regular singular point, in this case, we have expansions of the form

$$A(z) := (z - z_0)a(z) = a_0 + a_1(z - z_0) + a_2(z - z_0)^2 + \cdots,$$
$$B(z) := (z - z_0)^2 b(z) = b_0 + b_1(z - z_0) + b_2(z - z_0)^2 + \cdots.$$

In order to determine a solution in the Frobenius series form of (7.10) with $c_0 = 1$, we start by multiplying of Equation (7.11) by $(z - z_0)^2$ and rewriting it in an equivalent form:

$$(z - z_0)^2 u'' + (z - z_0)A(z)u' + B(z)u = 0.$$

Substituting the Frobenius series (7.10), with unknown exponent and coefficients, into this equation, we find, upon grouping the coefficients of the powers of $z - z_0$, that it can be written in the form:

(7.12) $$(z - z_0)^\alpha P(\alpha) + \sum_{n \geq 1}(z - z_0)^{\alpha+n} Q_n(\alpha) = 0,$$

where

(7.13) $$P(\alpha) = \alpha^2 + (a_0 - 1)\alpha + b_0$$

is the **indicial polynomial**, and

$$Q_n(\alpha) = P(\alpha + n)c_n + \sum_{k=0}^{n-1}\left((\alpha + k)a_{n-k} + b_{n-k}\right)c_k.$$

Now, let's identify the power coefficients in (7.12). By equating the coefficient of $(z - z_0)^\alpha$ to zero, we find that α must be a root of the indicial polynomial, which leads to the **indicial equation**:

$$\boxed{P(\alpha) = 0,}$$

a second-degree polynomial with two roots, $\alpha = \alpha_1$ and α_2, generally complex. These roots can be ordered such that

$$\operatorname{Re}\alpha_1 \geq \operatorname{Re}\alpha_2.$$

Nonresonant case. Suppose that the indicial polynomial (7.13) has two roots such that their difference is not an integer,

$$\text{(7.14)} \qquad \alpha_1 - \alpha_2 \notin \mathbb{Z},$$

which is the case we refer to as nonresonant.

If we take $\alpha = \alpha_1$ in our calculations, then we can equate the coefficients of $(z - z_0)^{\alpha+n}$ (for $n \geq 1$) to zero in (7.12) and find the relationships:

$$\text{(7.15)} \qquad P(\alpha_1 + n)c_n = -\sum_{k=0}^{n-1} \big((\alpha_1 + k)a_{n-k} + b_{n-k}\big)c_k,$$

for $n \in \mathbb{N}$.

However, $P(\alpha_1 + n) \neq 0$ for $n \in \mathbb{N}$. To see why, let's assume the contrary. If there existed an integer $n_0 = 1, 2, \ldots$ for which $P(\alpha_1 + n_0) = 0$, then, given that $P(\alpha)$ possesses only two roots, we would necessarily find $\alpha_2 = \alpha_1 + n_0$, which would directly contradict the relation in (7.14).

This implies that (7.15) gives us each coefficient c_n in terms of the previous ones $c_0 = 1, c_2, \ldots, c_{n-1}$. In other words, we can completely determine a solution in the form of a Frobenius series.

Now, if we take $\alpha = \alpha_2$ in our calculations, then we return to (7.15) with α_2 instead of α_1.

It is therefore clear that if $P(\alpha_2 + n) \neq 0$ for $n \in \mathbb{N}$. (which is equivalent to saying that $\alpha_1 - \alpha_2 \neq n$). Then, the method provides us with another solution in the form of a Frobenius series, this time with the exponent $\alpha = \alpha_2$.

Resonant Case. Let's focus on the case where $\alpha_1 = \alpha_2$. Following a similar approach as in the study of the Euler equation, we introduce the differential operator

$$L = (z - z_0)^2 \frac{\mathrm{d}^2}{\mathrm{d}z^2} + (z - z_0)A(z)\frac{\mathrm{d}}{\mathrm{d}z} + B(z).$$

We let L act on a Frobenius series (7.10) with a variable exponent α, where the coefficients $c_n = c_n(\alpha)$ depend on α except for $c_0 = 1$. This gives us:

$$Lu = (z - z_0)^\alpha P(\alpha) + \sum_{n\geq 1}(z - z_0)^{\alpha+n} Q_n(\alpha),$$

where $P(\alpha)$ and $Q_n(\alpha)$ are the same functions as those in (7.12). The indicial polynomial is given by:

$$\boxed{P(\alpha) = \alpha^2 + (a_0 - 1)\alpha + b_0,}$$

and

$$Q_n(\alpha) = \big(\alpha(\alpha - 1) + 2\alpha n + n(n - 1) + \alpha a_0 + na_0 + b_0\big)c_n$$
$$+ \sum_{k=0}^{n-1}\big((\alpha + k)a_{n-k} + b_{n-k}\big)c_k$$

$$= P(\alpha + n)c_n + \sum_{k=0}^{n-1} \left((\alpha + k)a_{n-k} + b_{n-k} \right)c_k.$$

For $\alpha = \alpha_1$, since there are no other roots, we can construct a solution in a recursive form by setting

$$Q_n(\alpha_1) = 0,$$

which yields the solution $u_1(z)$.

Now, consider any $\alpha \in \mathbb{C}$ such that

$$\operatorname{Re}\alpha > \operatorname{Re}\alpha_1 - 1.$$

For such α, we have $P(\alpha + n) \neq 0$ for $n \in \mathbb{N}$. This allows us to determine all coefficients $c_n(\alpha)$ using the recurrence relation:

$$P(\alpha + n)c_n = -\sum_{k=0}^{n-1} \left((\alpha + k)a_{n-k} + b_{n-k} \right)c_k,$$

for $n \in \mathbb{N}$.

The corresponding series satisfies:

$$Lu = (z - z_0)^\alpha P(\alpha).$$

Taking the derivative with respect to α in this equation yields:

$$L\left(\frac{\partial u}{\partial \alpha} \right) = (z - z_0)^\alpha P'(\alpha) + (z - z_0)^\alpha \log(z - z_0)P(\alpha).$$

Since

$$P(\alpha) = (\alpha - \alpha_1)^2, \quad P'(\alpha) = 2(\alpha - \alpha_1),$$

both vanish at $\alpha = \alpha_1$, we find that:

$$L\left(\frac{\partial u}{\partial \alpha} \bigg|_{\alpha = \alpha_1} \right) = 0.$$

Therefore, we conclude that

$$\frac{\partial u}{\partial \alpha} \bigg|_{\alpha = \alpha_1}$$

serves as a second solution of the form (7.1):

$$u_1(z) = u_0(z) \log(z - z_0) + (z - z_0)^{\alpha_1} \sum_{n=1}^{\infty} c_n'(z - z_0)^n,$$

where $u_0(z)$ is the Frobenius series solution corresponding to α_1.

We now introduce Fuchs' theorem for Equation (7.11).

Nonresonant Fuchs' Theorem

Let α_1 and α_2 be the roots of the indicial polynomial, ordered such that $\operatorname{Re}\alpha_1 \geq \operatorname{Re}\alpha_2$, satisfying $\alpha_1 - \alpha_2 \notin \mathbb{N}_0$, then there exist two linearly independent solutions in the form of Frobenius series:

$$u_1(z) = (z - z_0)^{\alpha_1} \sum_{n=0}^{\infty} c_n(z - z_0)^n,$$

$$u_2(z) = (z - z_0)^{\alpha_2} \sum_{n=0}^{\infty} c_n'(z - z_0)^n,$$

with initial conditions $c_0 = c_0' = 1$.

Theorem

Resonant Fuchs' Theorem

Let α_1 and α_2 be the roots of the indicial polynomial, satisfying $\operatorname{Re}\alpha_1 \geq \operatorname{Re}\alpha_2$, then

(1) In the case where $\alpha_1 - \alpha_2 = 0$, we have two linearly independent solutions in the form of:

$$u_1(z) = (z - z_0)^{\alpha} \sum_{n=0}^{\infty} c_n(z - z_0)^n,$$

$$u_2(z) = u_1(z)\log(z - z_0) + (z - z_0)^{\alpha} \sum_{n=0}^{\infty} c_n'(z - z_0)^n,$$

where $\alpha := \alpha_1 = \alpha_2$ and the initial conditions are $c_0 = 1$ and $c_0' = 0$.

(2) If $\alpha_1 - \alpha_2 \in \mathbb{N}$, then we have two linearly independent solutions of the form:

$$u_1(z) = (z - z_0)^{\alpha_1} \sum_{n=0}^{\infty} c_n(z - z_0)^n,$$

$$u_2(z) = c u_1(z)\log(z - z_0) + (z - z_0)^{\alpha_2} \sum_{n=0}^{\infty} c_n'(z - z_0)^n,$$

where $c_0 = 1$, $c_0' = 1$, and c is a constant that can be zero.

Theorem

In all cases, nonresonant and resonant, the radii of convergence of the power series involved in the solutions u_1 and u_2 are at least the distance to the nearest singularity.

§7.4. Bessel Equation

ESSEL functions are mathematical functions that play a pivotal role in solving complex differential equations, particularly those involving circular or cylindrical symmetry. Their significance spans various scientific and engineering domains, where they provide essential solutions to a wide range of problems. In the realm of electromagnetics, Bessel functions are instrumental in analyzing the behavior of electromagnetic waves and radiation patterns from antennas. Engineers and physicists rely on these functions to design efficient antennas, microwave devices, and radar systems.

In the field of acoustics, Bessel functions are used to model sound wave propagation within cylindrical or spherical environments. This has practical applications in designing musical instruments, predicting underwater acoustics, and optimizing architectural acoustics in concert halls and auditoriums. Vibrations and oscillations in circular membranes (like drumheads) and cylindrical structures (such as vibrating strings) are analyzed using Bessel functions. Heat conduction problems in cylindrical or spherical objects also benefit from Bessel functions. Engineers utilize them to describe temperature distributions, aiding in the design of heat exchangers, thermal insulation systems, and cooling mechanisms.

In quantum mechanics, Bessel functions appear when dealing with systems exhibiting cylindrical symmetry. They are employed to solve the Schrödinger equation for particles confined within cylindrical potentials, such as in quantum dots or carbon nanotubes.

Optics relies on Bessel functions to understand the diffraction patterns produced by circular apertures and optical elements with circular symmetry. They are also crucial for modeling laser beams and optical fibers. In the context of wave diffraction and scattering, Bessel functions are valuable tools when waves encounter cylindrical or circular objects. Applications abound in radar, sonar, and imaging systems. Fluid dynamics problems with circular or cylindrical symmetry frequently employ Bessel functions to describe fluid flow phenomena, such as flows in pipes or around cylinders.

Signal processing utilizes Bessel functions to analyze and filter signals exhibiting cylindrical or spherical symmetry, making them indispensable for tasks like image processing and pattern recognition.

Finally, Bessel functions serve as fundamental mathematical tools in physics and engineering, aiding in solving a broad spectrum of differential equations. Their orthogonality properties are particularly useful for tackling boundary value problems and representing solutions using Fourier–Bessel series.

Let's examine the Bessel equation:

$$(7.16) \qquad z^2 u'' + z u' + (z^2 - \nu^2) u = 0, \quad \nu \in \mathbb{R}, \quad \nu \geq 0.$$

In this case, we have

$$a(z) = \frac{1}{z},$$

$$b(z) = 1 - \frac{v^2}{z^2},$$

making $z_0 = 0$ a regular singular point. The indicial polynomial reduces to:

$$P(\alpha) = \alpha^2 - v^2 = (\alpha - v)(\alpha + v).$$

Considering that we assume $v > 0$, the solutions α_1 and α_2 of the indicial equation with $\operatorname{Re}\alpha_1 \geq \operatorname{Re}\alpha_2$ are $\alpha_1 = v$ and $\alpha_2 = -v$.

§7.4.1. The solution u_1 and its Frobenius Series

The corresponding Frobenius series for

$$\alpha_1 = v$$

takes the form:

$$u_1 = c_0 z^v + c_1 z^{v+1} + \cdots.$$

When introduced into (7.16), it yields:

$$P(v+1)c_1 = 0$$

so that

$$((v+1)^2 - v^2)c_1 = 0,$$

and, consequently, $c_1 = 0$. From

$$P(v+n)c_n + c_{n-2} = 0$$

we get

$$c_n = -\frac{c_{n-2}}{n(n+2v)}.$$

For the corresponding Frobenius series, all coefficients of odd order vanish:

$$\boxed{c_1 = c_3 = \cdots = c_{2n+1} = \cdots = 0,}$$

while coefficients of even order satisfy the recurrence relation:

$$c_{2k} = -\frac{c_{2(k-1)}}{4k(k+v)}.$$

This leads to:

$$\boxed{c_{2k} = \frac{(-1)^k c_0}{4^k k!(v+k)\cdots(v+2)(v+1)}.}$$

Taking $c_0 = 2^{-v}$, we find the following solution in series form:

(7.17)
$$\boxed{\begin{aligned} J_v(z) &= \sum_{n=0}^{\infty} \frac{(-1)^n}{n!\Gamma(v+1)(v+n)\cdots(v+2)(v+1)}\left(\frac{z}{2}\right)^{2n+v} \\ &= \sum_{n=0}^{\infty} \frac{(-1)^n}{n!\Gamma(n+v+1)}\left(\frac{z}{2}\right)^{2n+v} \end{aligned}}$$

This equation defines the **Bessel function of the first kind** of order v.

Here, we use Euler's gamma function, to be discussed later.

Note that the series defining $J_\nu(z)$ converges across the entire complex plane due to the absence of singularities outside the origin in the coefficients $a(z)$ and $b(z)$.

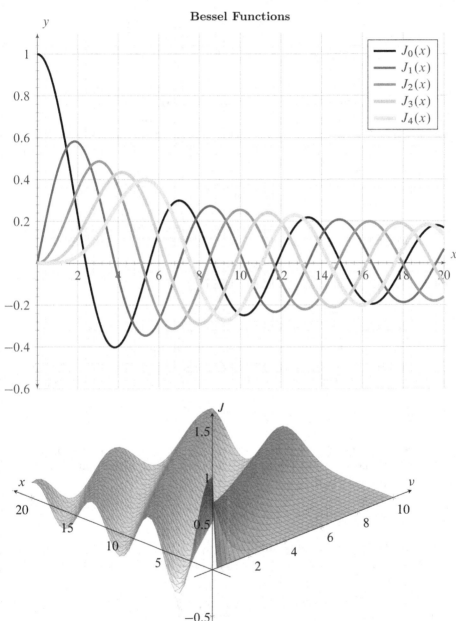

Graph of $F(x, \nu) = J_\nu(x)$ as function of (x, ν)

§7.4.2. The solution u_2

According to the statement of Fuchs' Theorem, to determine the solution u_2 associated with the other root $\alpha_2 = -\nu$ of the indicial polynomial, we have to distinguish the following cases according to the values it takes $\alpha_1 - \alpha_2 = 2\nu$:

(1) $2\nu \notin \mathbb{N}_0$.
(2) $2\nu \in \mathbb{N}_0$. In this case, there are two possible variants:
    (a) $\nu \in \mathbb{N}_0$.
    (b) $\nu \in \mathbb{N}_0 + \frac{1}{2} := \{\frac{1}{2}, \frac{3}{2}, \frac{5}{2}, \ldots\}$.

Case 1: $2\nu \notin \mathbb{N}_0$

If $\alpha_1 - \alpha_2 = 2\nu \notin \mathbb{N}_0$, the first alternative of the Fuchs Theorem assures us that the second solution u_2 will be a Frobenius series with exponent $\alpha_2 = -\nu$. In fact, if we repeat the previous calculation of the Frobenius series for $\alpha = \alpha_1 = \nu$, but now with $\alpha = \alpha_2 = -\nu$, we obtain the solution u_2 as the Bessel function of order $-\nu$:

$$J_{-\nu}(z) = \sum_{n=0}^{\infty} \frac{(-1)^n}{n!\,\Gamma(n-\nu+1)} \left(\frac{z}{2}\right)^{2n-\nu}.$$

Consequently, a general solution of the Bessel equation will be:

$$u(x) = C_1 J_\nu(x) + C_2 J_{-\nu}(x),$$

where C_1 and C_2 are two arbitrary constants.

Case 2a: $\nu \in \mathbb{N}_0$

Now, when $\alpha_1 - \alpha_2 = 2\nu$, this difference becomes a non-negative integer number $0, 2, 4, \ldots$, and we encounter two possibilities:

- If $\nu = 0$, then $\alpha_1 = \alpha_2 = 0$, leading us to alternative (2) of the Fuchs Theorem. Consequently, there exists a solution u_2 of the form:

$$u_2(z) = J_0(z)\log z + \sum_{n=0}^{\infty} c_n z^n.$$

- If $\nu = m$ with $m \in \mathbb{N}$, then $\alpha_1 = m$ and $\alpha_2 = -m$, placing us in alternative (3) of the Fuchs Theorem. Thus, a solution u_2 can be represented as:

(7.18) $$u_2(z) = c J_m(z)\log z + z^{-m}\sum_{n=0}^{\infty} c_n z^n,$$

where c could potentially be zero.

In both cases, the second solution $u_2(x)$ can be expressed using the Neumann functions, often referred to as:

(7.19) $$N_\nu(z) = \frac{J_\nu(z)\cos(\nu\pi) - J_{-\nu}(z)}{\sin \nu\pi}.$$

Also known as the Bessel functions of the second kind, they can alternatively be denoted as $Y_\nu(x)$. For $\nu \in \mathbb{N}_0$, the denominator in (7.19) vanishes, and the definition (7.19) is not applicable. However, the following notable identity holds for Bessel functions of integer order, which we omit proving:

$$J_{-m}(z) = (-1)^m J_m(z),$$

for $m \in \mathbb{N}_0$. By applying l'Hôpital's rule, we can define Neumann functions $N_m(z)$ using the limit:

$$N_m(z) = \lim_{\nu \to m} N_\nu(z),$$

for $m \in \mathbb{N}_0$, and consequently obtain the desired second solution for the Bessel equation. These functions can be expanded in the form:

$$(7.20) \quad N_m(z) = \frac{2}{\pi}\left[\log\left(\frac{z}{2}\right) + \gamma\right]J_m(z) - \frac{1}{\pi}\sum_{k=0}^{m-1}\frac{\Gamma(m-k)}{\Gamma(k+1)}\left(\frac{z}{2}\right)^{-m+2k}$$

$$- \frac{1}{\pi}\sum_{k=0}^{\infty}\frac{(-1)^m}{\Gamma(m+k-1)\Gamma(k+1)}\left(1 + \frac{1}{2} + \frac{1}{3} + \cdots + \frac{1}{k} + 1 + \frac{1}{2} + \frac{1}{3}\right.$$

$$\left. + \cdots + \frac{1}{m+k}\right)\left(\frac{z}{2}\right)^{m+2k},$$

where

$$\gamma = \lim_{n\to\infty}\left(1 + \frac{1}{2} + \frac{1}{3} + \cdots + \frac{1}{n} - \log n\right) \approx 0.5772156649$$

represents the so-called Euler–Mascheroni constant. Importantly, this formula indicates that the constant c in the expression (7.18) is not zero.

In summary, the general solution of the Bessel equation when ν is an integer $m \in \mathbb{N}_0$ can be expressed in the form:

$$u(x) = C_1 J_m(x) + C_2 N_m(x),$$

where C_1 and C_2 are two arbitrary constants.

The expressions (7.17) and (7.20) for the functions $J_m(x)$ and $N_m(x)$ allow us to find their asymptotic behavior as $x \to 0$, which is given by:

$$J_m(x) \sim \frac{1}{m!}\frac{x^m}{2^m},$$

$$N_m(x) \sim \begin{cases} \dfrac{2}{\pi}\log\dfrac{x}{2}, & \text{if } m = 0, \\[2mm] -\dfrac{(m-1)!\, 2^m}{\pi}\dfrac{1}{x^m}, & m \in \mathbb{N}. \end{cases}$$

The asymptotic behavior as $x \to \infty$ is deduced from other characterizations of these functions and turns out to be:

$$\begin{cases} J_m(x) \sim \sqrt{\dfrac{2}{\pi x}}\cos\left(x - \dfrac{m\pi}{2} - \dfrac{\pi}{4}\right), \\[2mm] N_m(x) \sim \sqrt{\dfrac{2}{\pi x}}\sin\left(x - \dfrac{m\pi}{2} - \dfrac{\pi}{4}\right). \end{cases}$$

Neumann Functions

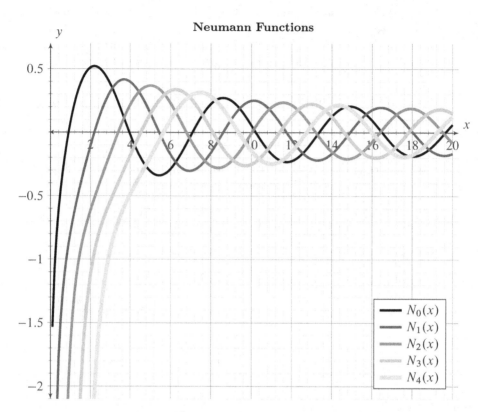

Case 2b: $v \in \mathbb{N}_0 + \frac{1}{2}$. Spherical Bessel and Neumann Functions

For $v = l + \frac{1}{2}$, $l \in \mathbb{N}_0$, the difference between the roots $\alpha_1 = v$ and $\alpha_2 = -v$ is a positive integer $\alpha_1 - \alpha_2 = 2l + 1$. We are once again in alternative (3) of the Fuchs Theorem, which leads to:

$$u_2(z) = c J_{l+\frac{1}{2}}(x) \log z + \frac{1}{z^{l+\frac{1}{2}}} \sum_{n=0}^{\infty} c_n z^n,$$

where c could be zero. In this case, it can be verified that $c = 0$, and therefore, the solution u_2 does not contain a logarithmic term and is reduced to a Frobenius series that coincides with $J_{-(l+\frac{1}{2})}(x)$. Indeed, $J_{-(l+\frac{1}{2})}(x)$ is a solution of the Bessel equation and is not proportional to $J_{l+\frac{1}{2}}(x)$ because $J_{l+\frac{1}{2}}(0) = 0$, while $J_{-(l+\frac{1}{2})}(x)$ diverges at $x = 0$.

In summary, a general solution of the Bessel equation when

$$v = l + \frac{1}{2},$$

for $l \in \mathbb{N}_0$, can be written in the form:

$$u(x) = C_1 J_{l+\frac{1}{2}}(x) + C_2 J_{-(l+\frac{1}{2})}(x),$$

where C_1 and C_2 are two arbitrary constants. An alternative form, using Neumann functions, is:

$$u(x) = C_1 J_{l+\frac{1}{2}}(x) + C_2 N_{l+\frac{1}{2}}(x).$$

A modified form of the functions of Bessel and Neumann of semi-integer order is their spherical versions:

$$\boxed{\begin{aligned} j_l(x) &= \sqrt{\frac{\pi}{2x}} J_{l+\frac{1}{2}}(x), \\ n_l(x) &= \sqrt{\frac{\pi}{2x}} N_{l+\frac{1}{2}}(x), \end{aligned}}$$

which appear in important applications. It should be noted that using the negative index Bessel functions, we have

$$n_l(x) = (-1)^{l+1} j_{-l}(x).$$

It is immediate to verify that if u is a solution of the Bessel equation of order v:

$$x^2 u'' + x u' + (x^2 - v^2)u = 0,$$

then

$$v = \frac{u}{\sqrt{x}}$$

satisfies:

$$x^2 v'' + 2x v' + \left(x^2 - v^2 + \frac{1}{4}\right)v = 0,$$

which for $v = l + \frac{1}{2}$ reduces to:

$$x^2 v'' + 2x v' + (x^2 - l(l+1))v = 0.$$

Therefore, this is the differential equation satisfied by the spherical functions of Bessel and Neumann. A general solution is given by:

$$v(x) = C_1 j_l(x) + C_2 n_l(x),$$

where C_1 and C_2 are two arbitrary constants. To conclude our discussion on these spherical functions, it is worth noting that their explicit forms are quite simple and can be determined using the following **Rayleigh's formulas:**

$$\boxed{\begin{cases} j_l(x) = (-x)^l \left(\dfrac{1}{x}\dfrac{d}{dx}\right)^l \dfrac{\sin x}{x}, \\ n_l(x) = -(-x)^l \left(\dfrac{1}{x}\dfrac{d}{dx}\right)^l \dfrac{\cos x}{x}. \end{cases}}$$

Initial Spherical Bessel Functions

$$j_0(x) = \frac{\sin x}{x},$$

$$j_1(x) = \frac{\sin x}{x^2} - \frac{\cos x}{x},$$

$$j_2(x) = \left(\frac{3}{x^2} - 1\right)\frac{\sin x}{x} - \frac{3\cos x}{x^2},$$

$$j_3(x) = \frac{3}{x}\left(\frac{5}{x^2} - 2\right)\frac{\sin x}{x} - \left(\frac{15}{x^2} - 1\right)\frac{\cos x}{x},$$

$$j_4(x) = \left(\frac{105}{x^4} - \frac{45}{x^2} + 1\right)\frac{\sin x}{x} - \frac{5}{x}\left(\frac{21}{x^2} - 2\right)\frac{\cos x}{x},$$

$$j_5(x) = \frac{15}{x}\left(\frac{63}{x^4} - \frac{28}{x^2} + 1\right)\frac{\sin x}{x} - \left(\frac{945}{x^4} - \frac{105}{x^2} + 1\right)\frac{\cos x}{x},$$

$$j_6(x) = \left(\frac{10395}{x^6} - \frac{4725}{x^4} + \frac{210}{x^2} - 1\right)\frac{\sin x}{x} - \frac{21}{x}\left(\frac{495}{x^4} - \frac{60}{x^2} + 1\right)\frac{\cos x}{x}.$$

Spherical Bessel Functions

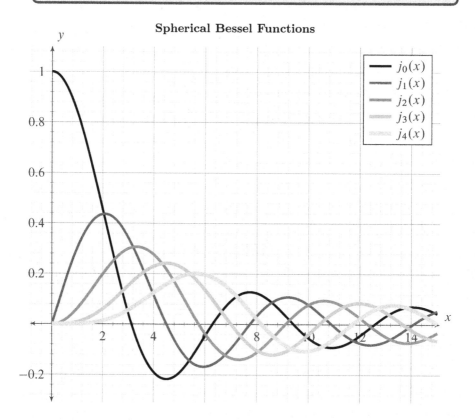

Initial Spherical Neumann Functions

$$n_0(x) = -\frac{\cos x}{x},$$

$$n_1(x) = -\frac{\cos x}{x^2} - \frac{\sin x}{x},$$

$$n_2(x) = -\left(\frac{3}{x^2} - 1\right)\frac{\cos x}{x} - \frac{3\sin x}{x^2}$$

$$n_3(x) = -\frac{3}{x}\left(\frac{5}{x^2} - 2\right)\frac{\cos x}{x} - \left(\frac{15}{x^2} - 1\right)\frac{\sin x}{x},$$

$$n_4(x) = -\left(\frac{105}{x^4} - \frac{45}{x^2} + 1\right)\frac{\cos x}{x} - \frac{5}{x}\left(\frac{21}{x^2} - 2\right)\frac{\sin x}{x},$$

$$n_5(x) = -\frac{15}{x}\left(\frac{63}{x^4} - \frac{28}{x^2} + 1\right)\frac{\cos x}{x} - \left(\frac{945}{x^4} - \frac{105}{x^2} + 11\right)\frac{\sin x}{x},$$

$$n_6(x) = -\left(\frac{10395}{x^6} - \frac{4725}{x^4} + \frac{210}{x^2} - 1\right)\frac{\cos x}{x} - \frac{21}{x}\left(\frac{495}{x^4} - \frac{60}{x^2} + 1\right)\frac{\sin x}{x}.$$

Spherical Neumann Functions

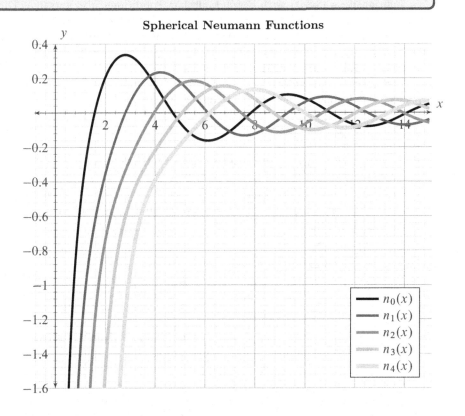

The Fourier–Bessel expansion is a mathematical tool used to represent functions with radial or cylindrical symmetry in terms of Bessel functions. It is a valuable technique in solving a wide range of physical problems where such symmetry arises, and it provides a means to approximate complex functions in a simpler and more manageable form.

Fourier–Bessel Expansions

For $\nu > -1$, consider the set of infinite zeros of $J_\nu(x) = 0$, denoted by $\{c_{\nu,n}\}_{n=1}^\infty$ and ordered strictly in increasing manner: $c_{\nu,1} < c_{\nu,2} < \cdots$. Then, the sequence $\{J_\nu(c_{\nu,n}x)\}_{n=1}^\infty$ forms a complete orthogonal set in $L^2([0,1], x\mathrm{d}x)$. For any function $f(x) \in L^2([0,1], x\mathrm{d}x)$, the expansion

$$f(x) = \sum_{n=1}^\infty c_n J_\nu(c_{\nu,n}x), \quad c_n = \frac{2}{(J_{\nu+1}(c_{\nu,n}))^2} \int_0^1 J_\nu(c_{\nu,n}x)f(x)x\mathrm{d}x,$$

converges strongly in $L^2([0,1], x\mathrm{d}x)$ and is referred to as the Fourier–Bessel expansion. It's worth noting that $(J_{\nu+1}(c_{\nu,n}))^2 = (J_\nu'(c_{\nu,n}))^2$.

Theorem

Fourier–Bessel series with Robin boundary conditions, also known as Dini series, or Fourier–Dini series, is a mathematical technique used to approximate functions defined on an interval with boundary conditions that involve the first derivative of the function. It leverages Bessel functions (typically of the first kind) and expansion coefficients determined by the Robin boundary conditions to provide a representation of the function that satisfies the specified conditions at the endpoints of the interval.

Dini Expansions

For $a \in \mathbb{R}$ let $\{c_{\nu,n}\}_{n=1}^\infty$ represent the set of infinite zeros of $aJ_\nu(x) + xJ_\nu'(x)$, arranged in strictly increasing order: $c_{\nu,1} < c_{\nu,2} < \cdots$. In this case, the sequence $\{J_\nu(c_{\nu,n}x)\}_{n=1}^\infty$ forms a complete orthogonal set in $L^2([0,1], x\mathrm{d}x)$. For any function in $L^2([0,1], x\mathrm{d}x)$, the following expansion is valid:

$$f(x) = \sum_{n=1}^\infty c_n J_\nu(c_{\nu,n}x),$$

$$c_n = \frac{2c_{\nu,n}^2}{(c_{\nu,n}^2 - \nu^2 + a^2)(J_\nu(c_{\nu,n}))^2} \int_0^1 J_\nu(c_{\nu,n}x)f(x)x\mathrm{d}x.$$

This expansion converges strongly in $L^2([0,1], x\mathrm{d}x)$ and is referred to as the Dini expansion.

Theorem

Fourier–Bessel series and Dini series exhibit pointwise convergence, similar to Fourier series.

Dirichlet's Theorem

The Fourier–Bessel series and Dini series corresponding to any piecewise C^1 function u in $[0, 1]$ converge pointwise in the interval $x \in [0, 1]$ to the average

$$\frac{1}{2}(u(x_+) + u(x_-))$$

for each x.

Theorem

Recurrence relations are essential tools for calculating Bessel and Neumann functions, particularly when dealing with higher-order Bessel and Neumann functions. These relations allow you to express Bessel and Neumann functions of different orders in terms of another one, simplifying their computation.

Recurrence Relations

Let's assume that

$$\operatorname{Re} v > -\frac{1}{2}.$$

In this case, the Bessel and Neumann functions satisfy the following differential relations:

$$(x^{\pm v} J_v(x))' = \pm x^{\pm v} J_{v \mp 1}(x), \quad (x^{\pm v} N_v(x))' = \pm x^{\pm v} N_{v \mp 1}(x),$$

which can be written as:

$$J_v'(x) = \mp J_{v \pm 1}(x) \pm \frac{v}{x} J_v(x), \quad N_v'(x) = \mp N_{v \pm 1}(x) \pm \frac{v}{x} N_v(x),$$

and further imply:

$$\frac{2v}{x} J_v(x) = J_{v-1}(x) + J_{v+1}(x), \quad \frac{2v}{x} N_v(x) = N_{v-1}(x) + N_{v+1}(x),$$

$$2J_v'(x) = J_{v-1}(x) - J_{v+1}(x), \quad 2N_v'(x) = N_{v-1}(x) - N_{v+1}(x).$$

Theorem

Generating functions for Bessel functions are mathematical tools that provide a compact and systematic way to express Bessel functions as power series. These generating functions are particularly useful for deriving various properties, relationships, and identities associated with Bessel functions.

Generating Functions

(1) **Generating Function:**

$$\exp\left(\frac{1}{2}x\left(t-\frac{1}{t}\right)\right) = \sum_{n=-\infty}^{\infty} J_n(x).$$

(2) **Jacobi–Anger Expansion:**

$$e^{iz\cos(\phi)} = \sum_{n=-\infty}^{\infty} i^n J_n(z)e^{in\phi},$$

$$e^{\pm iz\sin(\phi)} = J_0(z) + 2\sum_{n=1}^{\infty} J_{2n}(z)\cos 2n\phi \pm 2i\sum_{n=0}^{\infty} J_{2n+1}(z)\sin(2n+1)\phi.$$

Theorem

Further reading: For a comprehensive and in-depth exploration of Bessel functions and their various extensions, including spherical versions and practical applications, the book by V. B. Korenev [47] is an invaluable resource. This comprehensive volume delves into the theory, properties, and wide-ranging applications of Bessel functions, making it an essential reference for both novices and experts in the field. In addition to Korenev's work, readers seeking a more concise yet informative introduction to Bessel functions can turn to [12]. This booklet offers a clear and approachable entry point into the subject.

For those interested in exploring the asymptotic behavior of Bessel functions and related topics, *Asymptotics and Special Functions* by Frank W. J. Olver [54] is an indispensable reference. Olver's expertise in the field provides valuable insights into the behavior of Bessel functions in various limits.

For the classical and authoritative treatment of Bessel functions, *A Treatise on the Theory of Bessel Functions* by G. N. Watson [75] remains a cornerstone in the study of Bessel functions and their properties. Watson's work is a classic that continues to be a go-to resource for researchers and enthusiasts alike.

Other valuable references include [3, 6, 53, 65, 77].

Historical notes: The origins of Bessel functions can be traced back to Daniel Bernoulli's investigation of heavy chains in 1738, where the zeroth-order Bessel function first emerged. Leonhard Euler, in 1781, expanded the study of these functions, exploring their properties while delving into the vibrations of membranes.

In 1764, Euler also encountered them in the context of vibrating membranes. Subsequently, Joseph Fourier harnessed Bessel functions to unravel the intricacies of heat conduction within solid cylinders, while Poisson applied them to the realm of heat conduction in spheres in 1823.

However, the historical significance of Bessel functions extends beyond the domains of waves and heat equations. Their emergence was also intertwined with the understanding of the Kepler problem, a pursuit to elucidate planetary motion. According to Watson [75] Joseph-Louis Lagrange (1736-1813) unraveled the formulation and solution of Kepler's Problem as early as 1770. Lagrange sought to express the radial coordinate and the eccentric anomaly as functions of time. While Lagrange managed to formulate these expressions with trigonometric functions of time, his computation was limited to initial coefficients.

It was not until 1816 that Bessel ingeniously demonstrated that these coefficients in the expansion for r could be elegantly represented through integrals. Then, in 1824, Bessel presented a comprehensive exploration of these functions, now widely recognized as Bessel functions, marking a milestone in their formalization and nomenclature.

§7.4.3. Bessel Meets Sturm and Liouville

We examine the Sturm–Liouville operator defined as follows:

$$\rho(x) = x, \qquad p(x) = x, \qquad q_v(x) = \frac{v^2}{x}, \qquad v \geq 0.$$

This leads to the corresponding operator:

$$L_v u = -\frac{1}{x}\frac{d}{dx}\left(x\frac{du}{dx}\right) + \frac{v^2}{x^2}u = -\frac{d^2u}{dx^2} - \frac{1}{x}\frac{du}{dx} + \frac{v^2}{x^2}u.$$

The eigenvalue problem $L_v u = \lambda u$ gives rise to the following differential equation:

$$-\frac{d^2u}{dx^2} - \frac{1}{x}\frac{du}{dx} + \left(\frac{v^2}{x^2} - \lambda\right)u = 0.$$

By introducing the variable transformation:

$$z = \sqrt{\lambda}x$$

the equation transforms into the Bessel equation:

$$\frac{d^2u}{dz^2} + \frac{1}{z}\frac{du}{dz} + \left(1 - \frac{v^2}{z^2}\right)u = 0$$

with a general solution of the form:

$$u(z) = c_1 J_v(z) + c_2 N_v(z),$$

where J_v represents Bessel functions and N_v represents Neumann functions. Consequently, the general solution of the eigenvalue equation $L_v u = \lambda u$ is:

(7.21) $$u(x) = c_1 J_v(\sqrt{\lambda}x) + c_2 N_v(\sqrt{\lambda}x).$$

For Bessel and Neumann functions near the origin, their behavior is as follows:

$$J_\nu(x) \sim \frac{1}{\Gamma(\nu+1)} \left(\frac{x}{2}\right)^\nu, \qquad N_\nu(x) \sim \begin{cases} \dfrac{2}{\pi} \log x & \nu = 0, \\[2mm] -\dfrac{\Gamma(\nu)}{\pi(2x)^\nu} & \nu \neq 0, \end{cases}$$

for $x \to 0$.

Two types of intervals, namely those with $a > 0$ and those with $a = 0$, are of interest. In cases where $a > 0$, the operator is both regular and symmetric over the domains described in Chapter 3. However, when $a = 0$, the operator becomes singular due to properties such as $\rho(0) = 0$, $p(0) = 0$, or the nondifferentiability of $q_\nu(x)$ at $x = 0$. In this scenario, for $\nu > 0$, an appropriate domain over which the operator L_ν is symmetric can be represented by the following linear subspace of $L^2([0,b], x\,\mathrm{d}x)$:

$$\mathfrak{D}_\nu := \left\{ u \in C^\infty((0,b]) : u(b) = 0, \lim_{x\to 0}(x^\nu u(x)) = 0, \exists \lim_{x\to 0}\left(x^{1-\nu}\frac{\mathrm{d}u}{\mathrm{d}x}\right) \right\}.$$

The derivation of the symmetric nature involves considering that for functions u and v in $\mathfrak{D}_\nu$, the two limits

$$\lim_{x\to 0}\left(x^{1-\nu}\frac{\mathrm{d}\bar{u}}{\mathrm{d}x}(x)x^m v(x) - x^m \bar{u}(x)x^{1-\nu}\frac{\mathrm{d}v}{\mathrm{d}x}\right),$$

$$\lim_{x\to b}\left(x^{1-\nu}\frac{\mathrm{d}\bar{u}}{\mathrm{d}x}(x)x^\nu v(x) - x^m \bar{u}(x)x^{1-\nu}\frac{\mathrm{d}v}{\mathrm{d}x}\right),$$

are both equal to zero. Therefore, condition (4.9) is satisfied. Furthermore, considering the solution (7.21) from the eigenvalue equation and accounting for the behavior at the origin of the Bessel functions, we observe that the conditions at $x = 0$

$$\lim_{x\to 0}(x^m u(x)) = 0, \quad \exists \lim_{x\to 0}\left(x^{1-m}\frac{\mathrm{d}u}{\mathrm{d}x}\right),$$

are met when $c_2 = 0$. Additionally, J_ν is in $C^\infty((0,b])$ if $\nu \in \{0,1,2,\dots\}$. Therefore, we will only consider the case where m is a nonnegative integer number. By imposing the boundary condition $u(b) = 0$, we obtain the following condition to determine the eigenvalues λ_n:

$$J_\nu(\sqrt{\lambda}b) = 0.$$

Due to the symmetric nature of the operator, the eigenfunctions

$$u_n(x) = J_\nu\left(\sqrt{\lambda_n}x\right),$$

for $n \in \mathbb{N}$ form an orthogonal set:

$$\begin{aligned}(u_n, u_{n'}) &= \int_0^b J_\nu\left(\sqrt{\lambda_n}x\right) J_{n'}\left(\sqrt{\lambda_{n'}}x\right) x\,\mathrm{d}x \\ &= \frac{b^2}{2}\left(\frac{\mathrm{d}J_\nu}{\mathrm{d}x}\left(\sqrt{\lambda_n}b\right)\right)^2 \delta_{nn'}.\end{aligned}$$

Moreover, this set is complete in $L^2([0, b], x\mathrm{d}x)$. Therefore, any function $u \in L^2([0, b], x\mathrm{d}x)$ can be expanded as

$$u(x) = \sum_{n=1}^{\infty} c_n J_v\left(\sqrt{\lambda_n} x\right).$$

For $v = 0$, a suitable domain is

$$\mathfrak{D}_0 = \left\{u \in C^{\infty}((0, b]) : u(b) = 0, \exists \lim_{x \to 0} u(x), \lim_{x \to 0}\left(x\frac{\mathrm{d}u}{\mathrm{d}x}\right) = 0\right\}.$$

The inferred symmetric nature arises from observing that, for any pair of functions u and v in $\mathfrak{D}_0$, the limits

$$\lim_{x \to 0}\left(x\frac{\mathrm{d}\bar{u}}{\mathrm{d}x}(x)v(x) - \bar{u}(x)x\frac{\mathrm{d}v}{\mathrm{d}x}\right)$$

and

$$\lim_{x \to b}\left(x\frac{\mathrm{d}\bar{u}}{\mathrm{d}x}(x)v(x) - \bar{u}(x)x\frac{\mathrm{d}v}{\mathrm{d}x}\right)$$

are zero.

§7.5. Euler Gamma Function

ULER gamma function is an extension to the complex plane of the factorial of a positive integer, denoted as $n! := n(n-1)\cdots 1$. There are several equivalent ways to define it. In these notes, we will present three of them.

Firstly, there's Euler's original definition, which expresses the gamma function as an infinite limit. Then, we discuss Legendre's definition of the Euler gamma function as an integral, and finally, the Weierstrass definition as an infinite product.

§7.5.1. Euler's Definition as an Infinite Product

Euler's definition is given by the following limit:

$$\Gamma(z) := \lim_{n \to \infty} \frac{1 \cdot 2 \cdot 3 \cdots n}{z(z+1)\cdots(z+n)}n^z,$$

with

$$z \in \mathbb{C} \setminus \{0, -1, -2, \dots\},$$

and

$$n^z = e^{z \log n}.$$

From this definition, we can deduce the recurrence relation:

(7.22) $$\Gamma(z+1) = z\Gamma(z).$$

This difference equation characterizes Euler's Γ function. Unlike other significant functions in mathematical physics, this Γ function doesn't satisfy any differential equation. In fact, Otto Hölder proved in 1887 that for every $n \in \mathbb{N}_0$, there is no nonzero polynomial

$$P \in \mathbb{C}[X; Y_0, Y_1, \ldots, Y_n]$$

such that

$$P\left(z; \Gamma(z), \Gamma'(z), \ldots, \Gamma^{(n)}(z)\right) = 0$$

for all $z \in \mathbb{C} \setminus \{0, -1, -2, \ldots\}$.

From Euler's definition, we can compute the value of the function at $z = 1$:

$$\Gamma(1) = \lim_{n \to \infty} \frac{1 \cdot 2 \cdot 3 \cdots n}{2 \cdots (1+n)} n = 1,$$

and further, using (7.22):

$$\Gamma(2) = 1,$$
$$\Gamma(3) = 2\Gamma(2) = 2,$$

$$\vdots$$

$$\Gamma(n) = 1 \cdot 2 \cdot 3 \cdots (n-1) = (n-1)!.$$

Thus, Euler's Γ function extends the factorial function to the complex plane, and sometimes the notation $z! = \Gamma(z+1)$ is used.

§7.5.2. Legendre's Definition as an Integral

Legendre's definition of the gamma function is given by the integral:

$$(7.23) \qquad \boxed{\Gamma(z) := \int_0^\infty t^{z-1} e^{-t} \, dt, \quad \mathrm{Re}\, z > 0.}$$

Legendre introduced this definition, along with the name of gamma function and the Greek symbol Γ, around the year 1811.

The restriction on the real part of z is to prevent divergences of the integral at the origin $t = 0$. An integration by parts leads to the same recurrence relation as Euler's definition:

$$\Gamma(z+1) = z\Gamma(z).$$

For $\mathrm{Re}\, z > 0$, we can express it in other forms, like:

$$\Gamma(z) = 2 \int_0^\infty e^{-t^2} t^{2z-1} \, dt,$$
$$= \int_0^1 \left(\log \left(\frac{1}{t} \right) \right)^{z-1} dt.$$

For instance, from the latter, we deduce that

$$\Gamma\left(\frac{1}{2}\right) = \sqrt{\pi}.$$

To demonstrate the equivalence of both Euler's and Legendre's definitions, we introduce the function:

$$F(z,n) = \int_0^n \left(1 - \frac{t}{n}\right)^n t^{z-1} dt, \quad \operatorname{Re} z > 0,$$

for positive integers $n \in \mathbb{N}$. It can be shown that

$$\lim_{n\to\infty} F(z,n) = \Gamma(z).$$

By changing variables to $u = \frac{t}{n}$, we obtain an alternative expression:

$$F(z,n) = n^z \int_0^1 (1-u)^n u^{z-1} du.$$

This leads us to:

$$F(z,n) = \frac{1 \cdot 2 \cdot 3 \cdots n}{z \cdots (z+n)} n^z,$$

confirming the result.

While the definition (7.23) is valid only for $\operatorname{Re} z > 0$, the relationship (7.22) allows us to determine values for negative real parts. For example,

$$(-1/2)\Gamma(-1/2) = \Gamma(1 - 1/2) = \sqrt{\pi},$$

yielding

$$\Gamma(-1/2) = -2\sqrt{\pi}.$$

The same recurrence relation implies that

$$\lim_{z\to 0} z\Gamma(z) = \lim_{z\to 0} \Gamma(z+1) = \Gamma(1) = 1,$$

establishing $z = 0$ as a simple pole of the Euler gamma function. Similar reasoning applies to the other poles $z = -n$ (the negative integers) of the Γ function, rendering it a meromorphic extension in the complex plane with poles that are simple, resembling the behavior of the factorial function.

§7.5.3. Weierstrass Infinite Product

The Weierstrass infinite product for the gamma function, also known as the Weierstrass product theorem, is a fundamental result in complex analysis that provides an infinite product representation for the gamma function. It was developed by the German mathematician Karl Weierstrass and plays a crucial role in understanding the properties of Euler's Γ function, especially in the context of complex numbers.

$$\boxed{\frac{1}{\Gamma(z)} := z e^{\gamma z} \prod_{n=1}^{\infty} \left(1 + \frac{z}{n}\right) e^{-z/n},}$$

where we've used the Euler–Mascheroni constant γ.

Euler gamma function

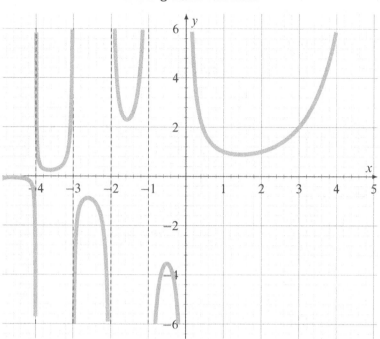

☞   **Observations:** Let's discuss other aspects of Euler's Γ function.

- **Euler's Reflection Formula:**

$$\Gamma(z)\Gamma(1-z) = \frac{\pi}{\sin \pi z}.$$

The Euler reflection formula is a powerful tool for understanding and calculating the values of the gamma function, especially for complex numbers. It provides a bridge between the values of the gamma function at different points and trigonometric functions, offering insights into the behavior of this important mathematical function.

- **Stirling Formula:** For $|z| \to \infty$ with $-\pi < \arg z < \pi$ we have

$$\Gamma(z) \sim \sqrt{2\pi} z^{z-1/2} e^{-z}.$$

The Stirling formula is an asymptotic approximation used to estimate the factorial of a positive integer, especially for large values of that integer. It provides an asymptotic representation of the factorial function and is named after the Scottish mathematician James Stirling, who contributed to its development. In statistics and probability theory, the Stirling formula is used to approximate the factorial in various distribution functions, such as the Poisson distribution and the normal distribution. In the field of statistical physics, this formula is commonly expressed as:

$$n! \sim \sqrt{2\pi n}\left(\frac{n}{e}\right)^n, \quad n \to \infty.$$

- **Factorial Variants:** Let's expand upon the concept of factorial by introducing various factorial variants:

  (1) Double Factorial: $n!!$ is defined as the product of all positive integers less than or equal to n that are congruent to n modulo 2.

  (2) Triple Factorial: $n!!!$ is defined as the product of all positive integers less than or equal to n that are congruent to n modulo 3.

  (3) q-Multiple Factorial: $(n)!^{(q)}$ is defined as the product of all positive integers less than or equal to n that are congruent to n modulo q.

These definitions lead to intriguing values for the gamma function:

$$\Gamma\left(n+\frac{1}{2}\right) = \frac{(2n-1)!!}{2^n}\sqrt{\pi} = \frac{(2n)!}{4^n n!}\sqrt{\pi},$$

$$\Gamma\left(n+\frac{1}{3}\right) = \frac{(3n-2)!!!}{3^n}\Gamma\left(\frac{1}{3}\right),$$

$$\Gamma\left(n+\frac{1}{4}\right) = \frac{(4n-3)!!!!}{4^n}\Gamma\left(\frac{1}{4}\right),$$

$$\Gamma\left(n+\frac{1}{q}\right) = \frac{(qn-(q-1))!^{(q)}}{q^n}\Gamma\left(\frac{1}{q}\right).$$

In general, this relationship can be expressed as:

(7.24)
$$\Gamma\left(n+\frac{p}{q}\right) = \Gamma\left(\frac{p}{q}\right)\frac{1}{q^n}\prod_{k=1}^{n}(kq+p-q).$$

It's worth noting that from the first equation and the reflection formula, we obtain:

$$\Gamma\left(\frac{1}{2}-n\right) = \frac{(-2)^n}{(2n-1)!!}\sqrt{\pi} = \frac{(-4)^n n!}{(2n)!}\sqrt{\pi}.$$

- **Multiplication Formulas:** The duplication formula for the gamma function, also known as the Legendre duplication formula, is given by:

$$\Gamma(z)\,\Gamma\left(z+\frac{1}{2}\right) = 2^{1-2z}\sqrt{\pi}\,\Gamma(2z).$$

Additionally, there is the multiplication theorem, sometimes referred to as Gauss's multiplication formula, which is expressed as:

$$\Gamma(z)\,\Gamma\left(z+\frac{1}{k}\right)\Gamma\left(z+\frac{2}{k}\right)\cdots\Gamma\left(z+\frac{k-1}{k}\right) = (2\pi)^{\frac{k-1}{2}}k^{\frac{1-2kz}{2}}\,\Gamma(kz).$$

This theorem applies for integer values of k belonging to the set of natural numbers, and it plays a fundamental role in the theory of the gamma function.

- **Bohr–Mollerup Log-Convexity Characterization:** In 1922, Danish mathematicians Harald Bohr and Johannes Mollerup established the Bohr–Mollerup theorem. This theorem is a pivotal result in the theory of special functions, particularly for the gamma function theory. It asserts that, among all functions that extend the factorial function to the positive real numbers, only the gamma function exhibits a unique property: log-convexity. This means that the natural logarithm of the gamma function is convex when considered over the positive real axis. To be more precise, the Bohr–Mollerup theorem characterizes the gamma function as the sole positive function $\Gamma(x)$ with a domain on positive numbers that satisfies three essential properties:

  - $\Gamma(1) = 1$,
  - $\Gamma(x + 1) = x\Gamma(x)$ for $x > 0$,
  - Logarithmic convexity: For $x, y > 0$ and $t \in [0, 1]$, it holds that $\log \Gamma(tx + (1-t)y) \le t \log \Gamma(x) + (1-t) \log \Gamma(y)$. This condition is equivalent to $\Gamma(tx + (1 - t)y) \le \Gamma(x)^t \Gamma(y)^{1-t}$.

  Bohr and Mollerup co-authored a highly influential four-volume textbook, "Lærebog i Matematisk Analyse" (Textbook in Mathematical Analysis), which featured this theorem. Despite their belief that it was a well-established result by others, it was, in fact, an original and significant contribution that had not been previously established by anyone else.

  Harald Bohr was the younger brother of the famous physicist, one of the founders and pioneers of quantum physics, Niels Bohr.

  Moreover, he was a famous footballer. During the 1905 season, Harald Bohr had the remarkable opportunity to share the field with his brother, Niels, who served as a goalkeeper. Harald's exceptional talent led to his selection to represent the Denmark national team in the 1908 Summer Olympics, a historic occasion, as it marked the first inclusion of football as an official event. Harald Bohr's efforts earned him a well-deserved silver medal, solidifying his place in Danish football history. His prowess as a footballer was so substantial that when he defended his doctoral thesis, the audience was reportedly filled with more football enthusiasts than mathematicians, a testament to his extraordinary popularity in both realms.

  In a particular context, the natural representation of the gamma function is often considered to be its natural logarithm $\log(\Gamma)$. This perspective becomes evident when examining its Taylor series expansion, which takes the form:

$$\log \Gamma(z+1) = -\gamma z + \sum_{k=2}^{\infty} \frac{\zeta(k)}{k} (-z)^k \quad |z| < 1.$$

Here, $\zeta(k)$ represents the Riemann zeta function.

- **Wielandt's Characterization:** The Wielandt theorem provides another significant characterization of the gamma function, highlighting its uniqueness. This theorem states that the gamma function is the sole function that simultaneously satisfies the following properties:
  - $\Gamma(1) = 1$,
  - For z in the complex domain excluding the nonpositive integers $\mathbb{Z} \setminus \mathbb{N}$:

$$\Gamma(z+1) = z\Gamma(z),$$

  - For all integers n and all complex numbers z:

$$\lim_{n \to \infty} \frac{\Gamma(n+z)}{\Gamma(n)\, n^z} = 1.$$

Historical notes: Daniel Bernoulli and Christian Goldbach were the first to consider, back in the 1720s, the extension of the factorial beyond natural numbers. In particular, on October 6, 1729 Bernoulli wrote a letter to Goldbach where he found

$$x! = \lim_{n \to \infty} \left(n+1+\frac{x}{2}\right)^{x-1} \prod_{k=1}^{n} \frac{k+1}{k+x},$$

that extends the factorial function to $\mathbb{R} \setminus \{0, -1, -2, \dots\}$. Almost at the same time on Octuber 13, Euler also wrote to Goldbach communicating to him the discovery of his product formula and a few months later, at the beginning of 1730, he wrote to Goldbach about the formula

$$n! = \int_0^1 (-\log x)^n \mathrm{d}x$$

for $\operatorname{Re} n > -1$. Note that under the change $x = -\log x$ this becomes Legendre's definition (7.23).

Further reading: For a clear and insightful exploration of the Euler gamma function, we highly recommend the booklet *Euler's Gamma Function* by Emil Artin [5]. Artin's work offers a lucid and approachable perspective on this fundamental mathematical function, making it an excellent choice for those seeking a solid foundation.

In addition to Artin's booklet, there are several other valuable references that delve into the Euler gamma function and related topics. These resources provide a comprehensive understanding of the subject and its applications. We suggest considering the following: This authoritative text [3] covers a wide range of special functions, including the gamma function, and it is a valuable resource for those interested in in-depth mathematical exploration. This comprehensive guide [6] offers a wealth of information on special functions, including the gamma function, suitable for both students and researchers.

For those interested in the asymptotic analysis of special functions, Olver's work [54] is an invaluable reference. It provides deep insights into the behavior of functions like the gamma function as variables become large.

§7.6. Remarkable Lives and Achievements

Who was Frobenius? Ferdinand Frobenius was a renowned German mathematician known for his significant contributions to various areas of mathematics, including group theory, matrix theory, linear algebra, and differential equations. He was born on October 26, 1849, in Charlottenburg, which is now part of Berlin, Germany, and he passed away on August 3, 1917 in Berlin.

Frobenius displayed exceptional mathematical talent from an early age and gained recognition for his abilities while still in school. He pursued his studies at the University of Berlin (now Humboldt University) and embarked on his mathematical journey under the guidance of eminent mathematicians like Leopold Kronecker and Karl Weierstrass. In 1870, Frobenius submitted his doctoral dissertation on the theory of linear differential equations and was awarded his doctorate.

Frobenius's groundbreaking work in group theory led to the introduction of "Frobenius groups." These are groups with specific properties that find applications across diverse mathematical areas. He developed the Frobenius normal form for matrices, a significant result in linear algebra with applications in algebraic number theory, representation theory, and other fields. Frobenius made pivotal contributions to the theory of linear transformations, vector spaces, and matrices. His insights and techniques have left a lasting impact on the study of linear algebra. He also formulated the Frobenius method for solving linear differential equations with regular singular points, which is crucial for finding solutions in terms of power series and holds applications in various scientific disciplines. Furthermore, Frobenius worked on topics in algebraic number theory, including quadratic forms and reciprocity laws.

Throughout his career, Frobenius held academic positions at several universities, including the University of Berlin, University of Halle, and the Eidgenössische Polytechnikum (now ETH Zurich). He mentored numerous students who later became prominent mathematicians themselves.

Frobenius's profound contributions to algebra, group theory, matrix theory, and more have solidified his place as a celebrated figure in the history of mathematics. His name is synonymous with foundational concepts and results that continue to exert a lasting impact on various branches of mathematics.

Who was Hermite? Charles Hermite, born on December 24, 1822, in Dieuze, France, and passing away on January 14, 1901, was a distinguished French mathematician celebrated for his profound contributions to various mathematical domains, with a particular focus on number theory, algebra, and mathematical analysis.

From an early age, Hermite displayed a remarkable affinity for mathematics, garnering recognition from his educators. Despite financial challenges, he pursued his mathematical education at the Collège de Nancy and later at the Collège Henri IV in Paris. It was in Paris that he encountered the works of influential mathematicians such as Joseph Liouville and Adrien-Marie Legendre, significantly shaping his intellectual journey.

Hermite's legacy includes a pivotal accomplishment: his proof that not all quintic equations are solvable algebraically through radicals. This monumental result, known as the Abel–Ruffini theorem, had long been conjectured by earlier mathematicians and was independently proven by Hermite and Évariste Galois. This breakthrough marked a pivotal moment in comprehending polynomial equations and their solvability.

His contributions extended deeply into the realm of number theory, notably in the exploration of algebraic numbers. Hermite introduced the concept of the Hermite normal form for quadratic forms, a concept essential to lattice theory and the study of Diophantine equations. His research on Hermite's constant and Hermite's reciprocity law further solidified his influence within the landscape of number theory.

Hermite's mathematical explorations also delved into transcendental numbers, where he delivered a proof of the transcendence of the mathematical constant e (the base of the natural logarithm), a demonstration later refined by subsequent mathematicians.

In the domain of mathematical analysis, Hermite left a significant mark through his contributions to the theory of elliptic functions. He pioneered a theory of modular functions that laid the groundwork for deeper insights into this branch of mathematics.

During his later years, Hermite confronted health adversities, including blindness. Yet, he remained actively engaged intellectually and continued to engage in correspondence with fellow mathematicians. He played a pivotal role in mentoring and guiding the next generation of mathematical minds, including the renowned Henri Poincaré.

Hermite's life and work stand as a testament to the dedication and brilliance of a mathematician who made substantial contributions across diverse mathematical domains. His theorems, findings, and concepts remain subjects of study and application among mathematicians globally, contributing to the continuous advancement of mathematical knowledge.

Who was Airy? George Biddell Airy, born on

July 27, 1801, in Alnwick, Northumberland, England, and passing away on January 2, 1892, in Greenwich, London, was a prominent British mathematician and astronomer. He made significant contributions to various areas of science and mathematics, particularly in the fields of optics, astronomy, and mathematical physics.

Airy's early education and passion for mathematics led him to study at Trinity College, Cambridge. He excelled in his studies and was later appointed the Lucasian Professor of Mathematics at Cambridge, a prestigious position previously held by legendary scientists such as Isaac Newton.

One of Airy's most notable achievements was his work on the theory of optics, specifically in the field of diffraction. He formulated the mathematical theory that describes how light waves interact with obstacles and openings, laying the foundation for understanding various optical phenomena. The "Airy disk," a pattern created by the diffraction of light, is named after him.

Airy's contributions to astronomy were equally significant. He served as the Astronomer Royal of the Royal Observatory in Greenwich for over four decades. In this role, he implemented reforms that improved the accuracy of astronomical observations and calculations. His work included the precise measurement of the Earth's shape and the study of planetary motion.

In the realm of mathematical physics, Airy's name is associated with the "Airy equation," a second-order linear differential equation with applications in various scientific disciplines, including fluid dynamics, quantum mechanics, and signal processing. Airy functions, solutions to the Airy equation, have a range of applications in both theoretical and applied mathematics.

Who was Legendre? Adrien-Marie Legendre, born on September 18, 1752, in Paris, France, and passing away on January 10, 1833, in Auteuil, France, was a distinguished French mathematician renowned for his significant contributions across a spectrum of mathematical disciplines, with particular emphasis on number theory, analysis, and celestial mechanics.

Legendre's early fascination with mathematics propelled him to enroll at the Collège Mazarin in Paris, where he embarked on cultivating his mathematical prowess. His exceptional aptitude did not go unnoticed, swiftly earning him a reputation as a promising young mathematician. Collaborations with prominent mathematicians of his era, including Joseph-Louis Lagrange and Pierre-Simon Laplace, further solidified his position in the mathematical community.

Among Legendre's most enduring accomplishments stands his notable work in number theory. He is particularly celebrated for his formulation and proof of the Law of Quadratic Reciprocity, a fundamental theorem offering crucial insights into the solvability of quadratic equations. This theorem holds profound implications in both pure and applied mathematics, making it a cornerstone result.

In addition to his contributions to number theory, Legendre played a pivotal role in the advancement of mathematical analysis. His development of the method of least squares, a powerful tool for fitting mathematical models to empirical data, brought about a revolution in data analysis and statistics. This method remains widely employed across various scientific disciplines to this day.

Legendre's influence extended into celestial mechanics, where his contributions to the understanding of planetary orbits and gravitational interactions laid a strong foundation for subsequent developments in the field.

His mathematical legacy encompasses the creation of the Legendre polynomials and associated Legendre functions, solutions with diverse applications to mathematical equations, physics, and engineering. For more information, see [11, 33].

It's worth noting that the portrait of Adrien-Marie Legendre had been inaccurately displayed for over a century. It had been erroneously identified as a portrait of the politician Louis Legendre, who played a prominent role during the French Revolution. The correct portrait, as shown here, was rediscovered in 2008 within the collection of 73 watercolor caricatures at the *Institut de France* in Paris. For more details, you can refer to Peter Duren's article titled *Changing Faces: The Mistaken Portrait of Legendre*, which was published in AMS Notices **56** (2009), pages 1440-1443.

Who was Bessel? Friedrich Bessel, born on July 22, 1784, in Minden, Westphalia (now in Germany), and passing away on March 17, 1846, in Königsberg, Prussia (now Kaliningrad, Russia), was a distinguished German mathematician and astronomer renowned for his significant contributions to various fields, particularly in mathematics, astronomy, and physics.

Bessel's early fascination with mathematics led him to pursue studies at the University of Göttingen, where he received guidance from prominent mathematicians such as Carl Friedrich Gauss. Bessel's mathematical talents were soon recognized, and he began his journey of influential discoveries.

Bessel is notably celebrated for his work on the Bessel functions, which are solutions to Bessel's differential equation and have far-reaching applications in diverse scientific disciplines, including physics and engineering. His contributions to the field of astronomy are equally remarkable. Bessel made pioneering advancements in the field of stellar parallax, which allowed for the measurement of distances to stars with unprecedented accuracy. This achievement revolutionized the field of astronomy and laid the groundwork for subsequent developments.

In addition to his work in mathematics and astronomy, Bessel played a crucial role in advancing geodesy—the science of measuring the Earth's shape and dimensions. His collaboration with fellow scientist Carl Friedrich Gauss resulted in the establishment of a highly accurate geodesic survey that contributed significantly to the understanding of Earth's physical characteristics.

Who was Laguerre? Edmond Nicolas Laguerre was born on April 9, 1834, in Bar-le-Duc, France. He displayed an early aptitude for mathematics and went on to pursue advanced studies in the subject. He completed his doctoral thesis at the University of Paris in 1857, with a dissertation titled "On the lengths of curves contained in a surface." His early work focused on geometry and analysis.

Laguerre is most famously associated with the development of Laguerre polynomials and the Laguerre equation. Laguerre polynomials are a family of orthogonal polynomials that arise in the study of various mathematical problems, including differential equations, numerical analysis, and quantum mechanics. Laguerre polynomials have numerous applications, including in solving problems related to quantum mechanics, where they describe the behavior of particles in certain potentials.

Laguerre also made contributions to continued fractions, complex analysis, and algebra. He developed a method for finding the continued fraction expansion of a function and used it to prove properties of certain functions. His work in complex analysis included investigations into the behavior of functions with complex variables and their singularities.

Laguerre passed away on August 14, 1886, in Paris, France. His legacy lives on through the mathematical concepts and results that bear his name. Laguerre's work continues to influence various branches of mathematics and science, and his polynomials and equations are used in a wide range of applications.

Who was Fuchs? Lazarus Immanuel Fuchs was born on May 5, 1833, in Moschin, Prussia (now Mosina, Poland). He made important contributions to the theory of linear differential equations, particularly those with regular singular points. His work was foundational in the study of what are now called "Fuchsian differential equations" and their solutions, often involving hypergeometric functions.

Fuchs's work was instrumental in advancing the understanding of the structure of solutions to linear differential equations with polynomial coefficients. His investigations into the properties of these equations and their solutions have applications in various areas of mathematics, including algebraic geometry, number theory, and mathematical physics.

Fuchs held academic positions at various institutions during his career, including the University of Göttingen. He contributed significantly to mathematical research in his lifetime, and his work continues to be influential in the study of differential equations and related fields.

Fuchs passed away on July 26, 1902, in Berlin, Germany. His contributions to mathematics, particularly in the theory of differential equations, have left a lasting impact on the field and continue to be studied by mathematicians today.

Who was Dini? Ulisse Dini was an Italian mathematician known for his contributions to the field of analysis, particularly in the areas of calculus and mathematical analysis. He is best known for his work on the theory of functions of a real variable, where he made significant advancements in understanding the convergence of series and the behavior of functions.

Dini was born on November 14, 1845, in Pisa, Italy. He studied at the University of Pisa, where he later became a professor and spent most of his academic career. Dini's work was recognized both in Italy and internationally, and he made substantial contributions to the development of mathematical analysis.

One of Dini's most well-known results is "Dini's theorem," which concerns the convergence of a sequence of functions. The theorem establishes conditions under which a sequence of functions converges uniformly to a limit function. This theorem is essential in the study of real analysis and provides insights into the behavior of functions and their convergence properties.

Dini also worked on the theory of Fourier series, the theory of potential, and the theory of elliptic functions. His research contributions significantly influenced the development of modern analysis and calculus.

He was an influential figure in the Italian mathematical community and played a role in shaping mathematical education and research in Italy.

Dini died on October 28, 1918, in Florence, Italy. His legacy lives on through his mathematical contributions and his impact on the study of analysis and calculus. Dini's theorem and his work on the convergence of functions remain fundamental results in the field of mathematics.

Who was Euler? Born on April 15, 1707, in Basel, Switzerland, Leonhard Euler stands as an eminent mathematician and physicist whose extensive and diverse contributions have indelibly marked the fields of mathematics and science. Throughout his life, Euler delved into a remarkable array of topics, showcasing an extraordinary ability to explore, innovate, and synthesize complex ideas.

Euler's stature places him among the greatest mathematicians in history, and his life story deserves more than a mere bio of a few paragraphs. Hence, let us offer only a few general comments on his monumental contributions.

His achievements span a wide spectrum of areas, ranging from the foundational concepts of calculus to the intricate realm of number theory. Euler's insights extended to graph theory, a field in which he played a pivotal role, notably exemplified by his solution to the Seven Bridges of Königsberg problem.

One of his most iconic contributions is undoubtedly Euler's formula, $e^{i\pi} + 1 = 0$, a brilliant linkage of five of the most significant constants in mathematics. This formula, a masterpiece of mathematical elegance, serves as a testament to Euler's capacity to discern profound relationships among seemingly disparate mathematical concepts.

In the realm of special functions, Euler's work on the gamma function holds particular importance. The gamma function, denoted by $\Gamma(x)$, is an extension of the factorial function to complex and real numbers. Euler's contributions to the development of the gamma function laid the groundwork for its deeper understanding and its applications in diverse areas of mathematics, including complex analysis, number theory, and statistics.

Beyond his theoretical pursuits, Euler achieved significant breakthroughs in mathematical physics. His work on topics such as the three-body problem and the dynamics of rigid bodies laid the groundwork for major developments in classical mechanics.

Euler's legacy is characterized not merely by the sheer volume of his output, which comprises over 800 published papers, but also by his remarkable ability to bridge divergent fields. His research on differential equations, for instance, played an integral role in shaping both mathematics and the physical sciences.

As a polymath proficient in multiple languages, Euler's collaborations with contemporaries enriched the global scientific community. His skill in uncovering connections between diverse realms of mathematics and science continues to inspire generations of thinkers. His name remains synonymous with brilliance, innovation, and an unyielding pursuit of knowledge.

On September 18, 1783, in Saint Petersburg, Russia, Euler passed away, leaving a legacy that continues to shape mathematics, science, and our perception of the interconnectedness of the universe. For more information, see [60].

§7.7. Exercises

§7.7.1. Exercises with Solutions

(1) To solve the initial value problem of the classical harmonic oscillator
using the Frobenius method, consider the equation

(7.25)
$$u'' + \omega^2 u = 0,$$

with initial conditions

$$\begin{cases} u(0) = a_0, \\ u'(0) = a_1. \end{cases}$$

Solution: Given that $a(z) \equiv 0$ and $b(z) = \omega^2$ are entire functions analytic in $\mathbb{C}$, the solution will also be an entire function. The
power series expansion (7.4) can be written as

$$u'' = \sum_{n=0}^{\infty} (n+2)(n+1)c_{n+2} z^n.$$

Substituting this into Equation (7.25), we obtain

$$\sum_{n=0}^{\infty} \Big((n+2)(n+1)c_{n+2} + \omega^2 c_n \Big) z^n = 0,$$

which, together with the boundary conditions, implies $c_0 = a_0$, $c_1 = a_1$ and

$$(n+2)(n+1)c_{n+2} + \omega^2 c_n = 0, \qquad n \in \mathbb{N}_0.$$

This leads to

$$c_{2n} = a_0 \frac{(-1)^n}{(2n)!} \omega^{2n}, \qquad c_{2n+1} = a_1 \frac{(-1)^n}{(2n+1)!} \omega^{2n},$$

and the power series expansion becomes

$$u = a_0 \Big(\sum_{n=0}^{\infty} \frac{(-1)^n}{(2n)!} \omega^{2n} z^{2n} \Big) + \frac{a_1}{\omega} \Big(\sum_{n=0}^{\infty} \frac{(-1)^n}{(2n+1)!} \omega^{2n+1} z^{2n+1} \Big)$$

$$= a_0 \cos(\omega z) + \frac{a_1}{\omega} \sin(\omega z).$$

Therefore, the solution is a combination of trigonometric functions
defined in the complex plane $\mathbb{C}$, making it an entire function.

(2) Solve the Airy equation initial value problem

$$u'' = zu,$$

with initial conditions

$$\begin{cases} u(0) = a_0, \\ u'(0) = a_1. \end{cases}$$

Solution: In this case, $a(z) \equiv 0$ and $b(z) = -z$, making $z_0 = 0$ an ordinary point. Therefore, the solution will be an entire function that admits a power series expansion (7.4) around that point:

$$u(z) = \sum_{n=0}^{\infty} c_n z^n,$$

which is convergent across the entire complex plane. Substituting this series into the differential equation yields

$$\sum_{n=0}^{\infty} (n+2)(n+1)c_{n+2}z^n = \sum_{n=1}^{\infty} c_{n-1}z^n,$$

which can be rewritten as

$$2c_2 + \sum_{n=0}^{\infty} \left((n+3)(n+2)c_{n+3} - c_n \right) z^{n+1} = 0.$$

As a result, we find that $c_2 = 0$. Additionally, the boundary conditions lead to the following relationships:

$$c_0 = a_0,$$
$$c_1 = a_1.$$

Furthermore, the ordinary differential equation provides us with the recurrence relation:

$$c_{n+3} = \frac{1}{(n+3)(n+2)} c_n, \quad n \in \mathbb{N}_0.$$

Hence, for $n \in \mathbb{N}$, we have $c_{3n+2} = 0$, and we can express c_{3n} and c_{3n+1} as follows:

$$\boxed{c_{3n} = \frac{(3n-2)!!!}{(3n)!}a_0, \quad c_{3n+1} = \frac{(3n-1)!!!}{(3n+1)!}a_1,}$$

Utilizing the formulas:

$$(3n)! = (3n)!!!(3n-1)!!!(3n-2)!!!, \quad (3n+1)! = (3n+1)!!!(3n)!!!(3n-1)!!!,$$

we can derive alternative expressions:

$$\boxed{c_{3n} = \frac{1}{(3n)!!!(3n-1)!!!}a_0, \quad c_{3n+1} = \frac{1}{(3n+1)!!!(3n)!!!}a_1.}$$

Recognizing that $(3n)!!! = 3^n n!$, we can simplify these expressions to:

$$\boxed{c_{3n} = \frac{1}{3^n n!(3n-1)!!!}a_0, \quad c_{3n+1} = \frac{1}{3^n n!(3n+1)!!!}a_1.}$$

Finally, we can rewrite these expressions in terms of Euler's gamma function. Using (7.24), we find:

$$(3n-1)!!! = 3^n \frac{\Gamma\left(n+\frac{2}{3}\right)}{\Gamma\left(\frac{2}{3}\right)}, \quad (3n+1)!!! = 3^n \frac{\Gamma\left(n+\frac{4}{3}\right)}{\Gamma\left(\frac{4}{3}\right)},$$

leading to the expressions:

$$c_{3n} = \frac{\Gamma\left(\frac{2}{3}\right)}{9^n n! \, \Gamma\left(n+\frac{2}{3}\right)} a_0, \quad c_{3n+1} = \frac{\Gamma\left(\frac{4}{3}\right)}{9^n n! \, \Gamma\left(n+\frac{4}{3}\right)} a_1.$$

As a result, the solution can be expressed as:

$$u = a_0 \sum_{n=0}^{\infty} \frac{\Gamma\left(\frac{2}{3}\right)}{9^n n! \, \Gamma\left(n+\frac{2}{3}\right)} z^{3n} + a_1 \sum_{n=0}^{\infty} \frac{\Gamma\left(\frac{4}{3}\right)}{9^n n! \, \Gamma\left(n+\frac{4}{3}\right)} z^{3n+1}.$$

The so-called Airy functions, often used in literature, are defined by

$$\mathrm{Ai}(z) = 3^{-2/3} \sum_{n=0}^{\infty} \frac{z^{3n}}{9^n n! \, \Gamma\left(n+\frac{2}{3}\right)} - 3^{-4/3} \sum_{n=0}^{\infty} \frac{z^{3n+1}}{9^n n! \, \Gamma\left(n+\frac{4}{3}\right)},$$

$$\mathrm{Bi}(z) = 3^{-1/6} \sum_{n=0}^{\infty} \frac{z^{3n}}{9^n n! \, \Gamma\left(n+\frac{2}{3}\right)} + 3^{-5/6} \sum_{n=0}^{\infty} \frac{z^{3n+1}}{9^n n! \, \Gamma\left(n+\frac{4}{3}\right)}.$$

These functions allow to write the solution of the Airy ODE:

$$u = \alpha_0 \mathrm{Ai}(z) + \alpha_1 \mathrm{Bi}(z).$$

The constants α_0, α_1 are obtained from the initial conditions

$$\begin{bmatrix} \alpha_0 \\ \alpha_1 \end{bmatrix} = \frac{1}{2} \begin{bmatrix} 3^{\frac{2}{3}} \Gamma\left(\frac{2}{3}\right) & -3^{\frac{4}{3}} \Gamma\left(\frac{4}{3}\right) \\ 3^{\frac{1}{6}} \Gamma\left(\frac{2}{3}\right) & 3^{\frac{5}{6}} \Gamma\left(\frac{4}{3}\right) \end{bmatrix} \begin{bmatrix} a_0 \\ a_1 \end{bmatrix}.$$

(3) Consider the following second-order ODE in complex plane

$$\frac{d^2 u}{dz^2} + \frac{1+z}{6z} \frac{du}{dz} + \frac{1+2z}{6z^2} u = 0$$

and answer the following questions:

(a) Classify the points of the plane according to whether they are regular, regular singular or irregular.

(b) For the origin, find the indicial polynomial and its two roots.

(c) Find the difference between the two roots and deduce how the two linearly independent solutions corresponding to each root of the indicial polynomial are.

(d) Find, for the first solution, its Frobenius series and the recurrence relation. Determine the first three nonzero coefficients.

(e) Find, for the second solution, the recurrence relation between its coefficients.

Solution:

(a) In this case, $a(z) = \frac{1+z}{6z}$ and $b(z) = \frac{1+2z}{6z^2}$, so $a(z)$ and $b(z)$ are meromorphic functions with a single singularity in $z = 0$, a simple pole and a double pole, respectively. Therefore, all the points of the plane except the origin are ordinary points of the ODE. However, $z = 0$ is a regular singular point.

(b) We have $a_0 = \frac{1}{6}$ and $b_0 = \frac{1}{6}$, so the indicial polynomial is

$$P(\alpha) = \alpha(\alpha - 1) + a_0\alpha + b_0 = \alpha^2 - \frac{5}{6}\alpha + \frac{1}{6}.$$

The roots of the indicial polynomial are

$$\alpha = \frac{\frac{5}{6} \pm \sqrt{\frac{25}{36} - \frac{4}{6}}}{2} = \begin{cases} \alpha_1 = \frac{1}{2}, \\ \alpha_2 = \frac{1}{3}. \end{cases}$$

(c) Since $\alpha_1 - \alpha_2 = \frac{1}{6}$ is not an integer, there are no resonances. Then, the two linearly independent solutions are written in the form of Frobenius series as follows

$$u_1(z) = \sum_{n=0}^{\infty} c_{1,n} z^{n+\frac{1}{2}}, \quad u_2(z) = \sum_{n=0}^{\infty} c_{2,n} z^{n+\frac{1}{3}},$$

with coefficients $c_{1,n}$ and $c_{2,n}$ to be determined recursively by the ODE itself.

(d) We write the ODE in the following way:

$$6z^2 \frac{d^2u}{dz^2} + (z + z^2)\frac{du}{dz} + (1 + 2z)u = 0$$

and calculate the first two derivatives, deriving term-by-term,

$$\frac{du_1}{dz} = \sum_{n=0}^{\infty} \left(n + \frac{1}{2}\right) c_{1,n} z^{n-\frac{1}{2}},$$

$$\frac{d^2u_1}{dz^2} = \sum_{n=0}^{\infty} \left(n + \frac{1}{2}\right)\left(n - \frac{1}{2}\right) c_{1,n} z^{n-\frac{3}{2}}.$$

Hence,

$$6z^2 \frac{d^2u_1}{dz^2} = \sum_{n=0}^{\infty} 6\left(n + \frac{1}{2}\right)\left(n - \frac{1}{2}\right) c_{1,n} z^{n+\frac{1}{2}}$$

$$= -\frac{3}{2}c_{1,0}z^{\frac{1}{2}} + \sum_{n=0}^{\infty} 6\left(n + \frac{3}{2}\right)\left(n + \frac{1}{2}\right) c_{1,n+1} z^{n+\frac{3}{2}},$$

$$(z + z^2)\frac{du_1}{dz} = \sum_{n=0}^{\infty} \left(n + \frac{1}{2}\right) c_{1,n} z^{n+\frac{1}{2}} + \sum_{n=0}^{\infty} \left(n + \frac{1}{2}\right) c_{1,n} z^{n+\frac{3}{2}}$$

$$= \frac{1}{2}c_{1,0}z^{\frac{1}{2}} + \sum_{n=0}^{\infty}\left(\left(n+\frac{3}{2}\right)c_{1,n+1} + \left(n+\frac{1}{2}\right)c_{1,n}\right)z^{n+\frac{3}{2}},$$

$$(1+2z)u_1 = \sum_{n=0}^{\infty} c_{1,n}z^{n+\frac{1}{2}} + \sum_{n=0}^{\infty} 2c_{1,n}z^{n+\frac{3}{2}}$$

$$= c_{1,0}z^{\frac{1}{2}} + \sum_{n=0}^{\infty}(c_{1,n+1} + 2c_{1,n})z^{n+\frac{3}{2}}.$$

Therefore, the ODE is written in terms of Frobenius series as follows:

$$\sum_{n=0}^{\infty}\left(6\left(n+\frac{3}{2}\right)\left(n+\frac{1}{2}\right)c_{1,n+1} + \left(n+\frac{3}{2}\right)c_{1,n+1} + \left(n+\frac{1}{2}\right)c_{1,n}\right.$$
$$\left. + c_{1,n+1} + 2c_{1,n}\right)z^{n+\frac{3}{2}} = 0.$$

That implies

$$6\left(n+\frac{3}{2}\right)\left(n+\frac{1}{2}\right)c_{1,n+1} + \left(n+\frac{3}{2}\right)c_{1,n+1} + \left(n+\frac{1}{2}\right)c_{1,n} + c_{1,n+1} + 2c_{1,n} = 0,$$

so

$$2((2n+3)(3n+2)+1)c_{1,n+1} + (2n+5)c_{1,n} = 0$$

and, finally, we can write the recurrence as

$$\boxed{c_{1,n+1} = -\frac{2n+5}{2((2n+3)(3n+2)+1))}c_{1,n},}$$

with $n \in \mathbb{N}$. Thus, $c_{1,0}$ is arbitrary and $c_{1,1} = -\frac{5}{14}c_{1,0}$, $c_{1,2} = -\frac{7}{82}c_{1,1} = \frac{35}{1148}c_{1,0}$.

(e) Proceed as above

$$\frac{du_2}{dz} = \sum_{n=0}^{\infty}\left(n+\frac{1}{3}\right)c_{2,n}z^{n-\frac{2}{3}},$$

$$\frac{d^2u_2}{dz^2} = \sum_{n=0}^{\infty}\left(n+\frac{1}{3}\right)\left(n-\frac{2}{3}\right)c_{2,n}z^{n-\frac{5}{3}}.$$

Hence,

$$6z^2\frac{d^2u_2}{dz^2} = \sum_{n=0}^{\infty}6\left(n+\frac{1}{3}\right)\left(n-\frac{2}{3}\right)c_{2,n}z^{n+\frac{1}{3}}$$

$$= -\frac{4}{3}c_{2,0}z^{\frac{1}{3}} + \sum_{n=0}^{\infty}6\left(n+\frac{4}{3}\right)\left(n+\frac{1}{3}\right)c_{2,n+1}z^{n+\frac{4}{3}},$$

$$(z+z^2)\frac{du_2}{dz} = \sum_{n=0}^{\infty}\left(n+\frac{1}{3}\right)c_{2,n}z^{n+\frac{1}{3}} + \sum_{n=0}^{\infty}\left(n+\frac{1}{3}\right)c_{2,n}z^{n+\frac{4}{3}}$$

$$= \frac{1}{3}c_{2,0}z^{\frac{1}{3}} + \sum_{n=0}^{\infty}\left(\left(n+\frac{4}{3}\right)c_{2,n+1} + \left(n+\frac{1}{3}\right)c_{2,n}\right)z^{n+\frac{4}{3}},$$

$$(1+2z)u_2 = \sum_{n=0}^{\infty} c_{2,n}z^{n+\frac{1}{3}} + \sum_{n=0}^{\infty} 2c_{2,n}z^{n+\frac{4}{3}}$$

$$= c_{2,0}z^{\frac{1}{3}} + \sum_{n=0}^{\infty}(c_{2,n+1} + 2c_{2,n})z^{n+\frac{4}{3}}.$$

Therefore, the ODE is written in terms of Frobenius series as follows:

$$\sum_{n=0}^{\infty}\left(6\left(n+\frac{4}{3}\right)\left(n+\frac{1}{3}\right)c_{2,n+1} + \left(n+\frac{4}{3}\right)c_{2,n+1} + \left(n+\frac{1}{3}\right)c_{2,n}\right.$$

$$\left. + c_{2,n+1} + 2c_{2,n}\right)z^{n+\frac{4}{3}} = 0.$$

Hence,

$$6\left(n+\frac{4}{3}\right)\left(n+\frac{1}{3}\right)c_{2,n+1} + \left(n+\frac{4}{3}\right)c_{2,n+1} + \left(n+\frac{1}{3}\right)c_{2,n} + c_{2,n+1} + 2c_{2,n} = 0.$$

So that,

$$\boxed{c_{2,n+1} = -\frac{3n+7}{2(3n+4)(3n+1)+3}c_{2,n},}$$

for $n \in \mathbb{N}_0$.

(4) For the ODE in the complex plane

$$3z^2\frac{d^2u}{dz^2} + (2z^2 - z)\frac{du}{dz} + u = 0.$$

(a) Classify the points of the complex plane, according to whether they are ordinary, regular singular, or irregular singular. What type of point is the origin $z = 0$?

(b) Find the indicial polynomial in $z = 0$, find its two roots α_1 and α_2, $\alpha_1 \geq \alpha_2$.

(c) Find the Frobenius series at the origin $u_1(z)$ for the largest root α_1.

(d) Find the Frobenius series at the origin $u_2(z)$ for the smaller root α_2; express this solution in terms of exponentials and radicals.

Solution:

(a) The ODE can be written as follows

$$\frac{d^2u}{dz^2} + a(z)\frac{du}{dz} + b(z)u = 0,$$

with

$$a(z) := -\frac{1}{3z} + \frac{2}{3},$$

$$b(z) := \frac{1}{3z^2}.$$

The functions $a(z)$ and $b(z)$ are holomorphic in $\mathbb{C} \setminus \{0\}$, so **all points of the complex plane are regular up to the origin.** In the origin, $a(z)$ has a single pole and $b(z)$ a double pole, so **the origin is a regular singular point.**

(b) As we know, if

$$a(z) = \frac{a_0}{z} + a_1 + a_2 z + \cdots,$$

$$b(z) = \frac{b_0}{z^2} + \frac{b_1}{z} + b_0 + \cdots,$$

then the indicial polynomial in the origin is

$$P(\alpha) = \alpha(\alpha - 1) + a_0\alpha + b_0.$$

In this case $a_0 = -\frac{1}{3}$ and $b_0 = \frac{1}{3}$, and hence $P(\alpha) = \alpha(\alpha - 1) - \frac{\alpha}{3} + \frac{1}{3}$, i.e.,

$$\boxed{P(\alpha) = \alpha^2 - \frac{4}{3}\alpha + \frac{1}{3}.}$$

The corresponding roots are

$$\frac{\frac{4}{3} \pm \sqrt{\frac{16}{9} - \frac{4}{3}}}{2} = \begin{cases} 1, \\ \frac{1}{3}. \end{cases}$$

So the roots are $\boxed{\alpha_1 = 1}$ and $\boxed{\alpha_2 = \frac{1}{3}.}$

(c) We are now looking for a solution in the form

$$u_1(z) = z \sum_{n=0}^{\infty} c_n z^n = \sum_{n=0}^{\infty} c_n z^{n+1}.$$

We calculate the first two derivatives

$$\frac{du_1}{dz} = \sum_{n=0}^{\infty}(n+1)c_n z^n, \qquad \frac{d^2u_1}{dz^2} = \sum_{n=0}^{\infty}(n+1)n c_n z^{n-1},$$

so that

$$3z^2\frac{d^2u_1}{dz^2} = \sum_{n=0}^{\infty} 3(n+1)n c_n z^{n+1},$$

$$(2z^2 - z)\frac{du_1}{dz} = \sum_{n=0}^{\infty} 2(n+1)c_n z^{n+2} - \sum_{n=0}^{\infty}(n+1)c_n z^{n+1}$$

$$= \sum_{n=0}^{\infty}(2n c_{n-1} - (n+1)c_n)z^{n+1}.$$

Therefore,

$$3z^2 \frac{d^2u_1}{dz^2} + (2z^2 - z)\frac{du_1}{dz} + u_1 = \sum_{n=0}^{\infty} \big(3(n+1)nc_n$$
$$+ 2nc_{n-1} - (n+1)c_n + c_n\big)z^{n+1}$$
$$= \sum_{n=0}^{\infty} n\big((3n+2)c_n + 2c_{n-1}\big)z^{n+1}.$$

Consequently, we must require

$$\boxed{c_n = -\frac{2}{3n+2}c_{n-1}, \quad n \in \mathbb{N}.}$$

In particular, $c_1 = -\frac{2}{5}c_0$, $c_2 = \frac{2^2}{8\times5}c_0$, $c_3 = -\frac{2^3}{11\times8\times5}c_0$, and

$$\boxed{c_n = (-1)^n \frac{2^{n+1}}{(3n+2)!!!}.}$$

The Frobenius series (power series in this case) is

$$\boxed{u_1(z) = c_0 \sum_{n=0}^{\infty} (-1)^n \frac{2^{2n+1}}{(3n+2)!!!} z^{n+1}.}$$

(d) The requested solution is

$$u_2(z) = z^{\frac{1}{3}} \sum_{n=0}^{\infty} c_n z^n = \sum_{n=0}^{\infty} c_n z^{n+\frac{1}{3}}.$$

We calculate the first two derivatives

$$\frac{du_2}{dz} = \sum_{n=0}^{\infty} \left(n + \frac{1}{3}\right)c_n z^{n-\frac{2}{3}},$$
$$\frac{d^2u_2}{dz^2} = \sum_{n=0}^{\infty} \left(n + \frac{1}{3}\right)\left(n - \frac{2}{3}\right)c_n z^{n-\frac{5}{3}}.$$

Hence,

$$3z^2 \frac{d^2u_2}{dz^2} = \sum_{n=0}^{\infty} 3\left(n + \frac{1}{3}\right)\left(n - \frac{2}{3}\right)c_n z^{n+\frac{1}{3}},$$
$$(2z^2 - z)\frac{du_2}{dz} = \sum_{n=0}^{\infty} 2\left(n + \frac{1}{3}\right)c_n z^{n+\frac{4}{3}} - \sum_{n=0}^{\infty} \left(n + \frac{1}{3}\right)c_n z^{n+\frac{1}{3}}$$
$$= -\frac{c_0}{3}z^{\frac{1}{3}} + \sum_{n=1}^{\infty} \left(2\left(n - \frac{2}{3}\right)c_{n-1} - \left(n + \frac{1}{3}\right)c_n\right)z^{n+\frac{1}{3}}.$$

Therefore,

$$3z^2\frac{d^2u}{dz^2} + (2z^2 - z)\frac{du}{dz} + u = \left(-3 \times \frac{1}{3} \times \frac{2}{3} - \frac{1}{3} + 1\right)c_0 z^{\frac{1}{3}}$$

$$+ \sum_{n=1}^{\infty}\left(\left(3\left(n + \frac{1}{3}\right)\left(n - \frac{2}{3}\right) - \left(n + \frac{1}{3}\right) + 1\right)c_n\right.$$

$$\left. + 2\left(n - \frac{2}{3}\right)c_{n-1}\right)z^{n+\frac{1}{3}}$$

$$= \sum_{n=1}^{\infty}(3n - 2)\left(nc_n + \frac{2}{3}c_{n-1}\right)z^{n+\frac{1}{3}},$$

so that, for $n \in \mathbb{N}$, we have

$$\boxed{c_n = -\frac{2}{3n}c_{n-1},}$$

and, consequently,

$$\boxed{c_n = \left(-\frac{2}{3}\right)^n \frac{1}{n!}c_0.}$$

Hence, the solution is

$$\boxed{u_2(x) = c_0 z^{\frac{1}{3}}\sum_{n=0}^{\infty}\left(-\frac{2z}{3}\right)^n \frac{1}{n!} = c_0 z^{\frac{1}{3}}\exp\left(-\frac{2z}{3}\right).}$$

(5) For the ODE

$$z(z^2 + 2)\frac{d^2u}{dz^2} + 2\frac{du}{dz} - 2zu = 0.$$

(a) Classify the points of the complex plane according to its role with respect the ODE: regular, regular singular, and irregular singular.

(b) For $z_0 = 0$, find the indicial polynomial $P(\alpha)$ of the ODE and its roots. Discuss the application of Fuchs theorem. Will terms with roots appear in the first series solution u_1? And what about logarithms in the second series solution $u_2(z)$?

(c) For the first solution $u_1(z)$, as a Frobenius series with respect to $z = 0$, find c_1 and the recurrence relationship between the coefficients.

(d) Discuss the recurrence. Find the value of odd coefficients c_{2n+1} and show that the series ends. Find the polynomial $u_1(z)$. What about the radius of convergence, is it bigger than the distance to the nearest singularity?

(e) Find the second solution and discuss the convergence radius of the corresponding power series.

Solution:

(a) We write the ODE as follows

$$\frac{d^2u}{dz^2} + a(z)\frac{du}{dz} + b(z)u = 0,$$

with

$$a(z) := \frac{2}{z(z^2 + 2)},$$

$$b(z) := -\frac{2}{z^2 + 2}.$$

We see that the functions $a(z)$ and $b(z)$ are meromorphic for $\mathbb{C}$. The sets of its singularities, which are simple poles, are $\{0, i\sqrt{2}, -i\sqrt{2}\}$ for $a(z)$ and $\{i\sqrt{2}, -i\sqrt{2}\}$ for $b(z)$. Therefore, all points in $\mathbb{C} \setminus \{0, i\sqrt{2}, -i\sqrt{2}\}$ are regular points. On the other hand, the regular single points make up the set $\{0, i\sqrt{2}, -i\sqrt{2}\}$, and there are no irregular singular points.

(b) To analyze local aspects around $z = 0$, we write the ODE in the form

$$z^2\frac{d^2u}{dz^2} + zA(z)\frac{du}{dz} + B(z)u = 0,$$

$$A(z) := za(z) = \frac{2}{z^2 + 2},$$

$$B(z) := z^2b(z) = -\frac{2z^2}{z^2 + 2}.$$

Where $A(z)$ and $B(z)$ present the following power series expansions

$$A(z) = 1 - \frac{z^2}{2} + \frac{z^4}{4} - \frac{z^6}{8} + \cdots,$$

$$B(z) = -z^2 + \frac{z^4}{2} - \frac{z^6}{4} + \frac{z^8}{8} + \cdots,$$

converging on the disk

$$D\left(0, \sqrt{2}\right) = \left\{z \in \mathbb{C} : |z| < \sqrt{2}\right\}.$$

Therefore, as

$$a_0 = 1,$$

$$b_0 = 0,$$

we have that the indicial polynomial

$$P(\alpha) = \alpha^2 + (a_0 - 1)\alpha + b_0$$

reduces to

$$P(\alpha) = \alpha^2.$$

There's only one double root:

$$\boxed{\alpha_1 = \alpha_2 = 0.}$$

We deduce that, on the one hand, $u_1(z)$ is a power series, and on the other hand, that logarithms will appear for the second series solution $u_2(z)$.

(c) For the Frobenius series for the root $\alpha_1 = 0$,

$$u_1 = \sum_{n=0}^{\infty} c_c z^n,$$

we find

$$-2zu_1(z) = -2z \sum_{n=0}^{\infty} c_n z^n = -\sum_{n=1}^{\infty} 2c_{n-1} z^n,$$

$$2u_1'(z) = \sum_{n=0}^{\infty} 2n c_n z^{n-1} = \sum_{n=0}^{\infty} 2(n+1)c_{n+1} z^n,$$

$$z(z^2 + 2)u_1''(z) = (z^3 + 2z) \sum_{n=0}^{\infty} n(n-1)c_n z^{n-2}$$

$$= \sum_{n=0}^{\infty} n(n-1)c_n z^{n+1} + 2 \sum_{n=0}^{\infty} n(n-1)c_n z^{n-1}$$

$$= \sum_{n=1}^{\infty} (n-1)(n-2)c_{n-1} z^n + 2 \sum_{n=0}^{\infty} (n+1)n c_{n+1} z^n.$$

Hence, adding together all these terms, we obtain

$$2c_1 + \sum_{n=1}^{\infty} \left((n-3)n c_{n-1} + 2(n+1)(n+1)c_{n+1} \right) z^n = 0$$

that leads to $\boxed{c_1 = 0}$ and to the recursion law

$$\boxed{c_{n+2} = -\frac{(n-2)(n+1)}{2(n+2)^2} c_n,}$$

for $n \in \{0, 1, 2, \ldots\}$.

(d) From the recurrence, since $c_1 = 0$, it follows that **all odd coefficients are zero,** $\boxed{c_{2n+1} = 0.}$ On the other hand, it also follows from the recurrence that $\boxed{c_4 = 0}$ and therefore that $\boxed{0 = c_6 = c_8 = c_{10} = \cdots .}$ Effectively, the series ends and the only coefficients not necessarily zero are c_0 and c_2, the latter, according to the recurrence is $c_2 = \frac{1}{4}$ and, therefore, choosing

$c_0 = 1$, we have

$$u_1(z) = 1 + \frac{z^2}{4}.$$

Obviously, as the power series collapses to a polynomial, the radius of convergence becomes infinity and is bigger than $\sqrt{2}$, the distance to the closest singularity.

(e) According to Fuchs Theorem, the second solution can be written in the form

$$u_2 = u_1 \log z + \sum_{n=1}^{\infty} c_n' z^n.$$

Let's use the notation

$$L := z(z^2 + 2)\frac{\mathrm{d}^2}{\mathrm{d}z^2} + 2\frac{\mathrm{d}}{\mathrm{d}z} - 2z.$$

Then, we compute

$$L(u_1 \log z) = L(u_1) \log z + z(z^2 + 2)\left(2\frac{1}{z}\frac{\mathrm{d}u_1}{\mathrm{d}z} - \frac{1}{z^2}u_1\right) + 2\frac{1}{z}u_1$$

with $u_1(z) = 1 + \frac{z^2}{4}$, to get

$$L(u_1 \log z) = L(u_1) \log z + \frac{3}{4}z^3 + z.$$

Therefore, we find

$$L(u_2) = L(u_1 \log z) + L\left(\sum_{n=1}^{\infty} c_n' z^n\right)$$

$$= L(u_1) \log z + \frac{3}{4}z^3 + z + L\left(\sum_{n=1}^{\infty} c_n' z^n\right).$$

Thus, as $L(u_1) = 0$ we find the coefficients c_n' from

$$L\left(\sum_{n=1}^{\infty} c_n' z^n\right) = -\frac{3}{4}z^3 - z.$$

As we have previously accomplished,

$$L\left(\sum_{n=1}^{\infty} c_n' z^n\right) = 2c_1' + \sum_{n=1}^{\infty} \left((n-3)n c_{n-1}' + 2(n+1)^2 c_{n+1}'\right)z^n,$$

where we have set $c_0' = 0$. Therefore, equating coefficients power by power

$$c_1' = 0, \qquad c_2' = -\frac{1}{8}, \qquad c_3' = 0, \qquad c_4' = -\frac{3}{128},$$

and all the odd coefficients are zero

$$\boxed{c_{2n+1}' = 0,}$$

and, for $n \geq 2$, we find the recurrence

$$\boxed{c_{2n+2}' = -\frac{(2n+1)(2n-2)}{2(2n+2)^2}c_{2n}'.}$$

Hence, as

$$\lim_{n \to \infty} \frac{c_{2n+2}'}{c_{2n}'} = \frac{1}{2}$$

according to d'Alembert's ratio convergence criterion the convergence radius is $\sqrt{2}$, and the second solution is

$$\boxed{u_2 = \left(\frac{z^2}{4}+1\right)\log z - \frac{1}{8}z^2 - \frac{3}{128}z^4 + \frac{5}{1536}z^6 + \cdots.}$$

For $n \geq 3$ we can write the explicit formula

$$c_{2n}' = \frac{(-1)^n}{3}\frac{(2n-1)!!(2n-4)!!}{2^{n-8}((2n)!!)^2}c_4'.$$

This formula simplifies for $n \geq 6$ to

$$c_{2n}' = \frac{(-1)^n}{3}\frac{(2n-1)!!}{2^{n-6}n(n-1)(2n-4)!!}c_4'.$$

(6) Consider the following ordinary differential equation in the complex plane:

$$\frac{d^2u}{dz^2} + \frac{du}{dz} + \frac{3+z}{16z^2}u = 0$$

and answer the following questions:
   (a) Classify the points in the complex plane as ordinary, regular singular, or irregular singular points.
   (b) For the origin, find the indicial polynomial and its two roots.
   (c) Determine the difference between the two roots and deduce how the two linearly independent solutions corresponding to each root of the indicial polynomial will be.
   (d) Find the recurrence relation between the coefficients of the first Frobenius series solution. Determine the first three nonzero coefficients.
   (e) Find the recurrence relation between the coefficients of the second solution.
   Solution:
   (a) In this case, $a(z) = 1$ and $b(z) = \frac{3+z}{16z^2}$, so $a(z)$ is an entire function and $b(z)$ is a meromorphic function with a single singularity at $z = 0$, a double pole. Therefore, all points in the complex plane except the origin are ordinary points of the ODE. However, $z = 0$ is a regular singular point.

(b) We have $a_0 = 0$ and $b_0 = \frac{3}{16}$, so the indicial polynomial is

$$P(\alpha) = \alpha(\alpha - 1) + a_0\alpha + b_0 = \alpha^2 - \alpha + \frac{3}{16}.$$

The roots of the indicial polynomial are

$$\alpha = \frac{1 \pm \sqrt{1 - \frac{12}{16}}}{2} = \begin{cases} \alpha_1 = \frac{3}{4}, \\ \alpha_2 = \frac{1}{4}. \end{cases}$$

(c) Since $\alpha_1 - \alpha_2 = \frac{1}{2}$ is not an integer, there are no resonances. Hence, two linearly independent solutions of the ODE are given by Frobenius series as follows:

$$u_1(z) = \sum_{n=0}^{\infty} c_{1,n} z^{n+\frac{3}{4}}, \quad u_2(z) = \sum_{n=0}^{\infty} c_{2,n} z^{n+\frac{1}{4}},$$

with coefficients $c_{1,n}$ and $c_{2,n}$ to be determined recursively from the ODE.

(d) Rewriting the ODE in the following form

$$16z^2 \frac{d^2u}{dz^2} + 16z^2 \frac{du}{dz} + (3 + z)u = 0$$

and calculating the first two derivatives term-by-term:

$$\frac{du_1}{dz} = \sum_{n=0}^{\infty} \left(n + \frac{3}{4}\right) c_{1,n} z^{n-\frac{1}{4}},$$

$$\frac{d^2u_1}{dz^2} = \sum_{n=0}^{\infty} \left(n + \frac{3}{4}\right)\left(n - \frac{1}{4}\right) c_{1,n} z^{n-\frac{5}{4}},$$

we get:

$$16z^2 \frac{d^2u_1}{dz^2} = \sum_{n=0}^{\infty} 16\left(n + \frac{3}{4}\right)\left(n - \frac{1}{4}\right) c_{1,n} z^{n+\frac{3}{4}}$$

$$= -3c_{1,0} z^{\frac{3}{4}} + \sum_{n=1}^{\infty} 16\left(n + \frac{3}{4}\right)\left(n - \frac{1}{4}\right) c_{1,n} z^{n+\frac{3}{4}},$$

$$16z^2 \frac{du_1}{dz} = \sum_{n=0}^{\infty} 16\left(n + \frac{3}{4}\right) c_{1,n} z^{n+\frac{7}{4}} = \sum_{n=1}^{\infty} 16\left(n - \frac{1}{4}\right) c_{1,n-1} z^{n+\frac{3}{4}},$$

$$(3 + z)u = \sum_{n=0}^{\infty} 3c_{1,n} z^{n+\frac{3}{4}} + \sum_{n=0}^{\infty} c_{1,n} z^{n+\frac{7}{4}}$$

$$= \sum_{n=0}^{\infty} 3c_{1,n} z^{n+\frac{3}{4}} + \sum_{n=1}^{\infty} c_{1,n-1} z^{n+\frac{3}{4}}$$

$$= 3c_{1,0}z^{\frac{3}{4}} + \sum_{n=1}^{\infty}(3c_{1,n} + c_{1,n-1})z^{n+\frac{3}{4}}$$

which leads to the following equation in terms of Frobenius series:

$$\sum_{n=1}^{\infty}\left(16\left(n+\frac{3}{4}\right)\left(n-\frac{1}{4}\right)c_{1,n} + 16\left(n-\frac{1}{4}\right)c_{1,n-1} + 3c_{1,n} + c_{1,n-1}\right)z^{n+\frac{3}{4}} = 0.$$

This implies:

$$16\left(n+\frac{3}{4}\right)\left(n-\frac{1}{4}\right)c_{1,n} + 16\left(n-\frac{1}{4}\right)c_{1,n-1} + 3c_{1,n} + c_{1,n-1} = 0,$$

which can be written as:

$$((4n+3)(4n-1)+3)c_{1,n} + (16n-3)c_{1,n-1} = 0.$$

Since $(4n+3)(4n-1)+3 = 8n(2n+1)$, we can express the recurrence relation as:

$$\boxed{c_{1,n} = -\frac{16n-3}{8n(2n+1)}c_{1,n-1},}$$

where $n \in \mathbb{N}$. The first three terms will be:

$$c_{1,1} = -\frac{13}{24}c_{1,0},$$

$$c_{1,2} = \frac{29}{80}\frac{13}{24}c_{1,0} = \frac{377}{1920}c_{1,0},$$

$$c_{1,3} = -\frac{45}{168}\frac{377}{1920}c_{1,0} = -\frac{377}{7168}c_{1,0},$$

and thus, the expression for $u_1(z)$ becomes:

$$u_1(z) = c_{1,0}z^{\frac{3}{4}}\left(1 - \frac{13}{24}z + \frac{377}{1920}z^2 - \frac{377}{7168}z^3 + \frac{4147}{294912}z^4 + \cdots\right).$$

(e) Proceeding as in the previous step,

$$\frac{du_2}{dz} = \sum_{n=0}^{\infty}\left(n+\frac{1}{4}\right)c_{2,n}z^{n-\frac{3}{4}},$$

$$\frac{d^2u_2}{dz^2} = \sum_{n=0}^{\infty}\left(n+\frac{1}{4}\right)\left(n-\frac{3}{4}\right)c_{2,n}z^{n-\frac{7}{4}}.$$

Hence,

$$16z^2\frac{d^2u_2}{dz^2} = \sum_{n=0}^{\infty}16\left(n+\frac{1}{4}\right)\left(n-\frac{3}{4}\right)c_{2,n}z^{n+\frac{1}{4}}$$

$$= -3c_{2,0}z^{\frac{1}{4}} + \sum_{n=1}^{\infty}16\left(n+\frac{1}{4}\right)\left(n-\frac{3}{4}\right)c_{2,n}z^{n+\frac{1}{4}},$$

$$16z^2 \frac{du_2}{dz} = \sum_{n=0}^{\infty} 16\left(n + \frac{1}{4}\right)c_{2,n} z^{n+\frac{5}{4}} = \sum_{n=1}^{\infty} 16\left(n - \frac{3}{4}\right)c_{2,n-1} z^{n+\frac{1}{4}},$$

$$(3+z)u = \sum_{n=0}^{\infty} 3c_{2,n} z^{n+\frac{1}{4}} + \sum_{n=0}^{\infty} c_{2,n} z^{n+\frac{5}{4}}$$

$$= \sum_{n=0}^{\infty} 3c_{2,n} z^{n+\frac{1}{4}} + \sum_{n=1}^{\infty} c_{2,n-1} z^{n+\frac{1}{4}}$$

$$= 3c_{2,0} z^{\frac{1}{4}} + \sum_{n=1}^{\infty} (3c_{2,n} + c_{2,n-1}) z^{n+\frac{1}{4}}.$$

Therefore, the ODE can be expressed in terms of Frobenius series as follows:

$$\sum_{n=1}^{\infty} \left(16\left(n + \frac{1}{4}\right)\left(n - \frac{3}{4}\right)c_{2,n} + 16\left(n - \frac{3}{4}\right)c_{2,n-1} + 3c_{2,n} + c_{2,n-1}\right) z^{n+\frac{1}{4}} = 0.$$

Hence,

$$16\left(n + \frac{1}{4}\right)\left(n - \frac{3}{4}\right)c_{2,n} + 16\left(n - \frac{3}{4}\right)c_{2,n-1} + 3c_{2,n} + c_{2,n-1} = 0.$$

Therefore,

$$((4n+1)(4n-3) + 3)c_{2,n} + (16n - 11)c_{2,n-1} = 0.$$

Since $(4n+1)(4n-3) + 3 = 8n(2n-1)$, we can express the recurrence relation as:

$$\boxed{c_{2,n} = -\frac{16n - 11}{8n(2n-1)} c_{2,n-1},}$$

with $n \in \mathbb{N}$.

§7.7.2. **Exercises**

(1) The Legendre equation for the polynomials of the same name $P_n(x)$
can be written as
$$LP_n = \lambda_n P_n, \quad \lambda_n = n(n+1),$$
where L is the operator of Sturm–Liouville
$$Lu := -D((1-x^2)Du) = -(1-x^2)D^2u + 2xDu.$$
Prove that L is symmetric over the domain
$$\mathfrak{D} := \left\{u \in C^\infty((-1,1)) : \lim_{x\to\pm1}(1-x^2)u'(x) = 0\right\} \subset L^2([-1,1], dx).$$
As a consequence, prove that Legendre's polynomials $\{P_n(x)\}_{n\geq0}$
form an orthogonal set in $L^2([-1,1], dx)$.

(2) Rodrigues' formula for Legendre's polynomials is:
$$P_n(x) = \frac{1}{2^n n!}D^n(x^2-1)^n, \quad n \in \mathbb{N}_0.$$
By using this expression:
   (a) Determine the first three polynomials.
   (b) Represent graphically in the interval $[-1,1]$ the polynomials P_2
   and P_3.
   (c) Show that $P_n(-x) = (-1)^n P_n(x)$, $n \in \mathbb{N}_0$.

(3) Through Rodrigues' formula for Legendre's polynomials:
   (a) Show that $P_n(x)$ is a polynomial of degree n.
   (b) Using the identity
   $$(x^2-1)^n = (x+1)^n(x-1)^n,$$
   in Rodrigues' formula and Leibniz's rule to differentiate, you
   get
$$P_n(x) = \frac{1}{2^n n!}\sum_{k=0}^{n}\frac{n!}{k!(n-k)!}[D^k(x+1)^n][D^{n-k}(x-1)^n].$$
   Use this formula to prove that
   $$P_n(1) = 1, \quad n \in \mathbb{N}_0,$$
   and deduce then that
   $$P_n(-1) = (-1)^n, \quad n \in \mathbb{N}_0.$$

(4) Developing by means of Newton's binomial formula, we have to
$$(x^2-1)^n = \sum_{k=0}^{n}\frac{n!}{k!(n-k)!}(-1)^{n-k}x^{2k}.$$
Making use of this identity in Rodrigues' formula, prove that
$$P_{2n}(0) = (-1)^n\frac{(2n)!}{2^{2n}(n!)^2}, \quad P_{2n+1}(0) = 0, \quad n \in \mathbb{N}_0.$$

(5) Making use of the identity

$$DP_{n+1} - DP_{n-1} = (2n+1)P_n, \quad n \geq 0,$$

determine the series expansion of Legendre polynomials

$$u(x) = \sum_{n=0}^{\infty} c_n P_n(x),$$

of the function

$$u(x) = \begin{cases} 1, & 0 < x < 1, \\ 0, & -1 < x < 0. \end{cases}$$

(6) Consider the Bessel equation

$$\frac{1}{x}(xu')' + (1 - \frac{m^2}{x^2})u = 0.$$

Multiplying by $2x^2 u'$, show that

$$2xu^2 = [(xu')^2 + (x^2 - m^2)u^2]'.$$

From here prove that

$$\int_0^b x J_m\big(\sqrt{\lambda_n}x\big)^2 dx = \frac{b^2}{2}\Big(J_m'\big(\sqrt{\lambda_n}b\big)\Big)^2,$$

with c_{mn} $(n \geq 1)$ being the zeros of the function $J_m(x)$, and

$$\lambda_n = (c_{mn}/b)^2.$$

(7) Show that the spherical Bessel function $j_l(x), n_l(x)$ fulfills

$$x^2 v'' + 2xu' + (x^2 - l(l+1))u = 0.$$

8. Cylindrical and Spherical BVPs

Contents

ELMHOLTZ equation for functions that satisfy a boundary value problem, with boundaries described naturally in given curvilinear coordinates—specifically cylindrical and spherical coordinates—is the focal point of this chapter. Building upon the knowledge and tools developed in previous chapters, we will delve into this topic for scenarios where cylindrical or spherical coordinates can be effectively utilized.

When working with cylindrical coordinates, a class of functions arises, involving Bessel and Neumann functions with integer indices. These functions play a pivotal role in describing phenomena governed by the Helmholtz equation within cylindrical geometries.

Transitioning to spherical coordinates, we encounter a distinct set of challenges. Here, the utilization of Legendre polynomials and their associated functions becomes indispensable in tackling the complexities of the Helmholtz equation. This mathematical framework empowers us to construct spherical harmonics, which hold significant importance across various scientific disciplines.

Furthermore, these harmonics give rise to tangible counterparts—tesseral harmonics—representing the real solutions within this intricate mathematical landscape.

§8.1. Cylindrical BVPs

APLACIAN operator is a key mathematical tool that takes center stage alongside its related eigenvalues. This helps us analyze problems related to boundaries and the Helmholtz equation in a detailed way. We'll focus on situations where boundaries can be easily described using cylindrical coordinates, which offers a unique perspective.

By using these cylindrical coordinates with the Helmholtz and MSV equations, something fascinating happens. We start seeing Bessel and Neumann functions emerge – two important math concepts that play a big role in our discussion.

We'll look at examples from quantum physics, fluid mechanics, sound theory and electrostatics to show how these ideas are used in real life. This highlights how the theory we're exploring is actually useful.

The figure below illustrates cylindrical coordinates, which are expressed as follows:

$$x = r\cos\theta,$$
$$y = r\sin\theta,$$
$$z = z.$$

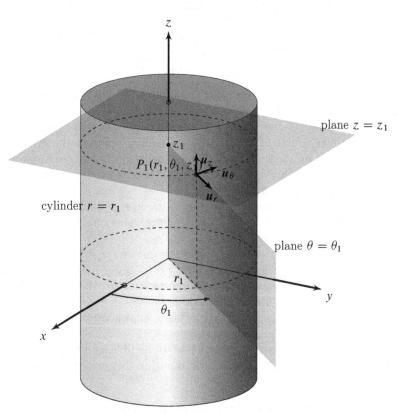

In this new coordinate system, the Helmholtz equation takes the form:

$$u_{rr} + \frac{1}{r}u_r + \frac{1}{r^2}u_{\theta\theta} + u_{zz} + k^2 u = 0.$$

To proceed, we introduce the technique of separating variables through factorization:

$$u(r, \theta, z) = V(r, \theta)Z(z).$$

Applying this separation of variables transforms the Helmholtz equation into two distinct equations:

$$Z'' + k^2 Z = \alpha^2 Z,$$
$$r^2 V_{rr} + r V_r + V_{\theta\theta} = -\alpha^2 r^2 V.$$

The solution to the first ordinary differential equation is given by:

(8.1)
$$Z_\alpha(z) = \begin{cases} Az + B, & \text{if } k^2 = \alpha^2, \\ A e^{i\sqrt{k^2-\alpha^2}\,z} + B e^{-i\sqrt{k^2-\alpha^2}\,z}, & \text{if } k^2 \neq \alpha^2. \end{cases}$$

For the second equation, we can also separate variables:

$$V(r, \theta) = R(r)\Theta(\theta).$$

This leads to the following equations:

$$r^2 R'' + r R' + \alpha^2 r^2 R = m^2 R,$$
$$\Theta'' = -m^2 \Theta.$$

The general solution for the second equation is:

(8.2)
$$\Theta_m(\theta) = \begin{cases} C\theta + D, & \text{if } m = 0, \\ C e^{im\theta} + D e^{-im\theta}, & \text{if } m \neq 0. \end{cases}$$

It remains to analyze the ODE in R, known as **radial equation**. There are two different cases for this one. First, we consider $\alpha = 0$ and the corresponding equation is

$$r^2 R'' + r R' - m^2 R = 0$$

whose solution is

(8.3)
$$R_{\alpha=0,m}(r) = \begin{cases} c_1 \log r + c_2, & m = 0, \\ c_1 r^m + c_2 r^{-m}, & m \neq 0. \end{cases}$$

When $\alpha \neq 0$, by changing the variable $\rho = \alpha r$, the corresponding ODE is reduced to

$$\rho^2 \frac{d^2 R}{d\rho^2} + \rho \frac{dR}{d\rho} + (\rho^2 - m^2) R = 0,$$

which is the well-known Bessel equation (7.16) in the variable ρ. Therefore, the solution is

(8.4)
$$R_{\alpha,m}(r) = E J_m(\alpha r) + F N_m(\alpha r).$$

§8.1.1. Polar Coordinates

Polar coordinates play a pivotal role as a simplification of cylindrical coordinates, offering a streamlined representation that proves especially valuable when addressing problems characterized by rotational symmetry. The shift from cylindrical to polar coordinates involves condensing the cylindrical system into a two-dimensional plane, effectively eliminating the z-axis component.

The Helmholtz equation in two dimensions, expressed within polar coordinates, reads as follows:

$$u_{rr} + \frac{1}{r}u_r + \frac{1}{r^2}u_{\theta\theta} + k^2 u = 0.$$

Remarkably, this equation mirrors the form that emerges when dealing with the Helmholtz equation in cylindrical coordinates, specifically in scenarios where solutions are sought that do not rely on the independent variable z. As a result, a simple transformation

$$\alpha^2 = k^2$$

is sufficient to derive the corresponding solutions.

§8.1.2. Quantum Particle in a Wedge

We now shift our attention to a quantum particle confined within a cylindrical wedge, characterized by a radius of a, a height of L, and an aperture angle

$$0 < \theta_0 < 2\pi,$$

as depicted in the figure.

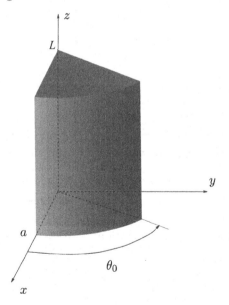

The stationary states of this system are described by square-integrable solutions that satisfy the following Dirichlet boundary problem:

$$\begin{cases} -\dfrac{\hbar^2}{2M}\Delta u = Eu, \\ u\big|_{\text{walls}} = 0. \end{cases}$$

Regarding the energy, we have

$$E = \frac{\hbar^2 k^2}{2M}.$$

We can formulate the corresponding boundary value problem for the Helmholtz equation in cylindrical coordinates as follows:

$$\begin{cases} \Delta u + k^2 u = 0, \\ \begin{cases} u\big|_{r=a} = 0, \\ u\big|_{\theta=0} = 0, \\ u\big|_{\theta=\theta_0} = 0, \\ u\big|_{z=0} = 0, \\ u\big|_{z=L} = 0. \end{cases} \end{cases}$$

By applying the MSV, solutions are obtained in the form:

$$R_{\alpha,m}(r)\Theta_m(\theta)Z_\alpha(z),$$

which must adhere to the boundary conditions and satisfy regularity criteria.

Firstly, our focus is directed toward the angular function Θ_m (8.2). The boundary conditions in the variable θ are given by:

$$\Theta_m(0) = \Theta_m(\theta_0) = 0,$$

leading, for $m \neq 0$, to a linear system for C and D as follows:

$$C + D = 0,$$

$$e^{im\theta_0}C + e^{-im\theta_0}D = 0.$$

For $m = 0$, this implies $\Theta = 0$.

For $m \neq 0$, a nontrivial solution exists if and only if:

$$\begin{vmatrix} 1 & 1 \\ e^{im\theta_0} & e^{-im\theta_0} \end{vmatrix} = 0,$$

which is equivalent to:

$$\sin m\theta_0 = 0.$$

This leads us to the conclusion that:

$$\boxed{m = j\frac{\pi}{\theta_0}, \quad j \in \mathbb{N}.}$$

Furthermore, we have $D = -C$ and:

$$\Theta_m(\theta) = 2C\sin(m\theta).$$

Moving on to the variable z, we impose the corresponding boundary conditions from (8.1) onto Z_α. If $k^2 = \alpha^2$, then it's clear that $Z = 0$. For $k^2 \neq \alpha^2$, a linear system for A and B emerges:

$$A + B = 0,$$
$$e^{i\sqrt{k^2-\alpha^2}L}A + e^{-i\sqrt{k^2-\alpha^2}L}B = 0.$$

A nontrivial solution exists if and only if:

$$\sin\sqrt{k^2-\alpha^2}L = 0.$$

Thus:

$$\boxed{k^2 = n^2\frac{\pi^2}{L^2} + \alpha^2,}$$

for $n \in \mathbb{N}$. Moreover, $B = -A$ and:

$$Z_\alpha(z) = 2A\sin\frac{\pi}{L}nz.$$

Lastly, we analyze the radial variable r. The radial function $R(r)$ must remain free of singularities at $r = 0$ and satisfy $R(a) = 0$. Considering (8.3), it's apparent that no nontrivial solutions exist for $\alpha = 0$, and thus, this case is dismissed. The focus turns to the case $\alpha \neq 0$. From (8.4), it requires that $F = 0$ (ensuring regularity at the origin), and the boundary condition mandates:

$$\boxed{J_m(\alpha a) = 0.}$$

Consequently, if $\{c_{m,\ell}\}_{\ell=1}^\infty$ represents the zeros of the Bessel function $J_m(x)$, the conclusion is:

$$\boxed{\begin{aligned} m &= j\frac{\pi}{\theta_0}, \\ \alpha &= \frac{c_{m,\ell}}{a}, \end{aligned}}$$

for $j, \ell \in \mathbb{N}$. Given the asymptotic behavior, $x \to \infty$:

$$J_m(x) \sim \sqrt{\frac{2}{\pi x}}\cos\left(x - (2m+1)\frac{\pi}{4}\right)$$

it's observed that for zeros:

$$c_{m,\ell} \sim (2\ell+1)\frac{\pi}{2} + (2m+1)\frac{\pi}{4}.$$

Therefore, the permissible energies are:

$$E_{j,\ell,n} = \frac{\hbar^2}{2M}\left(\frac{\pi^2 n^2}{L^2} + \frac{c_{\frac{j\pi}{\theta_0},\ell}^2}{a^2}\right),$$

for $j, \ell, n \in \mathbb{N}$, and the corresponding eigenfunctions will be:

$$J_{\frac{j\pi}{\theta_0}}\left(c_{\frac{j\pi}{\theta_0},\ell}\,\frac{r}{a}\right)\sin\left(j\pi\frac{\theta}{\theta_0}\right)\sin\left(n\pi\frac{z}{h}\right).$$

The Wedge

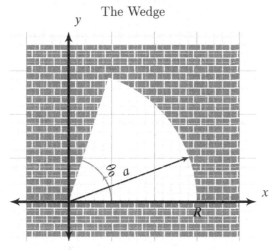

The issue of a two-dimensional quantum particle confined within a wedge, as depicted in the figure above, is resolved through polar coordinates by setting $n = 0$. As an illustration, for the values $a = 1$ and $\theta_0 = \pi/2$, the energies and stationary states are given by

$$\frac{\hbar^2}{2M}(c_{2j,\ell})^2, \quad J_{2j}(c_{2j,\ell}r)\sin(2j\theta),$$

for $j, \ell \in \mathbb{N}$. For instance, when $j = 1$, the initial two zeros of the Bessel function J_2 are approximately

$$c_{2,1} \approx 5.135622302, \qquad\qquad c_{2,2} \approx 8.417244140,$$

and for $j = 2$, the primary zero of the function J_4 is about

$$c_{4,1} \approx 7.588342435.$$

It is evident that these constitute the initial three zeros. Hence, the fundamental state, along with the first and second excited states, can be represented as

$$J_2(c_{2,1}r)\sin(2\theta), \qquad J_4(c_{4,1}r)\sin(4\theta), \qquad J_2(c_{2,2}r)\sin(2\theta).$$

Subsequently, we shall present the sequence formed by the squares of these functions.

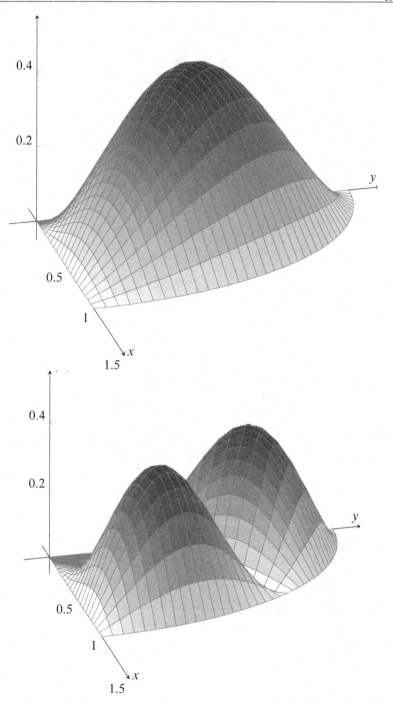

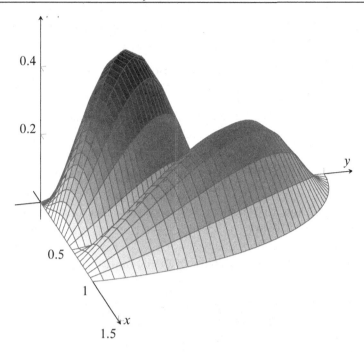

§8.1.3. Quantum Particle in a Cylinder

Let's examine the case of a quantum particle confined within a cylinder $(\theta_0 = 2\pi)$ of radius a. In this scenario, the conditions on θ can be expressed as follows:

$$\begin{cases} \Delta u + k^2 u = 0, \\ \begin{cases} u\big|_{r=a} = 0, \\ u\big|_{\theta=0} = u\big|_{\theta=2\pi}, \\ u\big|_{z=0} = 0, \\ u\big|_{z=L} = 0. \end{cases} \end{cases}$$

The periodic condition $u|_{\theta=0} = u|_{\theta=2\pi}$ implies that $m \in \mathbb{Z}$ and $\Theta_m(\theta) = C\,\mathrm{e}^{im\theta}$. Consequently, the admissible energies can be expressed as follows:

$$E_{m,\ell,n} = \frac{\hbar^2}{2M}\Big(\frac{\pi^2 n^2}{L^2} + \frac{(c_{m,\ell})^2}{a^2}\Big),$$

where $m \in \mathbb{N}_0$, $\ell, n \in \mathbb{N}$. The corresponding eigenfunctions take the form:

$$J_m\Big(c_{m,\ell}\frac{r}{a}\Big)\mathrm{e}^{\pm im\theta}\sin\Big(n\pi\frac{z}{h}\Big).$$

Additionally, there is a degeneracy for states with $|m| > 0$.

Moving forward, let's consider the case of a quantum particle confined within a quantum corral.

§8.1.4. Quantum Particle Trapped in a Corral

As previously mentioned, we set $n = 0$, resulting in the following expressions for the energies

$$E_{m,\ell} = \frac{\hbar^2}{2M} \frac{(c_{m,\ell})^2}{a^2},$$

and eigenfunctions:

$$J_m\left(c_{m,\ell} \frac{r}{a}\right) e^{\pm im\theta}.$$

Here, m belongs to $\mathbb{N}_0$ and ℓ belongs to $\mathbb{N}$. The first set of zeros are presented in Table 1 below.

TABLE 1. Zeros of Bessel functions

ℓ^m	0	1	2	3	4
1	2.4048	3.8317	5.1356	6.3802	7.5883
2	5.5201	7.0156	8.4172	9.7610	11.0647
3	8.6537	10.1735	11.6198	13.0152	14.3725
4	11.7915	13.3237	14.7960	16.2235	17.6160
5	14.9309	16.4706	17.9598	19.4094	20.8269
ℓ^m	5	6	7	8	9
1	8.7715	9.9361	11.0864	12.2251	13.3532
2	12.3386	13.5893	14.8213	16.0378	17.2489
3	15.7002	17.0038	18.2876	19.5545	20.9301
4	18.9801	20.3208	21.6415	22.9452	24.4929
5	22.2178	23.5861	24.9349	26.2668	27.9710

For $a = 1$, the fundamental state is

$$J_0(2.4048r).$$

The first excited state, which is degenerate, is

$$J_1(3.8317r)\cos(\theta - \beta).$$

The second excited state, also degenerate, is

$$J_2(5.1356r)\cos 2(\theta - \beta).$$

Finally, the third excited state is

$$J_0(5.1356r).$$

The plots of the first ten eigenstates are shown below.

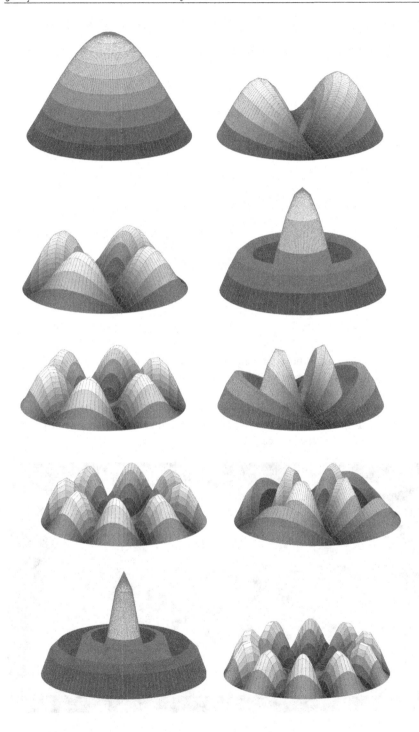

We present below the 16th excited state surface, which corresponds to $J_0(c_{0,4}r)$ with an energy of $\dfrac{\hbar^2}{2M}\dfrac{(c_{0,4})^2}{a^2}$.

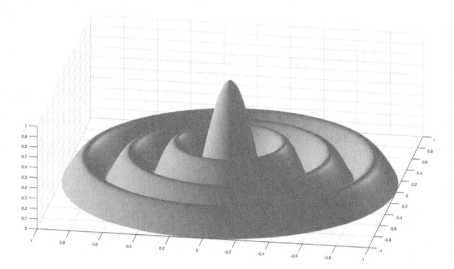

This can be compared to a renowned artwork depicting a solid-state surface quantum corral known as "The Well (Quantum Corral)." Crafted from gilded wood, the piece measures 3"× 13"× 12" (6 cm × 34 cm × 31 cm) and was created in 2009 by Julian Voss-Andreae. The artwork utilizes original experimental data to transform a human-engineered subatomic "quantum landscape" into an artistic creation.

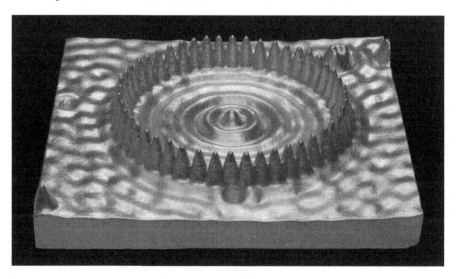

To enhance our comprehension of the probability density associated with the ten initial stationary states that were previously plotted, we will now explore an azimuthal perspective. This perspective will provide us with a more comprehensive view of how the probability density is distributed in these eigenstates.

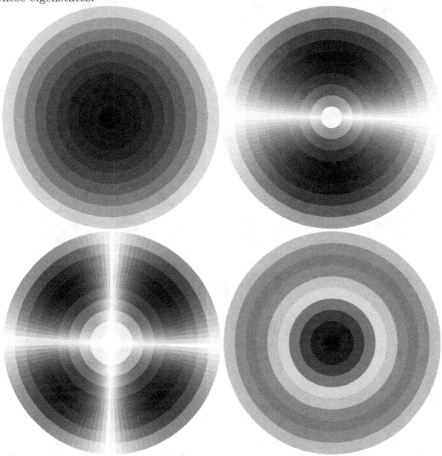

§8.1.5. Quantum Particle in an Annulus

Let's analyze the case of a quantum particle confined between two cylinders sharing the same axis and radii $0 < R_1 < R_2 < \infty$. In this scenario, the conditions on θ are given by:

$$
\begin{cases}
\Delta u + k^2 u = 0, \\
u\big|_{r=R_1} = 0, \\
u\big|_{r=R_2} = 0, \\
u\big|_{\theta=0} = u\big|_{\theta=2\pi}, \\
u\big|_{z=0} = 0, \\
u\big|_{z=L} = 0.
\end{cases}
$$

The periodic condition $u|_{\theta=0} = u|_{\theta=2\pi}$ implies that $m \in \mathbb{Z}$ and $\Theta_m(\theta) = C e^{im\theta}$.

While the discussion in the z direction remains unchanged, when we consider the radial variable, the situation changes entirely. For $m = 0$, we can have $\alpha = 0$, which leads to the following system of equations:

$$
\begin{bmatrix} \log R_1 & 1 \\ \log R_2 & 1 \end{bmatrix} \begin{bmatrix} c_1 \\ c_2 \end{bmatrix} = \begin{bmatrix} 0 \\ 0 \end{bmatrix}.
$$

However, this implies that

$$
\begin{vmatrix} \log R_1 & 1 \\ \log R_2 & 1 \end{vmatrix} = 0,
$$

but

$$
\begin{vmatrix} \log R_1 & 1 \\ \log R_2 & 1 \end{vmatrix} = \log R_1 - \log R_2 = \log \frac{R_1}{R_2},
$$

which means that from

$$
\log \frac{R_1}{R_2} = 0,
$$

we deduce $R_1 = R_2$, which is absurd. Moving on to the next case, we have:

$$
\begin{bmatrix} R_1^m & R_1^{-m} \\ R_2^m & R_2^{-m} \end{bmatrix} \begin{bmatrix} c_1 \\ c_2 \end{bmatrix} = \begin{bmatrix} 0 \\ 0 \end{bmatrix}.
$$

This requires that

$$
\begin{vmatrix} R_1^m & R_1^{-m} \\ R_2^m & R_2^{-m} \end{vmatrix} = 0,
$$

or

$$
\frac{R_1^m}{R_2^m} = \frac{R_2^m}{R_1^m},
$$

which again implies $R_1 = R_2$, making it impossible. Therefore, $\alpha = 0$ is ruled out.

When $\alpha \neq 0$, we have:

$$\begin{bmatrix} J_m(\alpha R_1) & N_m(\alpha R_1) \\ J_m(\alpha R_2) & N_m(\alpha R_2) \end{bmatrix} \begin{bmatrix} E \\ F \end{bmatrix} = \begin{bmatrix} 0 \\ 0 \end{bmatrix}.$$

Hence,

$$\begin{vmatrix} J_m(\alpha R_1) & N_m(\alpha R_1) \\ J_m(\alpha R_2) & N_m(\alpha R_2) \end{vmatrix} = 0,$$

i.e., α must be one of the zeros $\{\alpha_{n,\ell}\}$ of:

$$J_m(\alpha R_1) N_m(\alpha R_2) - N_m(\alpha R_1) J_m(\alpha R_2) = 0.$$

By denoting

$$c = \alpha R_1, \quad \rho = \frac{R_2}{R_1},$$

we can write this as:

$$\boxed{J_m(c) N_m(\rho c) - N_m(c) J_m(\rho c) = 0.}$$

Consequently, the admissible energies are given by:

$$\boxed{E_{m,\ell,n} = \frac{\hbar^2}{2M} \left(\frac{\pi^2 n^2}{L^2} + \frac{(c_{m,\ell})^2}{R_1^2} \right),}$$

for $m \in \mathbb{N}_0$, $\ell, n \in \mathbb{N}$, and the corresponding eigenfunctions are:

$$\boxed{\left(N_m(c_{m,\ell}) J_m \left(c_{m,\ell} \frac{r}{R_1} \right) - J_m(c_{m,\ell}) N_m \left(c_{m,\ell} \frac{r}{R_1} \right) \right) e^{\pm im\theta} \sin n\pi \frac{z}{L}.}$$

Upon planar reduction to the annulus, the energies become:

$$\boxed{E_{m,\ell} = \frac{\hbar^2}{2M} \frac{(c_{m,\ell})^2}{R_1^2},}$$

for $m \in \mathbb{N}_0$, $\ell \in \mathbb{N}$, and the eigenfunctions are given by:

$$\boxed{\left(N_m(c_{m,\ell}) J_m \left(c_{m,\ell} \frac{r}{R_1} \right) - J_m(c_{m,\ell}) N_m \left(c_{m,\ell} \frac{r}{R_1} \right) \right) e^{\pm im\theta}.}$$

For illustration, consider setting $R_1 = 1$, $R_2 = 2$ (yielding $\rho = 2$). In this context, we need to determine the collection of zeros of

$$f_m(x) := J_m(x) N_m(2x) - N_m(x) J_m(2x).$$

The functions $f_m(x)$ are depicted below:

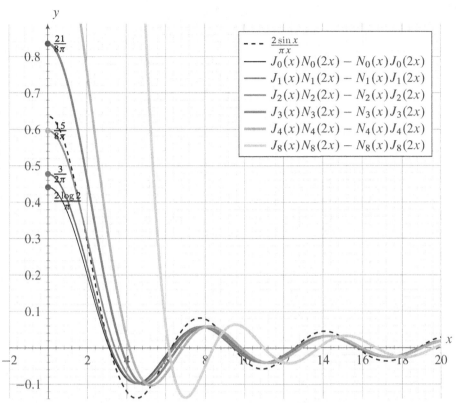

From the asymptotic behaviors of the Bessel and Neumann functions as x becomes large, and employing the trigonometric relation: $\sin(a - b) = \sin a \cos b - \sin b \cos a$, we deduce:

$$f_m(x) \sim \frac{2\pi}{x} \left(\cos\left(x - (2m+1)\frac{\pi}{4}\right) \sin\left(2x - (2m+1)\frac{\pi}{4}\right) \right.$$
$$\left. - \sin\left(x - (2m+1)\frac{\pi}{4}\right) \cos\left(2x - (2m+1)\frac{\pi}{4}\right) \right)$$
$$= \frac{2\sin x}{\pi x}$$

for $x \to \infty$. Thus, for a given m and sufficiently large x, much larger than $(2m+1)\frac{\pi}{4}$, the graph of $f_m(x)$ and $\frac{2\sin x}{\pi x}$ will practically overlap.

On the other hand, the asymptotic behavior for small x is given by:

$$f_0(x) \sim \frac{2\log 2}{\pi},$$
$$f_m(x) \sim \frac{4^m - 1}{m\pi 2^m},$$

for $m \in \mathbb{N}$.

TABLE 2. Zeros of $J_m(x)N_m(2x) - N_m(x)J_m(2x)$

$\ell^{\,m}$	0	1	2	3	4
1	3.1230	3.1966	3.4069	3.7289	4.1334
2	6.2734	6.3123	6.4278	6.6159	6.8712
3	9.4182	9.4445	9.5229	9.6522	9.8309
4	12.5614	12.5812	12.6404	12.7385	12.8747
5	15.7040	15.7199	15.7673	15.8462	15.9561
$\ell^{\,m}$	5	6	7	8	9
1	4.5950	5.0945	5.6179	6.1557	6.7015
2	7.1866	7.5550	7.9687	8.4206	8.9038
3	10.0564	10.3261	10.6371	10.9862	11.3703
4	13.0481	13.2574	13.5012	13.7779	14.0858
5	16.0964	16.2665	16.4656	16.6929	16.9474

The wave functions are

$$u_{m,\ell} = \big(N_m(c_{m,\ell})J_m(c_{m,\ell}r) - J_m(c_{m,\ell})N_m(c_{m,\ell}r)\big)\mathrm{e}^{\pm im\theta}$$

for $m \in \mathbb{N}_0$, where the initial zeros $c_{\ell,m}$ can be found in Table 2 above. The fundamental eigenstate is given by:

$$N_0(c_{0,1})J_0(c_{0,1}r) - J_0(c_{0,1})N_0(c_{0,1}r).$$

The first two excited states, both degenerated, with a two-dimensional eigensubspace are

$$(N_1(c_{1,1})J_1(c_{1,1}r) - J_0(c_{1,1})N_1(c_{1,1}r))\sin(\theta - \beta),$$
$$(N_2(c_{2,1})J_1(c_{2,1}r) - J_0(c_{2,1})N_1(c_{2,1}r))\sin(2(\theta - \beta)),$$

with corresponding plots

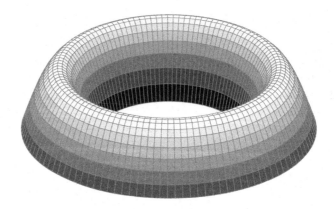

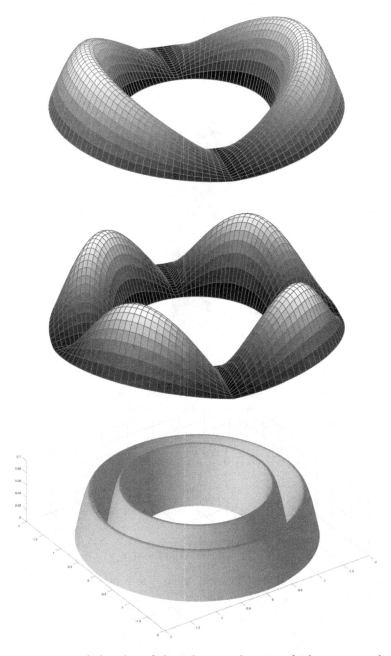

Above, we presented the plot of the 9th excited state, which corresponds to

$$N_0(c_{0,2})J_0(c_{0,2}r) - J_0(c_{0,2})N_0(c_{0,2}r)$$

and energy

$$\frac{\hbar^2}{2M}(c_{0,2})^2.$$

§8.1.6. Fluid in a Pipe

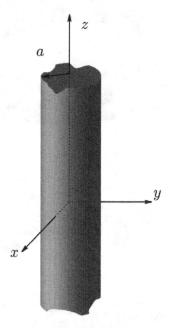

The velocity potential u of a stationary fluid in a circular section pipe, as shown in the figure, is governed by the following Neumann boundary value problem for the Laplace equation:

$$\begin{cases} \Delta u = 0, \\ \left.\dfrac{\partial u}{\partial r}\right|_{r=a} = 0. \end{cases}$$

Here, we have considered that the gradient of u in polar coordinates is given by:

$$\nabla u = u_r \boldsymbol{u}_r + \frac{1}{r} u_\theta \boldsymbol{u}_\theta + u_z \boldsymbol{u}_z,$$

where the orthonormal trihedron $\{\boldsymbol{u}_r, \boldsymbol{u}_\theta, \boldsymbol{u}_z\}$ is constructed in terms of the Cartesian trihedron $\{\boldsymbol{i}, \boldsymbol{j}, \boldsymbol{k}\}$ as follows

$$\boldsymbol{u}_r = \cos\theta\,\boldsymbol{i} + \sin\theta\,\boldsymbol{j}, \quad \boldsymbol{u}_\theta = -\sin\theta\,\boldsymbol{i} + \cos\theta\,\boldsymbol{j}, \quad \boldsymbol{u}_z = \boldsymbol{k}.$$

Hence, the unitary normal vector to the cylindrical pipe's surface is precisely $\boldsymbol{n} = \boldsymbol{u}_r$, leading to the normal derivative on that surface:

$$\frac{\partial u}{\partial \boldsymbol{n}} = \boldsymbol{n} \cdot \nabla u = u_r.$$

Let's apply these boundary conditions to the function:

$$R_{\alpha,m}(r)\Theta_m(\theta)Z_\alpha(z).$$

Considering the case $k = 0$ (thus leading to the Laplace equation), these functions must exhibit regularity at the origin ($r = 0$) and possess 2π-periodicity in θ, indicating that m takes values from the set of integers. Additionally, these functions are required to satisfy the homogeneous Neumann condition on the cylindrical surface with a radius of a, given by:

$$R'_{\alpha,m}(a) = 0.$$

When $\alpha = 0$, only the trivial solution with a constant value for u is feasible, representing a motionless fluid ($v = 0$). However, when $\alpha \neq 0$, the radial functions will take the form: $J_m(\alpha r)$.

Hence, the corresponding boundary condition is expressed as:

$$J'_m(\alpha a) = 0.$$

By denoting $\{c'_{m,\ell}\}_{\ell=0}^{\infty}$ the zeros of J'_m, we can deduce that:

$$\alpha_{m,\ell} = \frac{c'_{m,\ell}}{a},$$

where $m \in \mathbb{N}_0$ and $\ell \in \mathbb{N}$.

Considering the asymptotic behavior for large x:

$$J'_m(x) \sim -\sqrt{\frac{2}{\pi x}} \sin\left(x - (2m + 1)\frac{\pi}{4}\right),$$

it becomes evident that for large ℓ the zeros behave as:

$$c'_{m,\ell} \sim \ell\pi + (2m + 1)\frac{\pi}{4}.$$

Finally, the solutions take the form:

$$\boxed{J_m\left(c'_{m,\ell}\frac{r}{a}\right)\left(C\,e^{im\theta} + D\,e^{-im\theta}\right)\left(Ae^{\alpha_{m,\ell}z} + Be^{-\alpha_{m,\ell}z}\right).}$$

§8.1.7. Playing the Drum

A circular membrane with a radius of one meter, which is fixed at its edge, is initially at rest. Suddenly, at time $t = 0$, it is struck, imparting an instantaneous constant velocity of unit magnitude downward in a circle with a radius of one-third of a meter. The vertical oscillation amplitude $u(t, r, \theta)$, where r, θ are polar coordinates, is characterized by the following boundary problem and initial conditions for the wave equation:

$$\begin{cases} u_{tt} - \Delta u = 0, & 0 \leq r < 1, \quad 0 \leq \theta < 2\pi, \quad t > 0, \\ u|_{t=0} = 0, \\ u_t|_{t=0} = g(r) = \begin{cases} -1, & 0 \leq r < \frac{1}{3}, \\ 0, & \frac{1}{3} < r < 1, \end{cases} \\ u|_{r=1} = 0. \end{cases}$$

In polar coordinates, the wave equation is written as follows:

$$u_{tt} - \left(u_{rr} + \frac{1}{r}u_r + \frac{1}{r^2}u_{\theta\theta}\right) = 0,$$

and thus, the BVP for the associated Helmholtz equation is given by:

$$\begin{cases} w_{rr} + \dfrac{1}{r}w_r + \dfrac{1}{r^2}w_{\theta\theta} + k^2w = 0, \\ w|_{r=1} = 0. \end{cases}$$

Separation of variables,

$$w(r, \theta) = R(r)\Theta(\theta),$$

leads us to the following ordinary differential equations:

$$r^2 R'' + rR' + k^2 r^2 R = m^2 R,$$

$$\Theta'' = -m^2\Theta.$$

From the ODE for $\Theta(\theta)$, it follows that

$$\Theta(\theta) = Ae^{im\theta} + Be^{-im\theta},$$

where A and B are complex constants, and given the periodicity condition, we obtain that $\boxed{m \in \mathbb{Z}.}$ Thus, for a given $m \in \mathbb{Z}$, we consider the radial equation, which can be written as follows:

$$\begin{cases} R'' + \dfrac{1}{r}R' + \left(k^2 - \dfrac{|m|^2}{r^2}\right)R = 0, \\ R|_{r=1} = 0. \end{cases}$$

This is an Euler's differential equation when $k = 0$ or a Bessel equation when $k \neq 0$, modulo dilations. The boundary condition implies that $k \neq 0$, so the possible solutions are linear combinations of the Bessel function $J_{|m|}(kr)$ and the Neumann function $N_{|m|}(kr)$. The condition of regularity at $r = 0$ rules out the Neumann functions. Thus, the radial part becomes

$$\boxed{R(r) = J_{|m|}(kr).}$$

The boundary condition implies that $J_{|m|}(k) = 0$, i.e., k is a zero of the Bessel function $J_{|m|}(r)$. Let $\{c_{|m|,\ell}\}$ be the set of these zeros, so the possible wave vectors are

$$\boxed{k_{m,\ell} = c_{|m|,\ell}.}$$

Based on all that has been said, we conclude that the eigenfunctions and eigenvalues are:

$$\boxed{w_{m,\ell}(r, \theta) = J_{|m|}(c_{|m|,\ell}r)e^{im\theta}, \quad k_{m,\ell}^2 = c_{|m|,\ell}^2,}$$

for $m \in \mathbb{Z}$, $\ell \in \mathbb{N}$.

The set $\{w_{m,\ell}(r,\theta)\}_{\ell\in\mathbb{N},m\in\mathbb{Z}}$ forms an orthogonal basis for the Hilbert space

$$L^2(D(0,1), r\,dr\,d\theta)$$

of functions that are square-integrable over the unit disk. Note that $J_m(0) = 0$ for $m \in \mathbb{N}$ and

$$\int J_0(r)r\,dr = rJ_1(r).$$

Given a non-negative integer m, any function

$$f \in L^2([0,1], x\,dx)$$

can be represented as the following Fourier–Bessel expansion:

$$f(r) = \sum_{\ell=1}^{\infty} b_\ell J_m(c_{m,\ell}r),$$

$$b_\ell = \frac{2}{\left(J_{m+1}(c_{m,\ell})\right)^2} \int_0^1 J_m(c_{m,\ell}r) f(r)r\,dr,$$

where $c_{m,\ell}$, $\ell \in \mathbb{N}$, are the infinitely ordered increasing zeros of $J_m(r)$, i.e.,

$$J_m(c_{m,\ell}) = 0.$$

For numerical expressions of the zeros $c_{m,\ell}$, see Table 1.

Observe that the first initial condition for the membrane's height is zero, and thus it doesn't excite any mode. Also, the second initial condition is independent of θ, implying that necessarily $m = 0$. Therefore, only the eigenfunctions

$$w_{0,\ell}(r) = J_0(c_{0,\ell}r)$$

constructed in terms of the Bessel function J_0 and its infinite zeros come into play. This is known as an axisymmetric case.

We need to calculate the indicated Fourier–Bessel expansion of the initial condition $g(r)$:

$$g(r) = \sum_{\ell=0}^{\infty} A_\ell J_0(c_{0,\ell}r),$$

where the coefficients are given by:

$$A_\ell = \frac{2}{(J_1(c_{0,\ell}))^2} \int_0^1 J_0(c_{0,\ell}r) f(r)r\,dr = -\frac{2}{(J_1(c_{0,\ell}))^2} \int_0^{\frac{1}{3}} J_0(c_{0,\ell}r)r\,dr$$

$$= -\frac{2}{c_{0,\ell}^2(J_1(c_{0,\ell}))^2} \int_0^{\frac{c_{0,\ell}}{3}} J_0(x)x\,dx = -\frac{2}{c_{0,\ell}^2(J_1(c_{0,\ell}))^2} \left[xJ_1(x)\right]_0^{\frac{c_{0,\ell}}{3}}$$

$$= -\frac{2}{3c_{0,\ell}} \frac{J_1(\frac{c_{0,\ell}}{3})}{(J_1(c_{0,\ell}))^2}.$$

Hence,

$$g(r) = -\sum_{\ell=0}^{\infty} \frac{2}{3c_{0,\ell}} \frac{J_1(\frac{c_{0,\ell}}{3})}{(J_1(c_{0,\ell}))^2} J_0(c_{0,\ell}r).$$

TABLE 3. Initial 20 zeros of J_0

ℓ	$c_{0,\ell}$	ℓ	$c_{0,\ell}$
1	2.40483	11	33.77582
2	5.52008	12	36.91710
3	8.65373	13	40.05843
4	11.79153	14	43.19979
5	14.93092	15	46.34119
6	18.07107	16	49.48261
7	21.21164	17	52.62405
8	24.35247	18	55.76551
9	27.49348	19	58.90698
10	30.63461	20	62.04847

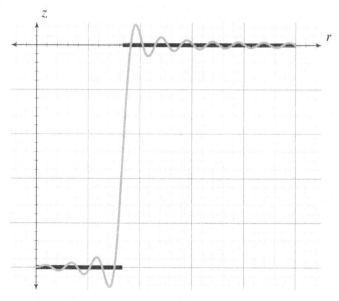

20 terms Fourier–Bessel for g

The eigenfunction expansion of the wave amplitude $u(t, r)$ (which is independent of θ since no longitudinal modes have been excited) on the circular membrane is given by:

$$u(t,r) = \sum_{\ell=1}^{\infty} v_\ell(t) J_0(c_{0,\ell} r).$$

For this expansion $u(t,r)$ to be a solution to the initial conditions of the wave equation on our circular membrane, the coefficients $v_\ell(t)$ must satisfy the following:

$$\begin{cases} v_\ell'' = -c_{0,\ell}^2 v_\ell, \\ v_\ell(0) = 0, \\ v_\ell'(0) = -\dfrac{2}{3c_{0,\ell}} \dfrac{J_1(\frac{c_{0,\ell}}{3})}{J_1(c_{0,\ell})}. \end{cases}$$

For the initial values of $c_{0,\ell}$ see Table 3. The solution of the ODE is

$$v_\ell(t) = a_\ell \cos c_{0,\ell} t + b_\ell \sin c_{0,\ell} t,$$

but since $v_\ell(0) = 0$, we deduce that $a_\ell = 0$, i.e., we can take

$$v_\ell(t) = b_\ell \sin c_{0,\ell} t.$$

However, the second initial condition implies that $v_\ell'(0) = b_\ell$.

$$v_\ell(t) = -\frac{2}{3c_{0,\ell}} \frac{J_1(\frac{c_{0,\ell}}{3})}{(J_1(c_{0,\ell}))^2} \sin c_{0,\ell} t$$

and, finally, the amplitude of the wave on the circular membrane is given by:

$$\boxed{u(t,r) = -\frac{2}{3} \sum_{\ell=1}^{\infty} \frac{1}{c_{0,\ell}} \frac{J_1(\frac{c_{0,\ell}}{3})}{(J_1(c_{0,\ell}))^2} J_0(c_{0,\ell} r) \sin c_{0,\ell} t.}$$

Thus, the solution is an *infinite* superposition of the steady waves

$$\boxed{J_0(c_{0,\ell} r) \sin c_{0,\ell} t.}$$

Observation: Back in 1966, the eminent mathematician Mark Kac posed a profound inquiry in his paper [44]. Kac's exploration delved into the fascinating realm of mathematical physics, specifically exploring the intriguing concept of whether the spectrum of vibrations produced by a drum is uniquely and unequivocally determined by the boundaries that enclose it, i.e., the drum's shape.

In simpler terms, Kac's question can be distilled into an intellectual puzzle: Imagine you stand blindfolded before an unfamiliar drum while someone beats it rhythmically. You are privy to the intricate symphony of frequencies emanating from the drum's vibrations, yet you remain oblivious to the drum's physical form. The question then arises: Can you, armed solely with this auditory information, reverse-engineer the precise contours and dimensions of the drum that orchestrates this auditory spectacle? Or does the world of drum shapes hide an enigmatic secret wherein multiple drum forms can conjure an identical symphony of frequencies?

Kac's work served as a catalyst, propelling these complex issues into the mathematical spotlight, drawing the attention of researchers to the intriguing field of isospectral geometry. This spark ignited a vibrant area of research that continues to thrive and evolve to this day.

More than two decades after Kac's groundbreaking paper, three accomplished mathematicians (Carolyn Gordon, David Webb, and Scott Wolpert) achieved a remarkable milestone by demonstrating that hearing the shape of a drum is, in fact, an elusive quest. This team managed to construct multiple examples of drums, each boasting distinct geometries, yet capable of producing identical sound frequencies. See for example their paper [32].

The breakthroughs made by these researchers took shape in a rather serendipitous manner. During a brief sojourn in Europe, one of the mathematicians, Carolyn Gordon found herself at Germany's Mathematical Research Institute of Oberwolfach, nestled amidst the serene beauty of the Black Forest. Her visit coincided with a pivotal juncture in the team's research on the audibility of shapes.

A pivotal turning point materialized when the researchers realized that an example previously dismissed by Gordon as unworkable held the key to demonstrating the existence of two differently shaped drums that produced indistinguishable sounds. Gordon reflects on this moment, saying, "We got ideas for other pairs that were much more complicated. We were making these huge paper constructions to represent drums of different shapes, and then trying to smash them." Yet, despite crafting these intricate paper "monstrosities," as one mathematician humorously referred to them, they discovered these elaborate constructions were ineffective. It was only when they revisited their original pair that they achieved the breakthrough they sought.

Effectively, their work accomplished what earlier generations of researchers deemed an insurmountable challenge. In 1882, Arthur Schuster, a German-born British physicist, expressed the sentiment that "To find out the different tunes sent out by a vibrating system is a problem which may or may not be solvable in certain special cases, but it would baffle the most skillful mathematician to solve the inverse problem and to find out the shape of a bell by means of the sounds which it is capable of sending out." The mathematicians of today have, in their own way, illuminated the intricate relationship between sound and geometry, proving that the shape of a drum remains a captivating enigma even in the face of scientific inquiry.

§8.1.8. Playing the Clarinet

A clarinet can be mathematically modeled as a cylindrical tube with radius a and length L. It has one open end at $z = 0$ and a closed end, the mouthpiece, at $z = L$. The difference in air pressure between its interior and the exterior, denoted as u, satisfies the following boundary value problem for the wave equation:

$$\begin{cases} u_{tt} = c^2 \Delta u, & r \in [0, a], \quad \theta \in [0, 2\pi], \quad z \in [0, L], \\ \begin{cases} u_r|_{r=a} = 0, \\ u|_{z=0} = 0, \\ u_z|_{z=L} = 0. \end{cases} \end{cases}$$

At the open end, the pressure is that of the external air; thus, the pressure difference u is zero. On the other hand, the normal velocity

$$-\frac{\partial \Psi}{\partial \mathbf{n}},$$

where Ψ is the velocity potential, of the air on the walls of the clarinet is zero for all t. Consequently, the normal acceleration is also zero, and therefore, according to the second law of Newton,

$$\left.\frac{\partial u}{\partial \mathbf{n}}\right|_{\text{walls}} = 0.$$

Substituting the form of the steady-state wave into the wave equation for the pressure difference, we obtain:

$$-\omega^2 \cos(\omega t) w(r, \theta, z) = c^2 \cos(\omega t) \Delta w(r, \theta, z).$$

Hence, the dispersion relation is

$$\boxed{\omega = \pm kc}$$

and the BVP for the Helmholtz equation is

$$\begin{cases} \Delta w + k^2 w = 0, \\ \begin{cases} w_r|_{r=a} = 0, \\ w|_{z=0} = 0, \\ w_z|_{z=L} = 0. \end{cases} \end{cases}$$

The separation of variables in cylindrical coordinates leads to

$$w(r, \theta, z) = R(r)\Theta(\theta)Z(z)$$

with separation constants α, m, and

$$Z_\alpha(z) = \begin{cases} Az + B, & \alpha^2 = \kappa^2, \\ A\sin(\sqrt{k^2 - \alpha^2}z) + B\cos(\sqrt{k^2 - \alpha^2}z), & \alpha^2 \neq \kappa^2, \end{cases}$$

$$\Theta_m(\theta) = \begin{cases} C\theta + D, & m = 0, \\ Ce^{im\theta} + De^{-im\theta}, & m \neq 0, \end{cases}$$

$$R_{\alpha,m}(r) = \begin{cases} E + F\log r, & \alpha = 0 = m, \\ Er^m + Fr^{-m}, & \alpha = 0 \neq m, \\ EJ_m(\alpha r) + FN_m(\alpha r), & \alpha \neq 0, \end{cases}$$

where A, B, C, D, E, and F are complex constants to be determined, J_m is the Bessel function of index m and N_m is the Neumann function of index m.

In addition to the boundary conditions, which we will now see how to impose, there must be single-valuedness in the angular variable when going through 2π radians. This implies that $m \in \mathbb{N}_0$ (negative integers can also appear, but as A and B are arbitrary, these are redundant). Note that for $m = 0$, we must impose $A = 0$. Since the stationary waves are bounded and the defined domain of the clarinet includes $r = 0$, we must impose in the radial part that $F = 0$. There are no logarithmic terms, negative powers, or Neumann functions, all of which are singular at the origin.

Let's now study the boundary conditions. For the radial part, we must impose $R'_{\alpha,m} = 0$, which leads to

$$\begin{cases} E a^m = 0, & \alpha = 0 \neq m, \\ \alpha E J'_m(\alpha r) = 0, & \alpha m \neq 0, \end{cases}$$

which implies that the case $\alpha = 0 = m$ is possible with $R_{0,0} = E$, that $\alpha = 0 \neq m$ leads to the trivial solution since $R = 0$, and the following equation:

$$\boxed{J'_m(\alpha a) = 0,}$$

for $\alpha \neq 0$. That is, α takes one of the following values:

$$\boxed{\alpha_{m,\ell} = \frac{c'_{m,\ell}}{a},}$$

where $\{c'_{m,\ell}\}_{\ell=1}^{\infty}$ **are the zeros of the function** $J'_m(x)$. Therefore,

$$R_{\alpha,m}(r) = \begin{cases} E, & \alpha = 0, \quad m = 0, \\ E J_m\left(\frac{c'_{m,\ell}}{a} r\right), & \alpha \neq 0, \quad m \in \mathbb{N}_0. \end{cases}$$

In the longitudinal part, we have two boundary conditions:

$$\begin{cases} Z_\alpha(0) = 0, \\ Z'_\alpha(L) = 0. \end{cases}$$

If

$$k^2 = \alpha^2,$$

this leads to

$$A = B = 0,$$

the trivial solution. In the case where

$$k^2 \neq \alpha^2,$$

we have

$$Z_\alpha(0) = 0$$

so that $B = 0$ and

$$\cos\sqrt{k^2 - \alpha^2}L = 0.$$

Therefore,

$$\sqrt{k^2 - \alpha^2}\, L = (2n + 1)\frac{\pi}{2}$$

for $n \in \mathbb{N}$. This implies that

$$k^2 = \begin{cases} \dfrac{(2n+1)^2\pi^2}{4L^2}, & \alpha = 0 = m, \\[3mm] \dfrac{(2n+1)^2\pi^2}{4L^2} + \dfrac{(c'_{m,\ell})^2}{a^2}, & \alpha \neq 0. \end{cases}$$

In this case,

$$Z_\alpha(z) = A \sin \frac{(2n+1)\pi}{2L} z.$$

In conclusion, using the dispersion relation for $\alpha \neq 0$:

$$\omega_{n,m,\ell} = c\sqrt{\frac{(2n+1)^2\pi^2}{4L^2} + \frac{(c'_{m,\ell})^2}{a^2}}$$

with $n \in \mathbb{N}$, $m, \ell \in \mathbb{N}_0$.

Therefore, the pure longitudinal steady waves are

$$\cos \frac{(2n+1)\pi}{2L} ct \ \sin \frac{(2n+1)\pi}{2L} z.$$

For $n \in \mathbb{N}$, $m, \ell \in \mathbb{N}_0$, the transverse-longitudinal steady waves are

$$\cos\left(\sqrt{\frac{(2n+1)^2\pi^2}{4L^2} + \frac{(c'_{m,\ell})^2}{a^2}}\ ct\right) J_m\left(\frac{c'_{\ell,m}}{a} r\right) e^{\pm im\theta} \sin \frac{(2n+1)\pi}{2L} z.$$

Further reading: For a deeper exploration of the relationship between music and mathematics, refer to [8]. For a more physical perspective, consult [7]. Additionally, you can explore the authoritative and historically significant treatment of sound theory by Rayleigh in [56].

§8.2. Spherical BVPs

ELMHOLTZ equation, a spectral problem for the Laplace operator, will be currently addressed with a focus on boundaries naturally expressed in spherical coordinates. In our earlier discussion in Chapter 1, we delved into how spherical coordinates offer an alternative perspective from the rectangular coordinates for describing locations in three-dimensional space.

Within the realm of solving these differential equations, Legendre polynomials emerge as a class of orthogonal polynomials, especially applicable when dealing with circular or spherical symmetries. Expanding this concept, associated Legendre polynomials come into play. They extend the notion of Legendre polynomials by accounting for angular dependencies in functions on spheres.

Here, we introduce spherical harmonics, which amalgamate associated Legendre polynomials with a complex exponential factor. Fields such as physics, chemistry, and engineering extensively employ spherical harmonics, particularly when addressing problems linked to partial differential equations on spheres and portraying angular aspects of wave functions in quantum mechanics. In disciplines like geophysics, they harness their potential for mapping the Earth's magnetic and gravitational fields, or in biology the shape of a migrating cell.

In spherical coordinates, the following relationships are established:

$$\begin{cases} x = r \sin\theta \cos\phi, \\ y = r \sin\theta \sin\phi, \\ z = r \cos\theta. \end{cases}$$

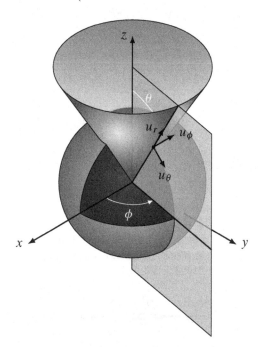

§8.2.1. Separation of Variables for the Helmholtz Equation

The Helmholtz equation in spherical coordinates is written as follows

$$r(ru)_{rr} + k^2 r^2 u + \frac{1}{\sin\theta}(\sin\theta\, u_\theta)_\theta + \frac{1}{\sin^2\theta} u_{\phi\phi} = 0.$$

Performing an initial separation of variables, $u(r,\theta,\phi) = R(r)Y(\theta,\phi)$, leads to the disentanglement of the equation into its radial and angular components as follows:

$$r(rR)'' + k^2 r^2 R = \lambda R,$$

$$\frac{1}{\sin\theta}(\sin\theta\, Y_\theta)_\theta + \frac{1}{\sin^2\theta} Y_{\phi\phi} = -\lambda Y.$$

The angular equation can be expressed in the following form:

$$\sin\theta(\sin\theta\, Y_\theta)_\theta + \lambda Y \sin^2\theta + Y_{\phi\phi} = 0.$$

This equation is amenable to separation of variables.

Hence, using second separation of variables, $Y(\theta,\phi) = P(\theta)\Phi(\phi)$, we arrive at the ensuing pair of ODEs:

$$\sin\theta(\sin\theta\, P_\theta)_\theta + \lambda \sin^2\theta\, P = m^2 P,$$

$$\Phi_{\phi\phi} = -m^2\Phi.$$

As a result, we find that: $\Phi(\phi) = C\,e^{im\phi}$.

To ensure continuity within the xz plane, it is necessary to assume that m belongs to the set of integers $\mathbb{Z}$.

§8.2.2. Associated Legendre Functions

On a different note, by introducing the variable $\xi := \cos\theta$, the equation for P takes on the following form:

$$(8.5) \qquad \boxed{\frac{d}{d\xi}\left((1-\xi^2)\frac{d}{d\xi}P\right) + \left(\lambda - \frac{m^2}{1-\xi^2}\right)P = 0.}$$

This equation represents the **associated** Legendre equation which reduces to the Legendre ODE when $m = 0$. Let us assume $\mathcal{P}(\xi)$ satisfies the Legendre ODE:

$$(8.6) \qquad \frac{d}{d\xi}\left((1-\xi^2)\frac{d\mathcal{P}}{d\xi}\right) + \lambda\mathcal{P} = 0.$$

Then

$$P(\xi) := (1-\xi^2)^{|m|/2}\frac{d^{|m|}\mathcal{P}}{d\xi^{|m|}}$$

is a solution to the associated Legendre equation. For $\ell \in \mathbb{N}_0$, the Legendre equation, as indicated by (8.6), possesses regular solutions at $\xi = \pm 1$ if and only if the following condition is met:

$$(8.7) \qquad\qquad \lambda = \ell(\ell + 1).$$

For $\ell \in \mathbb{N}_0$, these solutions are expressed by the Legendre polynomials, which have an unnormalized form:

$$\mathcal{P}_\ell(\xi) := \frac{d^\ell(\xi^2 - 1)^\ell}{d\xi^\ell}.$$

Consequently, Equation (8.5) will exhibit regular solutions at $\xi = \pm 1$ exclusively if condition (8.7) is satisfied. We introduce the set of pairs:

$$\mathbb{Q} := \{(\ell, m) \in \mathbb{Z}^2 : \ell \in \mathbb{N}_0, m \in \{-\ell, -\ell + 1, \ldots, \ell - 1, \ell\}\}.$$

In atomic physics and quantum chemistry, the collection of quantum numbers denoted as $\mathbb{Q}$ plays a crucial role. Among these quantum numbers, ℓ is referred to as the azimuthal quantum number, while m is known as the magnetic quantum number. These two quantum numbers, ℓ and m, provide valuable information about the spatial orientation and distribution of electron orbitals within an atom.

For $(\ell, m) \in \mathbb{Q}$, the solutions of (8.5) are given by:

$$P_\ell^{|m|}(\xi) := \frac{(-1)^m}{2^\ell \ell!} (1 - \xi^2)^{|m|/2} \frac{d^{|m|+\ell}(\xi^2 - 1)^\ell}{d\xi^{|m|+\ell}}.$$

The factor $(-1)^m$ is referred to as the Condon–Shortley phase. It ensures that the spherical harmonics are properly normalized and have the desired symmetry properties, especially under complex conjugation and parity transformations. The Condon–Shortley phase convention is widely used in quantum mechanics literature and ensures consistency in the representation of angular wave functions.

The associated Legendre equation corresponds to a Sturm–Liouville problem within the interval $[-1, 1]$, with the following coefficients:

$$\rho(\xi) = 1,$$
$$p(\xi) = 1 - \xi^2,$$
$$q(\xi) = \frac{m^2}{1 - \xi^2}.$$

Despite its singularity, it possesses a complete orthogonal set of eigenfunctions:

$$\int_{-1}^{1} P_\ell^{|m|}(\xi) P_{\ell'}^{|m|}(\xi) \, d\xi = 0, \qquad\qquad \ell \neq \ell'.$$

It's worth noting that these solutions are nontrivial if and only if:

$$|m| \leq \ell.$$

Furthermore, for: $m = 0$, we retrieve the Legendre polynomials $P_\ell(\xi)$. Drawing inspiration from the expression:

$$\frac{d^{\ell-m}}{d\xi^{\ell-m}}(\xi^2 - 1)^\ell = (-1)^m \frac{(\ell-m)!}{(\ell+m)!}(1 - \xi^2)^m \frac{d^{\ell+m}}{d\xi^{\ell+m}}(\xi^2 - 1)^\ell,$$

for any

$$(\ell, m) \in \mathbb{Q}$$

we define the associated Legendre functions

$$P_\ell^m(\xi)$$

as follows

$$P_\ell^{-|m|}(\xi) := (-1)^m \frac{(\ell - |m|)!}{(\ell + |m|)!} P_\ell^{|m|}(\xi).$$

Associated Legendre Functions

For values of

$$(\ell, m) \in \mathbb{Q},$$

the associated Legendre functions are expressed as

$$P_\ell^m(\xi) = \frac{(-1)^m}{2^\ell \ell!}(1 - \xi^2)^{m/2} \frac{d^{\ell+m}}{d\xi^{\ell+m}}(\xi^2 - 1)^\ell.$$

It is noteworthy that for even values of m, the associated Legendre functions take the form of polynomials. However, when m is odd, a factor of

$$\sqrt{1 - \xi^2}$$

multiplies a polynomial. Despite this distinction,

$$P_l^m(\cos\theta)$$

assumes the form of a trigonometric polynomial in both $\sin\theta$ and $\cos\theta$. This characteristic has led to the designation of these associated Legendre functions as "associated Legendre polynomials" on occasion.

Let's present the initial 16 associated Legendre functions:

Initial Associated Legendre Functions

$$P_0^0(x) = 1, \qquad\qquad P_1^1(x) = -\sqrt{1 - x^2},$$

$$P_1^0(x) = x, \qquad\qquad P_1^{-1}(x) = \tfrac{1}{2}\sqrt{1 - x^2},$$

$$P_2^2(x) = 3(1 - x^2), \qquad\qquad P_2^1(x) = -3x\sqrt{1 - x^2},$$

$$P_2^0(x) = \tfrac{1}{2}(3x^2 - 1), \qquad\qquad P_2^{-1}(x) = \tfrac{1}{2}x\sqrt{1 - x^2},$$

$$P_2^2(x) = \tfrac{1}{8}(1 - x^2), \qquad\qquad P_3^3(x) = -15(1 - x^2)\sqrt{1 - x^2},$$

$$P_3^2(x) = 15x(1 - x^2), \qquad\qquad P_3^1(x) = -\tfrac{3}{2}(5x^2 - 1)\sqrt{1 - x^2},$$

$$P_3^0(x) = \frac{1}{2}(5x^3 - 3x), \qquad\qquad P_3^{-1}(x) = \tfrac{1}{8}(5x^2 - 1)\sqrt{1 - x^2},$$

$$P_3^{-2}(x) = \tfrac{1}{8}x(1 - x^2), \qquad\qquad P_3^3(x) = \tfrac{1}{48}(1 - x^2)\sqrt{1 - x^2}.$$

---

§8.2.3. Spherical Harmonics

Spherical Harmonics

The solutions of the angular equation, known as *spherical harmonics*, can be expressed in the following form

$$Y_\ell^m(\theta, \varphi) = (-1)^m \sqrt{\frac{(2\ell+1)}{4\pi} \frac{(\ell-m)!}{(\ell+m)!}} \, P_\ell^m(\cos\theta) e^{im\phi},$$

for

$$(\ell, m) \in \mathbb{Q}.$$

Recall that

$$\mathbb{Q} := \{(\ell, m) \in \mathbb{Z}^2 : \ell \in \mathbb{N}_0, m \in \{-\ell, -\ell+1, \dots, \ell-1, \ell\}\}$$

Definition

We must mention that there exist alternative methods for introducing spherical harmonics. The one presented here is commonly used in the context of quantum mechanics. In various other fields of physics, different normalizations are adopted. For instance, in seismology, the normalization takes the form:

$$Y_\ell^m(\theta, \varphi) = \sqrt{(2\ell+1)\frac{4}{\pi}\frac{(\ell-m)!}{(\ell+m)!}} \, P_\ell^m(\cos\theta) e^{im\phi}.$$

In the domains of geodesy and geophysics:

$$Y_\ell^m(\theta, \varphi) = \sqrt{(2\ell+1)\frac{(\ell-m)!}{(\ell+m)!}} \, P_\ell^m(\cos\theta) e^{im\phi}.$$

And within the realm of magnetism:

$$Y_\ell^m(\theta, \varphi) = \sqrt{\frac{(\ell-m)!}{(\ell+m)!}} \, P_\ell^m(\cos\theta) e^{im\phi}.$$

These alternative normalizations cater to the specific requirements and conventions of each field.

The spherical harmonics satisfy the relationship:

(8.8)
$$\boxed{\bar{Y}_\ell^m(\theta, \varphi) = (-1)^m Y_\ell^{-m}(\theta, \varphi).}$$

Here, $\bar{Y}_\ell^m$ denotes the complex conjugate of the spherical harmonic.

Spherical Harmonics Expansions

The spherical harmonics form a complete orthonormal set in the Hilbert space

$$L^2(S^2) := \left\{ f = f(\theta, \phi) : \int_{S^2} |f(\theta, \phi)|^2 \, \mathrm{d}S < \infty \right\},$$

where $S^2 = \{(x, y, z) \in \mathbb{R}^3 : x^2 + y^2 + z^2 = 1\}$ represents the unit sphere in $\mathbb{R}^3$. The area element in this sphere is denoted by $\mathrm{d}S = \sin\theta \mathrm{d}\theta \mathrm{d}\phi$. The scalar product is defined as:

$$(f, g) := \int_{S^2} \bar{f}(\theta, \phi) g(\theta, \phi) \mathrm{d}S$$

$$= \int_0^{2\pi} \int_0^{\pi} \bar{f}(\theta, \phi) g(\theta, \phi) \sin\theta \mathrm{d}\theta \mathrm{d}\phi.$$

Consequently, we have an orthonormal basis

$$\{Y_\ell^m\}_{(\ell, m) \in \mathbb{Q}}.$$

This implies that they form an orthonormal set:

$$(Y_\ell^m, Y_{\ell'}^{m'}) = \delta_{\ell, \ell'} \delta_{m, m'}.$$

Furthermore, any function $f(\theta, \phi)$ in $L^2(S^2)$ can be expressed as an expansion:

$$f(\theta, \phi) = \sum_{(\ell, m) \in \mathbb{Q}} c_\ell^m Y_\ell^m(\theta, \phi).$$

Here, the coefficients c_ℓ^m are given by:

$$c_\ell^m = (Y_\ell^m, f) = \int_0^{2\pi} \int_0^{\pi} \bar{Y}_\ell^m(\theta, \phi) f(\theta, \phi) \sin\theta \mathrm{d}\theta \mathrm{d}\phi.$$

Theorem

Spherical harmonics provide a fundamental framework for understanding atomic orbitals in atomic physics and quantum chemistry. These orbitals, often referred to as atomic orbitals or waves, come in various types based on their angular momentum quantum number ℓ. The specific names for different types of orbitals are as follows:

- s orbitals (*sharp*, $\ell = 0$),
- p orbitals (*principal*, $\ell = 1$),
- d orbitals (*diffuse*, $\ell = 2$),
- f orbitals (*fundamental*, $\ell = 3$).

For higher angular momentum values, such as $\ell \in \{4, 5, 6, 7, \dots\}$, additional letters are employed, following an alphabetical sequence of g, h, i, and k. Notably, the letter 'j' is often skipped in this sequence due to the lack of distinction for it in some languages.

Spherical harmonics are used in the multipole expansion of electromagnetic fields, which is useful in solving problems involving radiation, scattering, and antenna design. They are also used in the analysis of the temperature fluctuations in the cosmic microwave background, which provides critical information about the early universe's structure and evolution. They are employed to model the Earth's gravitational field, which helps in geodesy, satellite navigation systems, and understanding the Earth's internal structure. Spherical harmonics can be used to analyze and synthesize textures in computer graphics and image processing.

s wave: $l = 0$

$$Y_0^0 = \frac{1}{2}\frac{1}{\sqrt{\pi}}.$$

p wave: $l = 1$

$$Y_1^{-1} = \frac{1}{2}\sqrt{\frac{3}{2\pi}}\sin\theta e^{-i\phi} = \frac{1}{2}\sqrt{\frac{3}{2\pi}}\frac{(x-iy)}{r},$$

$$Y_1^0 = \frac{1}{2}\sqrt{\frac{3}{\pi}}\cos\theta = \frac{1}{2}\sqrt{\frac{3}{\pi}}\frac{z}{r},$$

$$Y_1^1 = -\frac{1}{2}\sqrt{\frac{3}{2\pi}}\sin\theta e^{i\phi} = -\frac{1}{2}\sqrt{\frac{3}{2\pi}}\frac{(x+iy)}{r}.$$

d wave: $l = 2$

$$Y_2^{-2}(\theta,\varphi) = \frac{1}{4}\sqrt{\frac{15}{2\pi}}\sin^2\theta e^{-2i\phi} = \frac{1}{4}\sqrt{\frac{15}{2\pi}}\frac{(x-iy)^2}{r^2},$$

$$Y_2^{-1}(\theta,\varphi) = \frac{1}{2}\sqrt{\frac{15}{2\pi}}\sin\theta\cos\theta e^{-i\phi} = \frac{1}{2}\sqrt{\frac{15}{2\pi}}\frac{(x-iy)z}{r^2},$$

$$Y_2^0 = \frac{1}{4}\sqrt{\frac{5}{\pi}}(3\cos^2\theta - 1) = \frac{1}{4}\sqrt{\frac{5}{\pi}}\frac{(2z^2 - x^2 - y^2)}{r^2},$$

$$Y_2^1 = -\frac{1}{2}\sqrt{\frac{15}{2\pi}}\sin\theta\cos\theta e^{i\phi} = -\frac{1}{2}\sqrt{\frac{15}{2\pi}}\frac{(x+iy)z}{r^2},$$

$$Y_2^2 = \frac{1}{4}\sqrt{\frac{15}{2\pi}}\sin^2\theta e^{2i\phi} = \frac{1}{4}\sqrt{\frac{15}{2\pi}}\frac{(x+iy)^2}{r^2}.$$

---

f wave: $l = 3$

$$Y_3^{-3} = \frac{1}{8}\sqrt{\frac{35}{\pi}}\sin^3\theta e^{-3i\phi} = \frac{1}{8}\sqrt{\frac{35}{\pi}}\frac{(x-iy)^3}{r^3},$$

$$Y_3^{-2} = \frac{1}{4}\sqrt{\frac{105}{2\pi}}\sin^2\theta\cos\theta e^{-2i\phi} = \frac{1}{4}\sqrt{\frac{105}{2\pi}}\frac{(x-iy)^2 z}{r^3},$$

$$Y_3^{-1} = \frac{1}{8}\sqrt{\frac{21}{\pi}}\sin\theta(5\cos^2\theta - 1)e^{-i\phi} = \frac{1}{8}\sqrt{\frac{21}{\pi}}\frac{(x-iy)(4z^2 - x^2 - y^2)}{r^3},$$

$$Y_3^0 = \frac{1}{4}\sqrt{\frac{7}{\pi}}(5\cos^3\theta - 3\cos\theta) = \frac{1}{4}\sqrt{\frac{7}{\pi}}\frac{z(2z^2 - 3x^2 - 3y^2)}{r^3},$$

$$Y_3^1 = -\frac{1}{8}\sqrt{\frac{21}{\pi}}\sin\theta(5\cos^2\theta - 1)e^{i\phi} = \frac{-1}{8}\sqrt{\frac{21}{\pi}}\frac{(x+iy)(4z^2 - x^2 - y^2)}{r^3},$$

$$Y_3^2 = \frac{1}{4}\sqrt{\frac{105}{2\pi}}\sin^2\theta\cos\theta e^{2i\phi} = \frac{1}{4}\sqrt{\frac{105}{2\pi}}\frac{(x+iy)^2 z}{r^3},$$

$$Y_3^3 = -\frac{1}{8}\sqrt{\frac{35}{\pi}}\sin^3\theta e^{3i\phi} = \frac{-1}{8}\sqrt{\frac{35}{\pi}}\frac{(x+iy)^3}{r^3}.$$

A useful approach is to construct a basis of real spherical harmonics, which are often referred to as tesseral harmonics. These tesseral harmonics are essentially the real and imaginary components of the complex spherical harmonics, up to some multiplicative constants. By employing the relation described in Equation (8.8), we can derive expressions for tesseral harmonics using their complex counterparts, the spherical harmonics. Note that m appears as a subindex in the notation of the tesseral harmonics.

Tesseral Harmonics I

$$Y_{\ell,m}(\theta,\phi) = \begin{cases} \dfrac{i}{\sqrt{2}}(Y_\ell^m(\theta,\phi) - (-1)^m Y_\ell^{-m}(\theta,\phi)), & m < 0, \\ Y_\ell^0(\theta,\phi), & m = 0, \\ \dfrac{i}{\sqrt{2}}(Y_\ell^m(\theta,\phi) + (-1)^m Y_\ell^{-m}(\theta,\phi)), & m > 0, \end{cases}$$

which are the real and imaginary parts

$$Y_{\ell,m}(\theta,\phi) = \begin{cases} \sqrt{2}(-1)^m \operatorname{Im} Y_\ell^{|m|}(\theta,\phi), & m < 0, \\ Y_\ell^0(\theta,\phi), & m = 0, \\ \sqrt{2}(-1)^m \operatorname{Re} Y_\ell^m(\theta,\phi), & m > 0. \end{cases}$$

Definition

This technique is particularly advantageous in applications where real-valued functions are desired, such as in physics and engineering, or when dealing with real-world data that naturally corresponds to real spherical harmonics. The tesseral harmonics provide a valuable tool for simplifying mathematical calculations, as they can facilitate the representation of real physical phenomena and simplify the analysis of real-world problems.

By employing the associated Legendre polynomials, we can represent tesseral harmonics as a combination of polynomial expressions in the cosine of the latitude angle θ and trigonometric functions on integer multiples of the longitude angle ϕ.

What makes tesseral harmonics particularly valuable is their ability to form a complete orthogonal set within the Hilbert space of square integrable functions on the unit sphere. This completeness, along with their orthogonality, enables tesseral harmonics to serve as a powerful tool for analyzing and understanding functions on the sphere.

As a Hilbert basis, the tesseral harmonics enable us to express solutions to boundary value problems associated with the Helmholtz equation. Particularly, they are wellsuited for BVPs where the boundaries are naturally described in spherical coordinates, and they exhibit spherical symmetry. These harmonics provide an efficient means of expanding solutions in this context, simplifying the analysis and solution of problems that involve spherical symmetry and real-valued solutions are required.

Tesseral Harmonics II

Tesseral harmonics form an orthonormal basis of $L^2(S^2)$, said cosine tesseral harmonics for $m > 0$, and sine tesseral harmonics for $m < 0$

$$Y_{\ell,m}(\theta,\phi) = \begin{cases} \sqrt{2}\sqrt{\frac{2\ell+1}{4\pi}\frac{(\ell-|m|)!}{(\ell+|m|)!}}\, P_\ell^{|m|}(\cos\theta)\sin|m|\phi, & m < 0, \\ \sqrt{\frac{2\ell+1}{4\pi}}\, P_\ell^m(\cos\theta), & m = 0, \\ \sqrt{2}\sqrt{\frac{2\ell+1}{4\pi}\frac{(\ell-m)!}{(\ell+m)!}}\, P_\ell^m(\cos\theta)\cos m\phi & m > 0. \end{cases}$$

Theorem

Here, we present various sets of tesseral harmonics using the s, p, d, and f wave notations, along with the distinct multipolar components of the orbitals. Additionally, we will depict the orbitals, or tesseral waves, as surfaces defined by the polar equation $r = \sqrt{4\pi}|Y_{\ell,m}(\theta,\phi)|$. These surfaces are not level surfaces.

---

s wave: $l = 0$

$$Y_{0,0} = s = \frac{1}{2}\sqrt{\frac{1}{\pi}}.$$

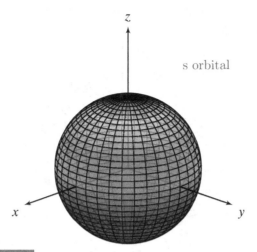

s orbital

p wave: $l = 1$

$$Y_{1,-1} = p_y = \sqrt{\frac{3}{4\pi}}\frac{y}{r} = \sqrt{\frac{3}{4\pi}}\sin\theta\sin\phi,$$

$$Y_{1,0} = p_z = \sqrt{\frac{3}{4\pi}}\frac{z}{r} = \sqrt{\frac{3}{4\pi}}\cos\theta,$$

$$Y_{1,1} = p_x = \sqrt{\frac{3}{4\pi}}\frac{x}{r} = \sqrt{\frac{3}{4\pi}}\sin\theta\cos\phi.$$

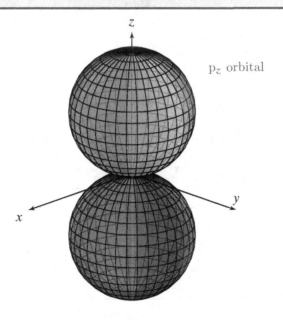

p_z orbital

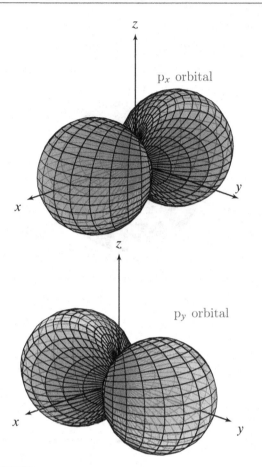

---

d wave: $l = 2$

$$Y_{2,-2} = \mathrm{d}_{xy} = \frac{1}{2}\sqrt{\frac{15}{\pi}}\frac{xy}{r^2} = \frac{1}{2}\sqrt{\frac{15}{\pi}}\sin^2\theta\cos\phi\sin\phi,$$

$$Y_{2,-1} = \mathrm{d}_{yz} = \frac{1}{2}\sqrt{\frac{15}{\pi}}\frac{yz}{r^2} = \frac{1}{2}\sqrt{\frac{15}{\pi}}\cos\theta\sin\theta\sin\phi,$$

$$Y_{2,0} = \mathrm{d}_{z^2} = \frac{1}{4}\sqrt{\frac{5}{\pi}}\frac{-x^2-y^2+2z^2}{r^2} = \frac{1}{4}\sqrt{\frac{5}{\pi}}(3\cos^2\theta-1),$$

$$Y_{2,1} = \mathrm{d}_{xz} = \frac{1}{2}\sqrt{\frac{15}{\pi}}\frac{zx}{r^2} = \frac{1}{2}\sqrt{\frac{15}{\pi}}\cos\theta\sin\theta\cos\phi,$$

$$Y_{2,2} = \mathrm{d}_{x^2-y^2} = \frac{1}{4}\sqrt{\frac{15}{\pi}}\frac{x^2-y^2}{r^2} = \frac{1}{4}\sqrt{\frac{15}{\pi}}\sin^2\theta(\cos^2\phi-\sin^2\phi).$$

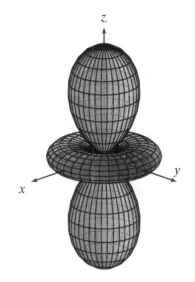

d_{z^2} orbital

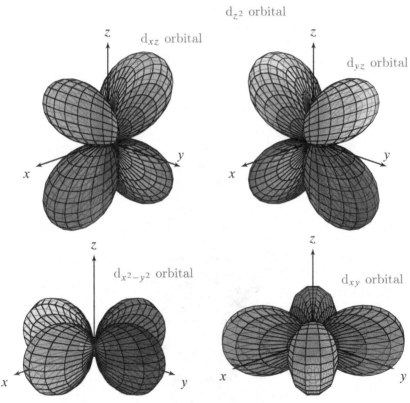

d_{xz} orbital

d_{yz} orbital

$d_{x^2-y^2}$ orbital

d_{xy} orbital

f wave: $l = 3$

$$Y_{3,-3} = f_{y(3x^2-y^2)} = \frac{1}{4}\sqrt{\frac{35}{2\pi}}\frac{y(3x^2-y^2)}{r^3} = \frac{1}{4}\sqrt{\frac{35}{2\pi}}\sin^3\theta\sin\phi(4\cos^2\phi - 1),$$

$$Y_{3,-2} = f_{xyz} = \frac{1}{2}\sqrt{\frac{105}{\pi}}\frac{xyz}{r^3} = \frac{1}{2}\sqrt{\frac{105}{\pi}}\cos\theta\sin^2\theta\sin\phi\cos\phi,$$

$$Y_{3,-1} = f_{yz^2} = \frac{1}{4}\sqrt{\frac{21}{2\pi}}\frac{y(4z^2-x^2-y^2)}{r^3} = \frac{1}{4}\sqrt{\frac{21}{2\pi}}(5\cos^2\theta - 1)\sin\theta\sin\phi,$$

$$Y_{3,0} = f_{z^3} = \frac{1}{4}\sqrt{\frac{7}{\pi}}\frac{z(2z^2-3x^2-3y^2)}{r^3} = \frac{1}{4}\sqrt{\frac{7}{\pi}}(5\cos^2\theta - 3)\cos\theta,$$

$$Y_{3,1} = f_{xz^2} = \frac{1}{4}\sqrt{\frac{21}{2\pi}}\frac{x(4z^2-x^2-y^2)}{r^3} = \frac{1}{4}\sqrt{\frac{21}{2\pi}}(5\cos^2\theta - 1)\sin\theta\cos\phi,$$

$$Y_{3,2} = f_{z(x^2-y^2)} = \frac{1}{4}\sqrt{\frac{105}{\pi}}\frac{z(x^2-y^2)}{r^3} = \frac{1}{4}\sqrt{\frac{105}{\pi}}\cos\theta\sin^2\theta(\cos^2\phi - \sin^2\phi),$$

$$Y_{3,3} = f_{x(x^2-3y^2)} = \frac{1}{4}\sqrt{\frac{35}{2\pi}}\frac{x(x^2-3y^2)}{r^3} = \frac{1}{4}\sqrt{\frac{35}{2\pi}}\sin^3\theta\cos\phi(1 - 4\sin^2\phi).$$

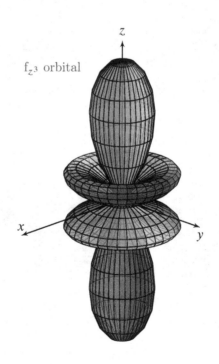

f_{z^3} orbital

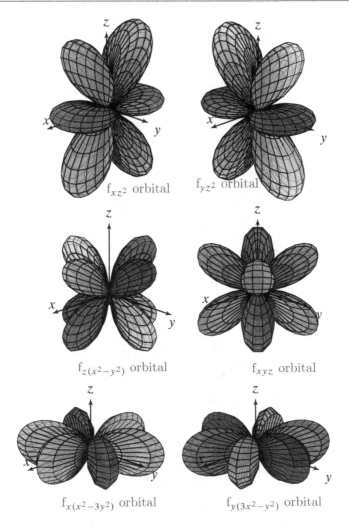

f_{xz^2} orbital

f_{yz^2} orbital

$f_{z(x^2-y^2)}$ orbital

f_{xyz} orbital

$f_{x(x^2-3y^2)}$ orbital

$f_{y(3x^2-y^2)}$ orbital

Historical notes: In 1867, a significant milestone occurred in the realm of mathematical physics when William Thomson (also known as Lord Kelvin) and Peter Guthrie Tait introduced what is now known as solid spherical harmonics within the pages of their seminal work, "Treatise on Natural Philosophy." It was within this groundbreaking treatise that they bestowed upon these functions the name "spherical harmonics," forever shaping the terminology used to describe these mathematical constructs.

Observation: The plots we've presented for the real spherical harmonics take the form of polar plots. In these plots, the distance of a given point on the surface from the origin corresponds to the absolute value of the corresponding spherical harmonics. Consequently, for a quantum particle, the larger this distance becomes, the greater the probability density is. However, these classical plots don't directly represent the probability density. Nowadays, due to the widespread availability of advanced plotting tools, it's becoming more common to depict the probability density on a unit sphere.

In the graphics below, regions that appear darker indicate a higher probability of finding the particle, while whiter regions indicate lower probability.

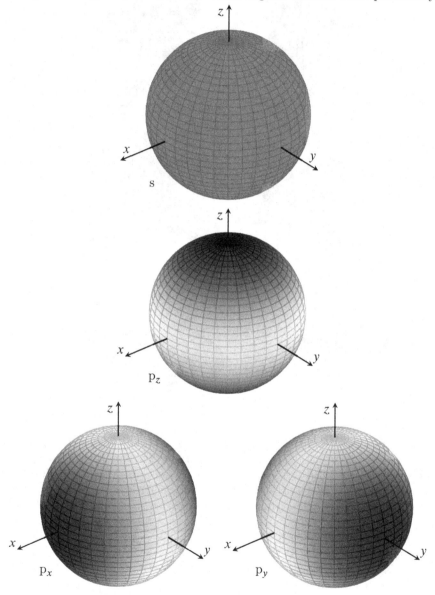

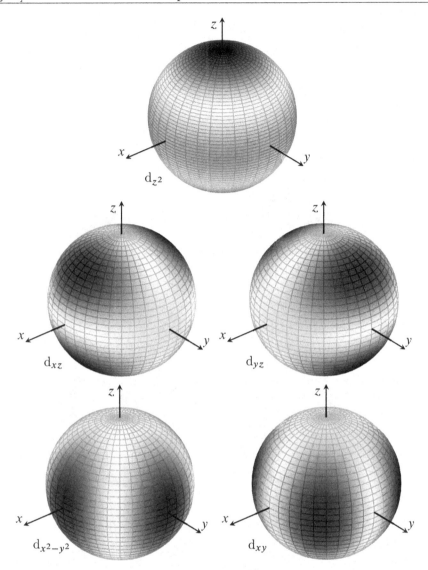

d_{z^2}

d_{xz}

d_{yz}

$d_{x^2-y^2}$

d_{xy}

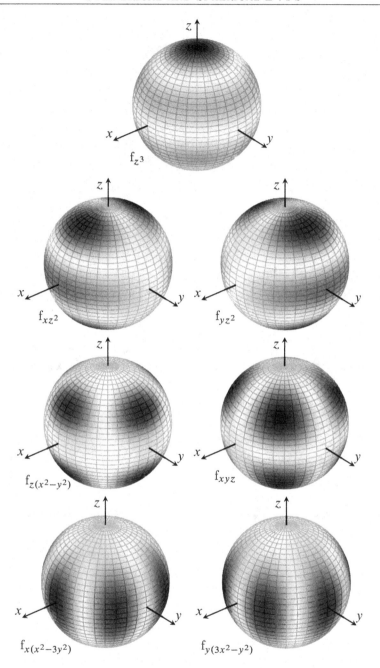

f_{z^3}

f_{xz^2} f_{yz^2}

$f_{z(x^2-y^2)}$ f_{xyz}

$f_{x(x^2-3y^2)}$ $f_{y(3x^2-y^2)}$

☞   **Observations:**

(1) These tesseral harmonics are aptly named because they vanish along
curves known as nodal curves, which correspond to $\ell - m$ parallels of

latitude and $2m$ meridians. These nodal curves effectively partition the surface of a sphere into quadrangles, each marked by right angles [77]. It's worth noting that the term "tessera" has Latin origins and signifies a small, square piece of stone or wood.

(2) Take note that:

$$\frac{1}{\sin\theta}(\sin\theta\, Y_\theta)_\theta + \frac{1}{\sin^2\theta}Y_{\phi\phi} = -\frac{L^2 Y}{\hbar^2}.$$

The differential operator L^2 is the squared orbital angular momentum operator in quantum mechanics. Similarly, the operator $L_z := -i\hbar\dfrac{\partial}{\partial\phi}$, is the vertical component of the angular momentum. The study of angular momentum has a rich and extensive literature, and we recommend interested readers to consult the following references for further details: [14, 24, 57].

§8.2.4. Cosmic Microwave Background

As previously mentioned, spherical harmonics play a pivotal role in modern cosmology and our comprehension of the cosmic microwave background (CMB). In the figure below (© ESA and the Planck Collaboration) presents a map (originally in color) illustrating recorded fluctuations in the CMB temperature, denoted as $\Delta T(\theta,\phi)$, in relation to the background temperature $T_0 = 2.7$ K. The darker regions (originally depicted in dark blue) correspond to directions in the sky where the CMB temperature is approximately 10^{-5} lower than the mean, $T_0 = 2.7$ K. These areas reflect conditions in the early universe. In contrast, the grayer regions (originally reddish and yellowish in color) represent hot (underdense) regions. The statistical properties of these fluctuations contain invaluable insights into both the background's evolution and the universe's initial conditions.

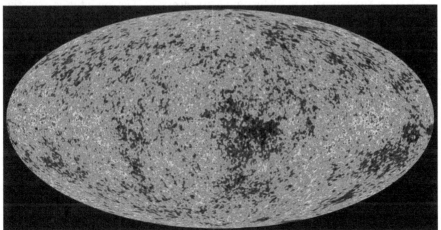

In this context, (θ, ϕ) denote the direction in the sky, specifically on the celestial sphere. The harmonic expansion of this map is defined as follows:

$$\Theta(\theta, \phi) = \frac{\Delta T(\theta, \phi)}{T_0} = \sum_{(\ell, m) \in \mathbb{Q}} c_{\ell, m} Y_{\ell, m}(\theta, \phi),$$

where:

$$c_{\ell, m} = \int Y_{\ell, m}(\theta, \phi) \Theta(\theta, \phi) \sin \theta \mathrm{d}\theta \mathrm{d}\phi.$$

Spherical harmonics with $\ell = 0$, $\ell = 1$, and $\ell = 2$ correspond to the monopole, dipole, and quadrupole, respectively, with m ranging from $-\ell$ to $+\ell$. These multipole moments, denoted as $c_{\ell, m}$, can be combined to create rotationally invariant angular power spectra.

For further information, refer to [22].

§8.2.5.   Periodic Table

The central field approximation, also referred to as the central field model or central field theory, serves as a foundational concept in atomic physics and quantum chemistry. It offers a simplified framework for understanding the behavior of electrons within many-electron atoms. In this approximation, each electron within an atom is primarily influenced by the positively charged atomic nucleus. The interaction between the electron and the nucleus is described by Coulomb's law, wherein the electron is attracted to the nucleus due to their opposite charges.

To account for the complexities arising from electron–electron repulsion, the central field model introduces the concept of an effective nuclear charge. This effective charge considers the shielding effect of inner-shell electrons, which reduces the net positive charge experienced by outermost electrons. Effective nuclear charge represents the average nuclear charge encountered by an electron in a multielectron atom.

Rather than directly modeling intricate electron–electron interactions, the central field approximation simplifies the problem by assuming that each electron moves independently within the effective field of the nucleus, accounting for modifications due to shielding effects.

Within this framework, electron energy levels in many-electron atoms are computed as if they were influenced by a solitary effective nucleus bearing the charge equivalent to the effective nuclear charge. These energy levels can be delineated using quantum numbers and orbitals, akin to the methodology applied in the hydrogen atom, albeit with necessary adaptations due to the effective nuclear charge.

The system is governed by a Schrödinger equation designed for a single electron, characterized by a spherically symmetric effective potential. Consequently, a separation of variables is readily achievable, spotlighting the pivotal roles played by spherical harmonics $Y_{\ell, m}(\theta, \phi)$ in describing the ground state of multielectronic atoms.

The central field approximation proves valuable in predicting electron configurations and the occupation of the 1-electron wave functions, known as atomic orbitals,

$$R_{n,\ell}(r)Y_{\ell,m}(\theta,\phi)\chi_s,$$

(χ_s takes into account the two possible spins of the electron) within many-electron atoms. It aids in understanding broad trends in the periodic table, such as the gradual increase in effective nuclear charge across a period and the sequential filling of electron shells.

However, it's essential to recognize that the central field approximation is a simplification. It does not fully encompass the intricacies of electron–electron interactions in many-electron systems. For precise calculations of electronic structure and properties in real atoms and molecules, more sophisticated models, such as the Hartree–Fock theory and density functional theory, are required. In Hartree theory, the one-electron wave functions, which are expressed in terms of spherical harmonics, are systematically represented using Slater determinants. Furthermore, the Pauli exclusion principle is integrated into the extended Hartree–Fock method. Consequently, in these improved approximation schemes, spherical harmonics play a pivotal and indispensable role.

For a more in-depth exploration of this topic, we recommend interested readers to consult chapter 6 in Bransden and Joachain work [13] and chapter 8 in Demtröder's book [21]. See also the authoritative monograph [68] by Sobelman; the second edition is an invaluable reference work on atomic physics.

Despite its simplifications, the central field approximation is a fundamental concept in atomic physics and quantum chemistry, as it forms the basis for understanding the electronic structure of atoms and the periodic table's organization. It provides a starting point for more sophisticated quantum mechanical models that take into account electron–electron interactions more accurately.

The periodic table is a tabular arrangement of chemical elements, organized by their atomic number, electron configuration, and recurring chemical properties. It is divided into rows and columns, with rows called "periods" and columns called "groups" or "families." The periodic table has 18 groups and seven periods. These divisions help chemists and scientists understand the relationships and trends among elements.

The periods are the horizontal rows in the periodic table, and there are seven periods in total. The period number indicates the energy level or shell in which the outermost electrons of an element are found. The first period (Period 1) contains only two elements, hydrogen (H) and helium (He), because these two elements have electrons filling the first energy level. The second period (Period 2) contains elements with electrons filling the second energy level, and so on. Each subsequent period adds another energy level for electrons, which results in elements in the same period sharing similar properties, as they have the same number of electron shells. For example, elements in Period 3 all have three energy levels, while those in Period 4 have four energy levels.

Periodic Table of Elements

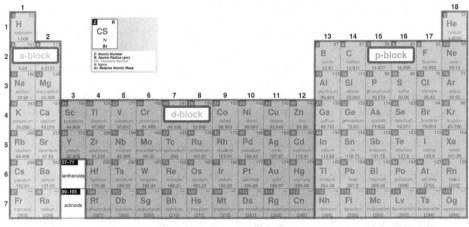

Groups, also known as families, are the vertical columns in the periodic table. Elements within the same group share similar chemical properties because they have the same number of valence electrons in their outermost electron shell. This outermost electron configuration is a key factor in determining an element's chemical behavior and reactivity. For example, group 1 is built up with the alkali metals, including elements like hydrogen (H), lithium (Li), and sodium (Na). They are highly reactive and have one valence electron. Then, group 2 consists of the alkaline earth metals, including elements like beryllium (Be) and magnesium (Mg). They are also reactive but less so than alkali metals and have two valence electrons.

In group 17 we find the halogens, including elements like fluorine (F) and chlorine (Cl). They are highly reactive nonmetals and have seven valence electrons. In group 18 we find the noble gases, including elements like helium (He) and neon (Ne). They are chemically inert and have full valence electron shells. Other groups, such as transition metals (groups 3–12), lanthanides, and actinides, have more complex electron configurations and exhibit a wide range of properties.

The periodic table includes the f-block elements, which are not numbered as part of the standard 18 groups. Instead, they form a separate region, consisting of 14 columns. Within this extension, the f-block elements are included, making a total of 32 columns. These f-block elements are found in the two rows at the bottom of the periodic table, known as the lanthanides and actinides series. The lanthanides encompass elements from cerium (Ce) to lutetium (Lu), while the actinides include elements from thorium (Th) to lawrencium (Lr). Due to their complex electron configurations and unique properties, the f-block elements are typically depicted separately from the main body of the periodic table.

Numbers 18 and 32 in the context of the periodic table can be explained by considering the quantum numbers, denoted as (ℓ, m), which characterize the behavior of electrons in atoms, particularly when it comes to the arrangement of electron orbitals. The explanation is rooted in the principles of quantum mechanics and the properties of spherical harmonics.

For each azimuthal quantum number, ℓ, there are $2\ell + 1$ possible values for the magnetic quantum number, m, within the range of $m \in \{-\ell, \ldots, \ell\}$. This fact stems from the properties of spherical harmonics, which describe the spatial distribution of electron probability density within an atom. Now, consider the sum of the first ℓ odd numbers. This sum is equal to ℓ^2. When you apply this to the context of electron orbitals and spherical harmonics, it tells us that up to $\ell = 3$, which corresponds to the d orbitals in the electron configuration, there are nine possible spherical harmonics. Taking into account the intrinsic property of electron spin, which has two possible values (spin-up and spin-down), you end up with a total of 18 electron eigenstates within each electron shell.

If you extend this concept to $\ell = 4$, corresponding to the f orbitals, you have $4^2 = 16$ possible spherical harmonics. When you include electron spin, this leads to a total of 32 eigenstates within the f-block of the periodic table, where electrons occupy these more complex orbitals. That is, for an azimuthal quantum number ℓ we get up to $2\ell^2$ configurations.

In summary, numbers 18 and 32 in the periodic table's context arise from the quantum mechanical principles that govern the distribution of electrons in various types of orbitals based on the angular quantum numbers and the consideration of electron spin, resulting in different eigenstates for each electron shell or block.

Understanding the periodic table's organization into periods and groups is essential for predicting an element's chemical behavior and its ability to form compounds with other elements. It also helps chemists and scientists make sense of the trends in properties across the periodic table, such as atomic size, electronegativity, and reactivity.

In the periodic table, elements that use s and p orbitals for their electron configurations are typically found in the s-block and p-block, respectively. Elements in the p-block, from groups 13 to 18, use both s and p orbitals as they fill their valence electron shells. The specific orbitals used depend on the element's position in the periodic table and the number of electrons it contains.

Elements that use d orbitals for their electron configurations are primarily found in the d-block of the periodic table. The d-block elements, also known as transition metals, are characterized by their use of d orbitals for filling their electron shells. Elements that use f orbitals for their electron configurations are found in the f-block of the periodic table. The f-block elements are part of the lanthanide and actinide series. These elements have partially filled 4f and 5f orbitals, respectively, in their electron configurations.

In the conventional notation used to describe electron configurations, there are no g orbitals with an azimuthal quantum number of $\ell = 5$. The orbitals up to and including the f orbitals are the primary determinants of electron behavior in chemical elements, encompassing the vast majority of known electron configurations. Mentioning g orbitals is typically theoretical or hypothetical, as they have not been observed or associated with any elements in the periodic table. In the current understanding of atomic structure, there are no elements known to have g orbitals.

However, if, in the future, g orbitals or similar higher-energy orbitals were to be confirmed and introduced, it would necessitate an extension of the periodic table. This extension would introduce a new region, often referred to as a "g-block," which would contain elements with electrons occupying these higher orbitals. Such an extension would also bring about the need for an additional set of groups, 50 in number, to accommodate the new elements and their unique properties within the periodic table.

§8.2.6. Addition Formulas and Multipolar Expansion

The spherical harmonics are known to satisfy a crucial set of addition formulas. Consider the unit vectors denoted as $\boldsymbol{u}(\theta, \phi)$, which point toward a specific location characterized by latitude θ and longitude ϕ. In terms of the Cartesian coordinate system represented by the basis vectors $\{\boldsymbol{i}, \boldsymbol{j}, \boldsymbol{k}\}$, these unit vectors can be expressed as:

$$\boldsymbol{u}(\theta, \phi) = \sin\theta\cos\phi\,\boldsymbol{i} + \sin\theta\sin\phi\,\boldsymbol{j} + \cos\theta\,\boldsymbol{k}.$$

Let us denote by γ the angle between $\boldsymbol{u}(\theta, \phi)$ and $\boldsymbol{u}(\theta', \phi')$. Then, as

$$\boldsymbol{u}(\theta, \phi) \cdot \boldsymbol{u}(\theta', \phi') = \cos\gamma,$$

we find

$$\cos\gamma = \cos\theta\cos\theta' + \sin\theta\sin\theta'\cos\left(\phi - \phi'\right).$$

The addition theorem for the spherical harmonics can then be stated as:

$$(8.9) \qquad \boxed{P_\ell(\cos\gamma) = \frac{4\pi}{2\ell+1} \sum_{m=-\ell}^{\ell} Y_\ell^m(\theta, \phi)\bar{Y}_\ell^m(\theta', \phi').}$$

In particular, when $\theta = \theta', \phi = \phi'$, this gives Unsöld's theorem

$$\boxed{\sum_{m=-\ell}^{\ell} Y_\ell^m \bar{Y}_\ell^m = \frac{2\ell+1}{4\pi}.}$$

which generalizes the identity $\cos^2\theta + \sin^2\theta = 1$ to two dimensions.

Let's consider a function $\rho(r, \theta, \phi)$ representing a charge or mass density, with support within a bounded domain $\Omega \subset \mathbb{R}^3$. We assume that it can be expanded in terms of spherical harmonics as follows:

$$r^2 \rho(r) = \sum_{(\ell,m) \in \mathbb{S}} \rho_\ell^m(r) Y_\ell^m(\theta, \phi).$$

Now, let's take a point r_0 outside of a ball $B(0, R)$ containing the bounded domain Ω, where the density ρ has its support, meaning $\Omega \subset B(0, R)$ (i.e., $r_0 > R$).

The corresponding potential energy is given by:

$$V(r_0) = \int_\Omega \frac{\rho(r)}{|r_0 - r|} \, dr.$$

By recalling (7.8), we can expand it in terms of spherical harmonics as follows:

$$V(r_0, \theta_0, \phi_0) = \int_0^R dr \int_{S^2} dS \sum_{(\ell,m) \in \mathbb{S}} \rho_\ell^m(r) Y_\ell^m(\theta, \phi) \sum_{\ell'} P_{\ell'}(\cos \gamma) \frac{r^{\ell'}}{r_0^{\ell'+1}}.$$

Then, utilizing the addition formula (8.9) and the fact that the spherical harmonics form an orthonormal basis in the Hilbert space $L^2(S^2, dS)$, we arrive at the following multipolar expansion for the potential energy:

$$\boxed{V(r_0, \theta_0, \phi_0) = 4\pi \sum_{\ell,m \in \mathbb{S}} \frac{1}{2\ell + 1} q_\ell^m \frac{1}{r_0^{\ell+1}} Y_\ell^m(\theta_0, \phi_0).}$$

Here, the so-called multipolar moments are defined as:

$$q_\ell^m := \int_0^R \rho_\ell^m(r) r^\ell dr.$$

It's important to note that for the multipolar moments, we have the relationship $q_\ell^{-m} = (-1)^m \bar{q}_\ell^m$. For more information, please refer to [34, 43].

The formulas for tesseral harmonics

$$Y_\ell^m = \begin{cases} \dfrac{1}{\sqrt{2}} \left(Y_{\ell,|m|} - iY_{\ell,-|m|} \right), & m < 0, \\[2mm] Y_{\ell,0}, & m = 0, \\[2mm] \dfrac{(-1)^m}{\sqrt{2}} \left(Y_{\ell,|m|} + iY_{\ell,-|m|} \right), & m > 0, \end{cases}$$

lead to the addition formula

$$\boxed{P_\ell(\cos \gamma) = \frac{4\pi}{2\ell + 1} \sum_{m=-\ell}^{\ell} Y_{\ell,m}(\theta, \phi) Y_{\ell,m}(\theta', \phi').}$$

Expressed in terms of tesseral harmonics, we can represent $r^2\rho$ as:

$$r^2\rho = \sum_{\ell,m} \rho_{\ell,m}(r)Y_{\ell,m}(\theta,\phi).$$

The corresponding tesseral multipolar moments are defined as:

$$q_{\ell,m} := \int_0^R \rho_{\ell,m}(r)r^\ell dr.$$

This leads us to the expression for the potential energy:

$$V(r_0,\theta_0,\phi_0) = 4\pi \sum_{(\ell,m)\in S} \frac{1}{2\ell+1}q_{\ell,m}\frac{1}{r_0^{\ell+1}}Y_{\ell,m}(\theta_0,\phi_0).$$

§8.2.7. Radial Equation

We differentiate between two cases based on whether k is zero or not.

- When $k = 0$, the radial equation takes the form:

$$r^2R'' + 2rR' - \ell(\ell+1)R = 0.$$

This equation belongs to the Euler ordinary differential equation type. By trying solutions in the form of r^α, we promptly deduce that the general solution is

(8.10)
$$R(r) = Ar^\ell + B\frac{1}{r^{\ell+1}}.$$

- When $k \neq 0$, the radial ordinary differential equation for R is:

$$r^2R'' + 2rR' + (k^2r^2 - \ell(\ell+1))R = 0.$$

Considering the change of variable $x = kr$, this radial equation transforms into the differential equation that corresponds to the spherical Bessel and Neumann functions:

$$j_\ell(x) := (-x)^\ell\left(\frac{1}{x}\frac{d}{dx}\right)^\ell\frac{\sin x}{x},$$
$$n_\ell(x) := -(-x)^\ell\left(\frac{1}{x}\frac{d}{dx}\right)^\ell\frac{\cos x}{x},$$

it is clear that the general solution of the radial equation for $k \neq 0$ is

(8.11)
$$R_\ell(r) = Aj_\ell(kr) + Bn_\ell(kr).$$

The initial spherical Bessel and Neumann functions are:

wave	s	p	d
j_ℓ	$\dfrac{\sin r}{r}$	$\dfrac{\sin r}{r^2} - \dfrac{\cos r}{r}$	$3\dfrac{\sin r}{r^3} - 3\dfrac{\cos r}{r^2} - \dfrac{\sin r}{r}$
n_ℓ	$-\dfrac{\cos r}{r}$	$-\dfrac{\cos r}{r^2} - \dfrac{\sin r}{r}$	$-3\dfrac{\cos r}{r^3} - 3\dfrac{\sin r}{r^2} + \dfrac{\cos r}{r}$

The asymptotic behavior of these functions as the argument approaches the origin, $r \to 0$, can be expressed as follows:

$$j_\ell(r) = \frac{r^\ell}{(2\ell + 1)!!} + O(r^{\ell+1}),$$

$$n_\ell(x) = -\frac{(2\ell - 1)!!}{r^{\ell+1}} + O(r^{-\ell}).$$

As for the asymptotic behavior of the spherical Bessel functions and Neumann functions as the argument approaches infinity, $r \to \infty$, they are given by:

$$j_\ell(r) = \frac{1}{r} \cos\left(r - (2\ell + 1)\frac{\pi}{2}\right) + O(r^{-2}),$$

$$n_\ell(r) = \frac{1}{r} \sin\left(r - (2\ell + 1)\frac{\pi}{2}\right) + O(r^{-2}).$$

§8.2.8. Quantum Particle in a Sphere

The stationary states of energy E for a free particle confined within a sphere of radius a and impenetrable walls are characterized by the solutions of the following boundary value problem:

$$\begin{cases} \Delta u + k^2 u = 0, \quad E = \frac{\hbar^2 k^2}{2M}, \\ u|_{r=a} = 0. \end{cases}$$

We must enforce both regularity at the origin and boundary value conditions on the solutions of the Helmholtz equation, given by:

$$u(r, \theta, \phi) = R_{k,\ell}(r) Y_{\ell,m}(\theta, \phi).$$

The requirement for regularity at the origin implies that B must be set to 0 in both (8.10) and (8.11). Conversely, the boundary value conditions for the (8.10) case lead to A being equal to 0, resulting in a trivial solution. For the (8.11) case, these conditions yield:

$$\boxed{j_\ell(ka) = 0.}$$

Consequently, if $\{c_{\ell,n}\}_{n=1}^{\infty}$ represents the increasing sequence of zeros of the spherical Bessel function j_ℓ, the energy values are:

$$\boxed{E_{\ell,n} = \frac{\hbar^2}{2M} \frac{c_{\ell,n}^2}{a^2}.}$$

The corresponding stationary states are given by:

$$u_{\ell,m,n}(r,\theta,\phi) = j_\ell \left(\frac{c_{\ell,n}}{a}r\right) Y_{\ell,m}(\theta,\phi).$$

The probability density $|u_{\ell,m,n}|^2$ does not vary with respect to the variable ϕ. As a result, we can represent the probability in the planes containing the z-axis, where ϕ is constant, using polar coordinates (r,θ). This representation remains consistent across all these planes.

The initial three zeros are

$$c_{0,1} = \pi, \qquad c_{1,1} \approx 4.493409458, \qquad c_{2,1} \approx 5.763459197.$$

Consequently, the energy of the fundamental state and the first excited state is approximately given by:

$$\frac{\hbar^2}{2Ma^2}9.869604404, \qquad \frac{\hbar^2}{2Ma^2}20.190728557, \qquad \frac{\hbar^2}{2Ma^2}33.217461915.$$

§8.2.9. Fluid Inside a Sphere

The velocity potential u for a stationary fluid within a sphere of radius a is defined by the following Neumann problem:

$$\begin{cases} \Delta u = 0, \\ u_r|_{r=a} = 0. \end{cases}$$

In this context, it is important to consider that the gradient of u is given by:

$$\nabla u = u_r \boldsymbol{u}_r + \frac{1}{r}u_\theta \boldsymbol{u}_\theta + \frac{1}{r\sin\theta}u_\phi \boldsymbol{u}_\phi.$$

The orthonormal basis $\boldsymbol{u}_r, \boldsymbol{u}_\theta, \boldsymbol{u}_\phi$ is defined as follows:

$$\begin{cases} \boldsymbol{u}_r = \sin\theta\cos\phi\,\boldsymbol{i} + \sin\theta\sin\phi\,\boldsymbol{j} + \cos\theta\,\boldsymbol{k}, \\ \boldsymbol{u}_\theta = \cos\theta\cos\phi\,\boldsymbol{i} + \cos\theta\sin\phi\,\boldsymbol{j} - \sin\theta\,\boldsymbol{k}, \\ \boldsymbol{u}_\phi = -\sin\phi\,\boldsymbol{i} + \cos\phi\,\boldsymbol{j}. \end{cases}$$

Hence, the unit normal vector to the spherical surface is

$$\boldsymbol{n} = \boldsymbol{u}_r,$$

which leads to the following relation:

$$\frac{\partial u}{\partial \boldsymbol{n}} = \boldsymbol{n} \cdot \nabla u = u_r.$$

In this case, we are dealing with a Helmholtz equation where $k = 0$, which reduces to the Laplace equation. The requirement of regularity at the origin entails that B must be set to 0, leading us to the possible eigenfunctions:

$$r^\ell Y_{\ell,m}(\theta,\phi).$$

The boundary condition implies $\ell = 0$, and as a result, the potential is constant. This signifies that the fluid is at rest.

§8.2.10.   Problems of Electrostatics and Fluid Mechanics

In electrostatics, it is common to encounter boundary value problems like:

$$\begin{cases} \Delta u = \rho, \\ a_i(u) = g_i, \quad i \in \{1, \ldots, n\}. \end{cases}$$

In such cases, ρ represents the charge density, u denotes the electric potential, consequently giving rise to the electric field

$$\boldsymbol{E} = -\nabla u,$$

while the boundary operators a_i solely operate on the variable r and

$$g_i = g_i(\theta, \phi).$$

A comparable problem also arises in the field of fluid mechanics:

$$\begin{cases} \Delta u = 0, \\ a_i(u) = g_i, \quad i \in \{1, \ldots, n\}. \end{cases}$$

Here, u represents the potential for fluid velocities, resulting in the velocity being given by

$$\boldsymbol{v} = \nabla u.$$

The Poisson equation in spherical coordinates is formulated as follows:

$$\frac{1}{r}(ru)_{rr} + \frac{1}{r^2 \sin\theta}(\sin\theta\, u_\theta)_\theta + \frac{1}{r^2 \sin^2\theta} u_{\phi\phi} = \rho(r, \theta, \phi).$$

To apply the eigenfunction expansion method (EEM), we express the Poisson equation in the form:

$$(A + B)u = r^2 \rho,$$

where

$$Au := r^2 u_{rr} + 2r u_r,$$

$$Bu := \frac{1}{\sin\theta}(\sin\theta\, u_\theta)_\theta + \frac{1}{\sin^2\theta} u_{\phi\phi}.$$

The operator B is symmetric in $C^\infty(S^2)$, and the corresponding eigenvalue problem can be solved as follows:

$$\boxed{BY_{\ell,m} = -\ell(\ell+1)Y_{\ell,m}.}$$

This holds for $(\ell, m) \in \mathcal{Q}$.

Since the spherical harmonics constitute a complete set in $L^2(S^2)$, we are able to expand expressions as follows:

$$\boxed{\begin{aligned} r^2 \rho(r, \theta, \phi) &= \sum_{l,m} \rho_{l,m}(r) Y_{l,m}(\theta, \phi), \\ g_i(\theta, \phi) &= \sum_{l,m} c_{i,l,m} Y_{l,m}(\theta, \phi). \end{aligned}}$$

Consequently, the eigenfunction expansion method seeks solutions in the form:

$$u(r, \theta, \phi) = \sum_{(l,m) \in \mathbb{Q}} v_{l,m}(r) Y_{l,m}(\theta, \phi),$$

where the functions $v_{l,m}(r)$ are subject to the following conditions:

$$\begin{cases} r^2 v_{l,m}'' + 2r v_{l,m}' - l(l+1) v_{l,m} = \rho_{l,m}, \\ a_i(v_{l,m}) = c_{i,l,m}. \end{cases}$$

In the homogeneous case where $\rho = 0$, the equation corresponds to the Laplace equation and possesses the solution:

$$v_{l,m}(r) = A_{l,m} r^l + \frac{B_{l,m}}{r^{l+1}}.$$

§8.2.11. Charged Conducting Sphere

Let's consider a conducting sphere with a radius of a, centered at the origin, and carrying a charge Q. This sphere has reached equilibrium while immersed in a constant electric field $E_0 \mathbf{k}$. It is well known that equilibrium is attained when the potential on the sphere remains constant, denoted as u_0.

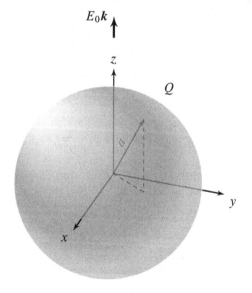

The electrostatic potential u is the solution to the following BVP for the Laplace equation:

$$\begin{cases} \Delta u = 0, \\ u|_{r=a} = u_0, \\ u = -E_0 z + V_0 + O\left(\dfrac{1}{r}\right), \quad r \to \infty. \end{cases}$$

The electric potential can be expressed using the following multipolar expansion:

$$u(r, \theta, \phi) = \sum_{\ell, m} \left(A_{\ell, m} r^\ell + B_{\ell, m} \frac{1}{r^{\ell+1}} \right) Y_{\ell, m}(\theta, \phi).$$

The boundary value condition at $r = a$ can be formulated as:

$$u|_{r=a} = u_0 \sqrt{4\pi} Y_{0,0}.$$

Additionally, considering $z = r \cos \theta$, the condition at infinity takes the form:

$$u = -E_0 r \sqrt{\frac{4\pi}{3}} Y_{1,0}(\theta, \phi) + V_0 \sqrt{4\pi} Y_{0,0} + O\left(\frac{1}{r}\right).$$

Hence, we can confine ourselves to an expansion:

$$u(r, \theta, \phi) = \left(A_{0,0} + B_{0,0} \frac{1}{r} \right) Y_{0,0}(\theta, \phi) + \left(A_{1,0} r + B_{1,0} \frac{1}{r^2} \right) Y_{1,0}(\theta, \phi).$$

Now, by enforcing the boundary value conditions at $r = a$, we can derive:

$$\begin{cases} A_{0,0} + B_{0,0} \dfrac{1}{a} = \sqrt{4\pi} u_0, \\ A_{1,0} a + B_{1,0} \dfrac{1}{a^2} = 0. \end{cases}$$

Similarly, from the conditions at $r \to \infty$:

$$\begin{cases} A_{0,0} = \sqrt{4\pi} V_0, \\ A_{1,0} = -\sqrt{\dfrac{4\pi}{3}} E_0. \end{cases}$$

Thus, the expression simplifies to:

$$u(r, \theta, \phi) = V_0 + \frac{(u_0 - V_0)a}{r} - E_0 \left(r - \frac{a^3}{r^2} \right) \cos \theta.$$

The potential is comprised of a monopolar term given by:

$$\frac{(u_0 - V_0)a}{r}$$

and a dipolar term

$$-E_0 \left(r - \frac{a^3}{r^2} \right) \cos \theta.$$

The monopolar moment is $4\pi\varepsilon_0(u_0 - V_0)a$ and the net charge is $Q = (u_0 - V_0)a$. Thus, the potential is

$$u(r, \theta, \phi) = V_0 + \frac{Q}{r} - E_0\left(r - \frac{a^3}{r^2}\right)\cos\theta.$$

Hence, the electric field $\boldsymbol{E} = -\nabla u$ is

$$\boldsymbol{E} := \left(\frac{Q}{r^2} + E_0\left(1 + 2\frac{a^3}{r^3}\right)\cos\theta\right)\boldsymbol{u}_r - E_0\left(1 - \frac{a^3}{r^3}\right)\sin\theta\ \boldsymbol{u}_\theta.$$

The polar moment of the sphere is given by:

$$p = 4\pi\varepsilon_0 a^3 E_0.$$

Therefore, the polarizability of the sphere is $\alpha = 4\pi\varepsilon_0 a^3$.

Since E_r is the normal component of the electric field on the sphere, the charge density on the surface of the sphere is:

$$\sigma(\theta) = \varepsilon_0 E_r = \varepsilon_0\left(\frac{Q}{a^2} + 3E_0\cos\theta\right).$$

The reference potential V_0 (which, therefore, could be taken as zero) is the potential in the xy plane in regions far from the origin, and in fact, it can be taken as the ground potential. Therefore, if we connect the sphere to the ground, $(u_0 - V_0)a = Q = 0$, then the sphere is neutral with total zero charge.

The points where the electric field vanishes are relevant in the study of equipotential surfaces, as the surface is no longer regular there. The component E_θ of $\boldsymbol{E} = E_r\boldsymbol{u}_r + E_\theta\boldsymbol{u}_\theta$ becomes zero only when:

- $r = a$, or
- $\theta = 0, \pi$.

Let's now analyze the zeros of E_r in each of these cases:

- The electric field becomes zero on the sphere $r = a$ whenever θ satisfies:

$$\frac{Q}{a^2} + 3E_0\cos\theta = 0$$

which is:

$$\theta^* = \arccos\frac{Q}{3a^2 E_0}.$$

So, if

$$\left|\frac{Q}{3a^2 E_0}\right| < 1,$$

then the electric field only becomes zero in a circle on the spherical surface with $\theta = \theta^*$. If

$$Q = \pm 3a^2 E_0$$

then the electric field only becomes zero at the north and south poles of the sphere, respectively. When:

$$\left| \frac{Q}{3a^2 E_0} \right| > 1,$$

the situation changes, and the electric field cannot be zero on the sphere.

- In this second case, the electric field becomes zero for $\theta = 0, \pi$ whenever r satisfies

$$P(r) := r^3 \pm \frac{Q}{E_0} r + 2a^3 = 0,$$

respectively. Let's analyze the roots of this equation. The function $P(r)$ is a cubic with:

$$P \to \pm\infty, \quad r \to \pm\infty$$

so it has at least one real root; and it has critical points at:

$$3r^2 \pm \frac{Q}{E_0} = 0.$$

The critical points will be real if and only if:

$$\frac{Q}{E_0} \lessgtr 0,$$

respectively. In this case, we'll have a minimum and a maximum located at:

$$r_\pm^* = \pm\sqrt{\left| \frac{Q}{3E_0} \right|}.$$

Otherwise, the function P is always increasing, and, therefore, it only has a single real root that is negative, since

$$P(0) = 2a^3 > 0.$$

In the case of two distinct critical points for $P(r)$, in the interval (r_-^*, r_+^*) the function is decreasing, and since

$$P(0) > 0,$$

we have

$$P\left(r_-^*\right) > 0.$$

Thus, the condition for two real roots to exist is:

$$P\left(r_+^*\right) = 0,$$

and for three real roots to exist:

$$P\left(r_+^*\right) < 0.$$

Evaluating the value of P at r_+^*, we obtain:

$$P\left(r_+^*\right) = -2\left(\left|\frac{Q}{3E_0}\right|\right)^{3/2} + 2a^3.$$

So, we'll have two real roots (one negative and one positive $(P(0) > 0)$) if:

$$\left|\frac{Q}{3E_0 a^2}\right| = 1$$

and three real roots (one negative and two positive $(P(0) > 0)$) if:

$$\left|\frac{Q}{3E_0 a^2}\right| > 1.$$

It turns out that if

$$\left|\frac{Q}{3E_0 a^2}\right| = 1,$$

then the value

$$r = a$$

is a root. On the other hand, if

$$\left|\frac{Q}{3E_0 a^2}\right| > 1 \Rightarrow r_+^* > a$$

and there is a root greater than a, and since:

$$P(a) = 3a^3\left(1 - \left|\frac{Q}{3E_0 a^2}\right|\right) < 0$$

there's only one root greater than a.

In summary, for $r > a$, the electric field becomes zero at a point if and only if:

$$\left|\frac{Q}{3E_0 a^2}\right| > 1$$

and this point lies on the z-axis.

The structure of the equipotential curves changes according to:

$$\left|\frac{Q}{3E_0 a^2}\right| \lessgtr 1.$$

The different cases are depicted graphically below.

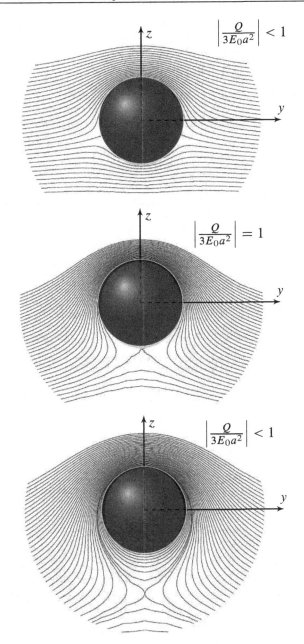

Recall that the electric field is orthogonal to these equipotential curves.

§8.2.12.   Fluid Around a Ball

Let us now consider a fluid in uniform motion in which we submerge a solid sphere of radius a. The velocity potential must satisfy the following boundary problem

$$\begin{cases} \Delta u = 0, \\ \begin{cases} u_r|_{r=a} = 0, \\ u = v_0 z + O\left(\dfrac{1}{r}\right), \quad r \to \infty. \end{cases} \end{cases}$$

Similar to the previous problem, we are searching for a solution in the form:

$$u(r, \theta, \phi) = \sum_{\ell, m} \left(A_{\ell,m} r^\ell + B_{\ell,m} \frac{1}{r^{\ell+1}} \right) Y_{\ell,m}(\theta, \phi).$$

Additionally, the boundary value conditions can be expressed as follows:

$$u_r|_{r=a} = 0,$$

$$u(r, \theta, \phi) = v_0 r \sqrt{\frac{4\pi}{3}} Y_{1,0}(\theta, \phi) + O\left(\frac{1}{r}\right), \qquad r \to \infty.$$

Hence, we confine ourselves to solutions of the form:

$$u(r, \theta, \phi) = \left(A_{1,0} r + B_{1,0} \frac{1}{r^2} \right) Y_{1,0}(\theta, \phi).$$

From the boundary conditions we derive that

$$\begin{cases} A_{1,0} = \sqrt{\dfrac{4\pi}{3}} v_0, \\ A_{1,0} - 2 B_{1,0} \dfrac{1}{a^3} = 0. \end{cases}$$

Therefore,

$$B_{1,0} = \frac{1}{2} \sqrt{\frac{4\pi}{3}} v_0 a^3.$$

Consequently, for the velocity potential, we find

$$\boxed{u(r, \theta, \phi) = v_0 \left(r + \frac{a^3}{2r^2} \right) \cos \theta}$$

and the velocity field is

$$\boxed{v = v_0 \left(\left(1 - \frac{a^3}{r^3} \right) \cos \theta \, u_r - \left(1 + \frac{a^3}{2r^3} \right) \sin \theta \, u_\theta \right).}$$

§8.3. Beyond Cylindrical and Spherical

SPHERICAL coordinates, alongside cylindrical and Cartesian coordinates, represent three coordinate systems where Helmholtz's equation admits separability. However, other coordinate systems also exhibit the separability property, as highlighted by [49, 50, 51]. An illuminating example is the domain of ellipsoidal harmonics, elucidated in works such as [19, 77]. The ellipsoidal coordinate system stands as one of the most comprehensive three-dimensional systems where the Helmholtz's equation exhibits separability. Despite this significance, the theory of ellipsoidal harmonics, much like its spherical counterpart, saw limited application historically, though it has garnered renewed attention in recent times. For recent applications of spherical and ellipsoidal harmonics in the fields of electroencephalography and magnetoencephalography, please refer to [20].

§8.3.1. Eisenhart's Coordinate Systems

The American mathematician Luther Pfahler Eisenhart (1876-1965) in the papers [25, 26] classified those systems of curvilinear coordinates having quadratic and linear relations for the coordinate surfaces in which the Helmholtz equation admits a separation of variables, see also his paper [27]. He found 11 coordinate systems, known as the Eisenhart coordinate systems. These systems have been studied in detail, and we refer the reader to [49, 50, 51]. These coordinate systems are indeed a fascinating aspect of mathematical physics. While they might not always yield profound results universally, their significance lies in their versatility and adaptability to specific problems in physics.

We plot (with Mathematica) here the corresponding ellipsoidal coordinate surfaces,

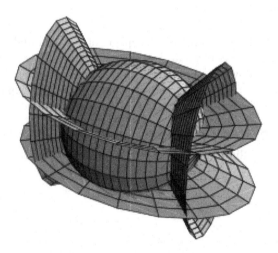

The depth of insight these coordinate systems provide can vary depending on the chosen coordinates and the equations under consideration. For example, when dealing with three-dimensional problems and equations related to the Helmholtz equation, there are precisely 11 sets of coordinates in the Eisenhart framework that can be employed.

However, it's crucial to note that the versatility of these coordinate systems goes beyond just these particular cases. Researchers and physicists can explore coordinates of different degrees, such as quartic coordinate surfaces, to address various scenarios.

One interesting observation is that the Laplace equation, which frequently arises in physics, often exhibits separable solutions when transformed into certain coordinate systems. An excellent example is bispherical coordinates, where the Laplace equation becomes separable, enabling more straightforward solutions to problems that might be challenging in other coordinate systems.

On the flip side, some more complex equations may not be as amenable to separation of variables, even when the Helmholtz equation exhibits separable solutions in a given coordinate system. This highlights the nuanced nature of these coordinate systems and their applications.

The 11 Eisenhart's coordinate systems, see for example [50, pages 1-48], are:

(1) Cartesian or rectangular coordinates.
(2) Circular-cylinder coordinates.
(3) Elliptic-cylinder coordinates.
(4) Parabolic-cylinder coordinates.
(5) Spherical coordinates
(6) Prolate spheroidal coordinates.
(7) Oblate spheroidal coordinates.
(8) Parabolic coordinates.
(9) Conical coordinates.
(10) Ellipsoidal coordinates.
(11) Paraboloidal coordinates.

Observe that all of them (with the exception of paraboloidal coordinates) are particular cases of the ellipsoidal system. As explained in chapter 3 of Miller's work [49], these 11 systems have a group-theoretical explanation based on the Euclidean group of translations and rotations. In [50] an exploration beyond the quadratic surfaces is given, and up to 40 cases are discussed.

§8.3.2.   Stäckel Determinant and Separation of Variables

A general curvilinear change of coordinates is described by the invertible transformation:

$$x = F_1(u_1, u_2, u_3), \qquad y = F_2(u_1, u_2, u_3), \qquad z = F_3(u_1, u_2, u_3),$$

and its inverse

$$u_1 = G_1(x, y, z), \qquad u_2 = G_2(x, y, z), \qquad u_3 = G_3(x, y, z).$$

If $r = (x, y, z)$ denotes the position vector, then by taking its partial derivatives with respect to the general curvilinear system $\{u_1, u_2, u_3\}$, we obtain the so-called local basis:

$$\left\{ \frac{\partial r}{\partial u_1}, \frac{\partial r}{\partial u_2}, \frac{\partial r}{\partial u_3} \right\}.$$

When these vectors are orthogonal, we refer to the system as an orthogonal curvilinear coordinate system. In such orthogonal systems, the Lamé coefficients or scale factors are given by:

$$h_i := \left| \frac{\partial r}{\partial u_i} \right|, \qquad i \in \{1, 2, 3\}.$$

To each orthogonal coordinate system, there is an associated Stäckel matrix, as discussed in [50]:

$$\mathcal{S} = \begin{bmatrix} \Phi_{11}(u_1) & \Phi_{12}(u_1) & \Phi_{13}(u_1) \\ \Phi_{21}(u_2) & \Phi_{22}(u_2) & \Phi_{23}(u_2) \\ \Phi_{31}(u_3) & \Phi_{32}(u_3) & \Phi_{33}(u_3) \end{bmatrix}$$

The Stäckel determinant is given by:

$$S = \begin{vmatrix} \Phi_{11}(u_1) & \Phi_{12}(u_1) & \Phi_{13}(u_1) \\ \Phi_{21}(u_2) & \Phi_{22}(u_2) & \Phi_{23}(u_2) \\ \Phi_{31}(u_3) & \Phi_{32}(u_3) & \Phi_{33}(u_3) \end{vmatrix}$$

The co-factors of this Stäckel matrix along the first column are:

$$M_{11} = \begin{vmatrix} \Phi_{22}(u_2) & \Phi_{23}(u_2) \\ \Phi_{32}(u_3) & \Phi_{33}(u_3) \end{vmatrix},$$

$$M_{21} = -\begin{vmatrix} \Phi_{12}(u_1) & \Phi_{13}(u_1) \\ \Phi_{32}(u_3) & \Phi_{33}(u_3) \end{vmatrix},$$

$$M_{31} = \begin{vmatrix} \Phi_{12}(u_1) & \Phi_{13}(u_1) \\ \Phi_{22}(u_2) & \Phi_{23}(u_2) \end{vmatrix}.$$

For the significance of the functions $\Phi_{i,j}$, see [50].

The necessary and sufficient conditions for the separability of the Helmholtz equation are given by the relations:

(8.12a) $$\qquad h_i^2 = \frac{S}{M_{i1}}, \quad i \in \{1, 2, 3\},$$

(8.12b) $$\qquad \frac{h_1 h_2 h_3}{S} = f_1(u_1) f_2(u_2) f_3(u_3).$$

Equation (8.12a) allows us to construct a Stäckel determinant related to scaling factors, and (8.12b) states that $\frac{h_1 h_2 h_3}{S}$ is a separable product. If (8.12) are satisfied, then the corresponding ordinary differential equations obtained from the separation of variables are:

$$\frac{1}{f_i(u_i)} \frac{d}{du_i} \left(f_i(u_i) \frac{dU_i}{du_i} \right) + U_i \sum_{j=1}^{3} \alpha_j \Phi_{ij} = 0, \quad i \in \{1, 2, 3\},$$

where $\alpha_1, \alpha_2, \alpha_3$ are the separation constants, and $\alpha_1 = k^2$. It's important to note that the Stäckel matrix is not unique.

Stäckel determinants and matrices were introduced by the German mathematician Paul Gustav Samuel Stäckel (1862-1919) in his Habilitation thesis on the separation of variables for the Hamilton–Jacobi equation in 1891 at Halle University and in papers [63, 64].

For Cartesian coordinates, the Stäckel matrix can be chosen as:

$$\mathscr{S} = \begin{bmatrix} 0 & -1 & -1 \\ 0 & 1 & 0 \\ 1 & 0 & 0 \end{bmatrix}.$$

For cylindrical coordinates:

$$\mathscr{S} = \begin{bmatrix} 0 & -\dfrac{1}{r^2} & -1 \\ 0 & 1 & 0 \\ 1 & 0 & 1 \end{bmatrix}.$$

And for spherical coordinates:

$$\mathscr{S} = \begin{bmatrix} 1 & \dfrac{1}{r^2} & 0 \\ 0 & 1 & -\dfrac{1}{\sin^2 \theta} \\ 0 & 0 & 1 \end{bmatrix}.$$

For an alternative exposition, please refer to [51, pp. 509-510].

§8.4. Exercises

§8.4.1. Exercises with Solutions

(1) Determine the standing waves in spherical coordinates, $u(t, r, \theta, \phi) = e^{-i\omega t} w(r, \theta, \phi)$ of the following BVP for wave equation

$$\begin{cases} u_{tt} = \Delta u, & r < a, \\ u|_{r=a} = 0. \end{cases}$$

Solution: Standing waves are obtained by considering the separation of variables $u(t, r, \theta, \phi) = e^{-i\omega t} w(r, \theta, \phi)$, where the following dispersion relation and the Helmholtz equation are satisfied

$$\omega = \pm k, \quad \Delta w + k^2 w = 0.$$

By imposing the condition of regularity at $r = 0$, it is clear that the admissible $w(r, \theta, \phi)$ are $j_l(kr)Y_{l,m}(\theta, \phi)$ with $j_l(ka) = 0$ and $\omega = \pm k$.

(2) Determine the regular solutions, in cylindrical coordinates, $u(r, \theta, z)$ of the following Neumann problem

$$\begin{cases} \Delta u = 0, & r < a, \\ \dfrac{\partial u}{\partial r}\bigg|_{r=a} = 0. \end{cases}$$

Solution: The functions that satisfy all requirements are those of the form $J_m(\alpha r)e^{im\theta + \alpha z}$ with $m \in \mathbb{Z}$, $J_m'(\alpha a) = 0$.

(3) Find the value of the constant a for which there is a solution to the following boundary value problem in spherical coordinates

$$\begin{cases} \Delta u = 1, & r < 1, \\ \dfrac{\partial u}{\partial r}\bigg|_{r=1} = a. \end{cases}$$

Solution: Applying the divergence theorem in the ball $B(0, 1)$, we have

$$\int_{B(0,1)} \Delta u \, d^3 x = \int_{S(0,1)} \nabla u \cdot dS$$

as $\Delta u = 1$ and $u_r dS = \nabla u \cdot dS$ we conclude

$$\frac{4}{3}\pi = a 4\pi$$

so that $a = 1/3$.

(4) Solve, by the method of separating variables in cylindrical coordinates, the following boundary value problem

$$\begin{cases} \Delta u = 0, & 1 < r < 2, \\ u|_{r=1} = u|_{r=2}. \end{cases}$$

Solution: The resolution by separation of variables in cylindrical coordinates of the Laplace equation leads to consider

$$u(r, \theta, z) = R_{\alpha,m}(r)\Theta_m(\theta)Z_\alpha(z),$$

with

$$
\begin{cases}
Z_\alpha(z) = \begin{cases} Az + B, & \alpha = 0, \\ Ae^{\alpha z} + Be^{-\alpha z}, & \alpha \neq 0, \end{cases} \\
\Theta_m(\theta) = \begin{cases} C\theta + D, & m = 0, \\ Ce^{im\theta} + De^{-im\theta}, & m \neq 0, \end{cases} \\
R_{\alpha,m}(r) = \begin{cases} c_1 \log r + c_2, & \alpha = m = 0, \\ c_1 r^m + c_2 r^{-m}, & \alpha = 0, m \neq 0, \\ E J_m(\alpha r) + F N_m(\alpha r), & \alpha \neq 0. \end{cases}
\end{cases}
$$

Not all of these functions are solutions to our problem. As we want the solution to be single-valued, we deduce that $\boxed{m \in \mathbb{N}_0.}$ The boundary condition, which is imposed on the $R_{\alpha,m}$, determines the following possibilities:

(a) $\alpha = m = 0 \Rightarrow c_2 = c_1 \log 2 + c_2$ and so $c_1 = 0$. Hence,

$$\boxed{R_{0,0} = 1.}$$

(b) $\alpha = 0, m \neq 0 \Rightarrow c_1 + c_2 = 2^m c_1 + 2^{-m} c_2$. Therefore,

$$\boxed{R_{0,m}(r) = (1 - 2^{-m})r^m - (1 - 2^m)r^{-m}.}$$

(c) $\alpha \neq 0 \Rightarrow E J_m(\alpha) + F N_m(\alpha) = E J_m(2\alpha) + F N_m(2\alpha)$. Thus,

$$\boxed{R_{\alpha,m}(r) = (N_m(\alpha) - N_m(2\alpha))J_m(\alpha r) - (J_m(\alpha) - J_m(2\alpha))N_m(\alpha r).}$$

(5) The solutions of the Helmholtz equation in spherical coordinates are of the form

$$[A_{lm} j_l(kr) + B_{lm} n_l(kr)]Y_{l,m}(\theta, \phi).$$

Determine the values of k for which there are nontrivial solutions with $l = 0$ of the boundary problem

$$
\begin{cases}
\Delta u + k^2 u = 0, & 1 < r < 2, \\
u|_{r=1} = 0, \\
\dfrac{\partial u}{\partial r}\bigg|_{r=2} = 0.
\end{cases}
$$

Hint: $j_0(r) = \dfrac{\mathrm{sen}\, r}{r}$, $n_0(r) = -\dfrac{\cos r}{r}$.

Solution: The solution has the form

$$u(r, \theta, \phi) = (Aj_0(kr) + Bn_0(kr))Y_{0,0},$$

and the boundary conditions impose

$$\begin{cases} Aj_0(k) + Bn_0(k) = 0, \\ Aj_0'(2k) + Bn_0'(2k) = 0. \end{cases}$$

For a nontrivial solution to exist, it is necessary that

$$\begin{vmatrix} j_0(k) & n_0(k) \\ j_0'(2k) & n_0'(2k) \end{vmatrix} = 0.$$

This is,

$$\begin{vmatrix} \sin k & -\cos k \\ 2k\cos 2k - \sin 2k & 2k\sin 2k + \cos 2k \end{vmatrix} = 0$$

which reduces to

$$\boxed{\tan k = 2k.}$$

The table of the first five zeros is

Initial zeros of $\tan k - 2k$				
1.1656	4.6042	7.7899	10.9499	14.1017

and the corresponding graph

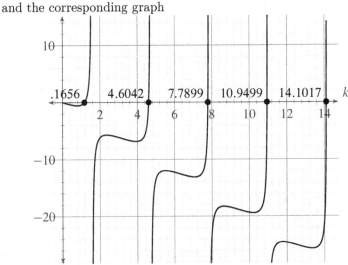

Graph $f(k) = \tan(k) - 2k$ and the initial five zeros

(6) Determine standing waves using the variable separation method

$$u(t, x) = e^{-i\omega t}w(x), \qquad\qquad \omega \geq 0,$$

of the next boundary problem in spherical coordinates

$$\begin{cases} u_{tt} = \Delta u, \quad r > 1, \\ u|_{r=1} = 0, \\ \lim_{r \to \infty} u = 0. \end{cases}$$

Solution: If we introduce the standing wave ansatz in the boundary value problem for the wave equation, then we get the following boundary problem for the Helmholtz equation:

$$\begin{cases} \Delta w + \omega^2 w = 0, \\ w|_{r=1} = 0, \\ \lim_{r \to \infty} w = 0. \end{cases}$$

Separation of variables into spherical coordinates leads to the following solutions of the Helmholtz equation:

$$R_l(r) Y_{l,m}(\theta, \phi),$$

where Y_{lm} are the spherical harmonics and R_l is a function that takes one form or another depending on whether ω is zero or not. When $\omega = 0$ we have $R_l(r) = a r^l + \frac{b}{r^{l+1}}$, but boundary conditions impose $a = 0$ and $a + b = 0$ then the solution is trivial. If $\omega \neq 0$ then the radial function is $R_l(r) = a j_l(\omega r) + b n_l(\omega r)$, with $j_l(r)$ and $n_l(r)$ are the spherical Bessel and spherical Neumann functions, respectively, which automatically satisfy the boundary condition at $r \to \infty$. The boundary condition in $r = 1$ imposes that the radial function is proportional to $n_l(\omega) j_l(\omega r) - j_l(\omega) n_l(\omega r)$. Then, the requested standing waves are

$$\boxed{a e^{-i\omega t} \left(n_l(\omega) j_l(\omega r) - j_l(\omega) n_l(\omega r) \right) Y_{l,m}(\theta, \phi).}$$

(7) Solve the following boundary problem for the Laplace equation by means of the eigenfunction expansion method in polar coordinates

$$\begin{cases} \Delta u = 0, \quad\quad 0 \leq r < 1, \quad 0 \geq 2\pi, \\ u|_{r=1} = \sin^2 \theta. \end{cases}$$

Solution: In polar coordinates, the problem takes the form

$$\begin{cases} r^2 u_{rr} + r u_r + u_{\theta\theta} = 0, \quad 0 \leq r < 1, \quad 0 \leq \theta < 2\pi, \\ u|_{r=1} = \sin^2 \theta. \end{cases}$$

To apply the EEM, we first solve the eigenvalue problem for $\dfrac{d^2}{d\theta^2}$ with periodic boundary conditions.

The eigenfunctions are

$$\{1, \sin\theta, \cos\theta, \sin 2\theta, \cos 2\theta, \dots\}$$

and the corresponding spectrum is $\{1^2, 2^2, 3^2, \dots\}$.

Therefore, we expand the solution as

$$u(r, \theta) = v_0(r) + \sum_{m=0}^{\infty} (v_m(r) \cos m\theta + w_m(r) \sin m\theta).$$

The functions $v(r), w(r)$ are characterized for being solutions of

$$r^2 v'' + r v' + \lambda_m v_m = 0.$$

Considering now the regularity of the solution for $r = 0$, we conclude that $v_0 = c_0$ and that

$$v_m(r) = c_m r^m, \quad w_m(r) = d_m r^m,$$

where the coefficients c_m, d_m are determined by the boundary condition

$$\sin^2 \theta = \frac{1}{2} - \frac{1}{2} \cos 2\theta.$$

Hence, all the d's are zero, as well as all the c's but for $c_0 = \frac{1}{2}$ and $c_1 = -\frac{1}{2}$. Therefore, the expansion of the solution is

$$\boxed{u(r, \theta) = \frac{1 - r^2 \cos 2\theta}{2}.}$$

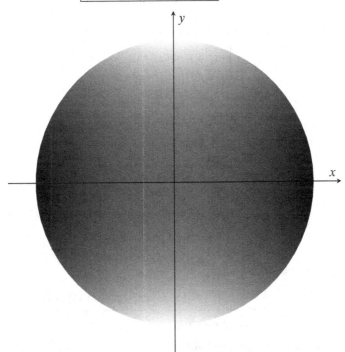

Color temperature on the circular plate

(8) Consider the Laplace equation in a cylindrical layer with Dirichlet boundary conditions

$$\begin{cases} \Delta u = 0, \quad 1 < r < 2, \\ \begin{cases} u|_{r=1} = 0, \\ u|_{r=2} = 0. \end{cases} \end{cases}$$

Determine, by means of the method of separation of variables the solutions $u(r, z)$ independent of θ.

Solution: We are solving the Laplace equation, so the solutions obtained by the method of separation of variables will be linear combinations of functions of the form $u = R(r)\Theta(\theta)Z(z)$, where

$$Z(z) = A e^{\alpha z} + B e^{-\alpha z},$$

$$\Theta(\theta) = C e^{im\theta} + D e^{-im\theta},$$

$$R(r) = \begin{cases} c_1 \log r + c_2, & \alpha = 0, m = 0, \\ c_1 r^m + \frac{c_2}{r^m}, & \alpha = 0, m \neq 0, \\ E J_m(\alpha r) + F N_m(\alpha r), & \alpha \neq 0. \end{cases}$$

As we don't want to have contributions dependent on θ, we put $m = 0$ and $\Theta = 1$. Therefore, the solutions are

$$u(r, z) = R(r)Z(z)$$

with

$$Z(z) = A e^{\alpha z} + B e^{-\alpha z},$$

$$R(r) = \begin{cases} c_1 \log r + c_2, & \alpha = 0, m = 0, \\ E J_0(\alpha r) + F N_0(\alpha r), & \alpha \neq 0. \end{cases}$$

Imposing the boundary conditions

$$R(r)|_{r=1} = R(r)|_{r=2} = 0$$

we see that $\alpha \neq 0$ and that we must have

$$\boxed{J_0(\alpha)N_0(2\alpha) = J_0(2\alpha)N_0(\alpha).}$$

The table of the initial zeros is

Initial zeros of $J_0(\alpha)N_0(2\alpha) - J_0(2\alpha)N_0(\alpha)$					
3.1230	6.2734	9.4182	12.5614	15.7040	18.8462

This transcendent equation determines the possible values of α. The solution sought is

$$\boxed{u = e^{\pm \alpha z}\left(N_0(\alpha)J_0(\alpha r) - J_0(\alpha)N_0(\alpha r)\right).}$$

The radial functions are:

$$R_1(r) := J_0(3.1230)N_0(3.1230r) - N_0(3.1230)J_0(3.1230r),$$

$$R_2(r) := J_0(6.2734)N_0(6.2734r) - N_0(6.2734)J_0(6.2734r),$$

$R_3(r) := J_0(9.4182)N_0(9.4182r) - N_0(9.4182)J_0(9.4182r),$

$R_4(r) := J_0(12.5614)N_0(12.5614r) - N_0(12.5614)J_0(12.5614r),$

$R_5(r) := J_0(15.7040)N_0(15.7040r) - N_0(15.7040)J_0(15.7040r).$

The graph of the function

$$J_0(\alpha)N_0(2\alpha) - J_0(2\alpha)N_0(\alpha)$$

and the initial five zeros is:

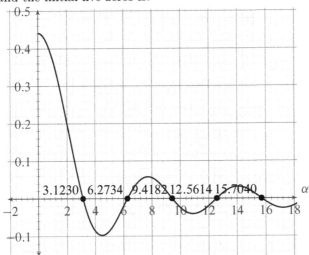

The radial functions have the following graphs, for $1 < r < 2$:

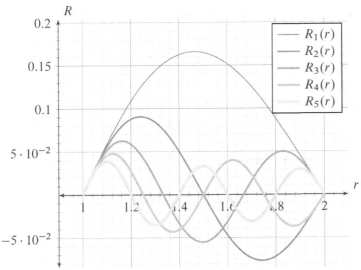

(9) Find the standing waves of the form $u(t, x) = \sin t\, w(r, \theta, \phi)$ for the wave equation in a spherical layer with the following Dirichlet

boundary conditions

$$\begin{cases} u_{tt} = \Delta u, & \dfrac{\pi}{2} < r < \pi, \\ \begin{cases} u|_{r=\frac{\pi}{2}} = \cos\theta, \\ u|_{r=\pi} = 0. \end{cases} \end{cases}$$

Hint: $Y_{1,0} = \sqrt{\dfrac{3}{4\pi}}\cos\theta.$

Solution: The conditions for $w(r,\theta,\phi)$ are

$$\begin{cases} \Delta w + w = 0, & \frac{\pi}{2} < r < \pi, \\ \begin{cases} w|_{r=\frac{\pi}{2}} = \cos\theta, \\ w|_{r=\pi} = 0. \end{cases} \end{cases}$$

This is a boundary value problem for the Helmholtz equation with $k^2 = 1$.

To apply the eigenfuncion expansion method, we write

$$\Delta + 1 = A + B,$$
$$A = r\frac{d}{dr}r\frac{d}{dr} + r^2,$$
$$B = \sin\theta\frac{d}{d\theta}\sin\theta\frac{d}{d\theta} + \frac{1}{\sin^2\theta}\frac{d^2}{d\phi^2},$$

and analyze the eigenvalue problem

$$BY = -\lambda Y$$

in $C^\infty(S^2)$. Let us remember that, in this case, the eigenfunctions are the spherical harmonics $\{Y_{l,m}(\theta,\phi)\}_{l=0,1,2,\dots}^{m=-l,\dots,l}$ that form an orthogonal basis, with eigenvalues $\lambda = -l(l+1)$. We only have one inhomogeneous term in a boundary condition, $g_1(\theta) = \cos\theta$, for which we have

$$g_1(\theta) = \sqrt{\frac{4\pi}{3}}Y_{1,0}(\theta,\phi).$$

The eigenfunction expansion of the solution reduces to

$$w(r,\theta,\phi) = v(r)Y_{1,0}(\theta,\phi),$$

where $v(r)$ satisfies

$$Av + \lambda v = 0$$

so that

$$v = c_1 j_1(r) + c_2 n_1(r),$$

where the first-order Bessel spherical functions are

$$j_1(r) = \frac{\sin r}{r^2} - \frac{\cos r}{r},$$

$$n_1(r) = -\frac{\cos r}{r^2} - \frac{\sin r}{r}.$$

Note that

$$j_1\left(\frac{\pi}{2}\right) = \frac{4}{\pi^2}, \quad n_1\left(\frac{\pi}{2}\right) = -\frac{2}{\pi},$$

$$j_1(\pi) = \frac{1}{\pi}, \qquad n_1(\pi) = \frac{1}{\pi^2}.$$

The boundary conditions at $r = \frac{\pi}{2}$ and $r = \pi$ are

$$v\left(\frac{\pi}{2}\right) = \sqrt{\frac{4\pi}{3}}, \quad v(\pi) = 0,$$

respectively. Hence, we are led to the following system for the coefficients c_1 and c_2,

$$\begin{bmatrix} \frac{4}{\pi^2} & -\frac{2}{\pi} \\ \frac{1}{\pi} & \frac{1}{\pi^2} \end{bmatrix} \begin{bmatrix} c_1 \\ c_2 \end{bmatrix} = \begin{bmatrix} \sqrt{\frac{4\pi}{3}} \\ 0 \end{bmatrix}.$$

We have that

$$\begin{bmatrix} \frac{4}{\pi^2} & -\frac{2}{\pi} \\ \frac{1}{\pi} & \frac{1}{\pi^2} \end{bmatrix}^{-1} = \frac{\pi^2}{2 + \pi^2} \begin{bmatrix} \frac{1}{2} & \pi \\ -\frac{\pi}{2} & 2 \end{bmatrix}$$

that implies

$$c_1 = \frac{1}{2} \frac{\pi^2}{2 + \pi^2} \sqrt{\frac{4\pi}{3}},$$

$$c_2 = -\frac{\pi}{2} \frac{\pi^2}{2 + \pi^2} \sqrt{\frac{4\pi}{3}}.$$

Hence, we find

$$u = \frac{1}{2} \frac{\pi^2}{2 + \pi^2} \sin t \cos \theta \left(j_1(r) - \pi n_1(r)\right)$$

and the solution is

$$\boxed{u(t, r, \theta) = \frac{1}{2} \frac{\pi^2}{2 + \pi^2} \frac{\sin t \cos \theta}{r^2} \left((1 + \pi r) \sin r + (\pi - r) \cos r\right).}$$

(10) The Schrödinger equation

$$\begin{cases} -\frac{\hbar}{2M} \Delta u = E u, \\ u|_{\text{walls}} = 0, \end{cases}$$

for a quantum particle of mass M confined within the region bounded by the given curve, determines its stationary states and energies $E = \frac{\hbar}{2M}k^2$, where k is the wave vector of that state.

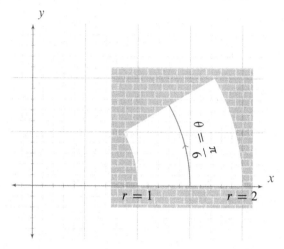

Confinement region

(a) Apply the method of separation of variables. For the factorized solution $u(r, \theta) = R(r)\Theta(\theta)$, find the corresponding boundary value problems for the ordinary differential equations (ODEs) that determine the radial modes $R(r)$ and the longitudinal modes $\Theta(\theta)$.

(b) What are the longitudinal modes like?

(c) Find the Bessel and Neumann functions involved in the radial part of the stationary states.

(d) Find the equation, in terms of Bessel and Neumann functions, that determines the possible values of k as the zeros of a function. Express the possible energies and stationary states in terms of these zeros.

Solution:

(a) The stationary Schrödinger equation is expressed as the following boundary value problem for the two-dimensional Helmholtz equation:

$$\begin{cases} \Delta u + k^2 u = 0, \quad 1 < r < 2, \quad 0 < \theta < \dfrac{\pi}{6}, \\ u|_{\text{wall}} = 0. \end{cases}$$

When written in polar coordinates, $x = r\cos\theta$ and $y = r\sin\theta$, it reads:

$$\begin{cases} r^2 u_{rr} + r u_r + k^2 r^2 u + u_{\theta\theta} = 0, \\ u|_{\text{wall}} = 0. \end{cases}$$

The separation of variables suggests to consider factoring the solutions expressed in polar coordinates in the form $u(r, \theta) = R(r)\Theta(\theta)$, which leads to

$$\begin{cases} r^2 R'' + r R' + k^2 r^2 R = m^2 R, \\ R(1) = 0, \\ R(2) = 0, \end{cases}$$

$$\begin{cases} \Theta'' = -m^2 \Theta, \\ \Theta(0) = 0, \\ \Theta(\tfrac{\pi}{6}) = 0. \end{cases}$$

(b) The ODE of the boundary problem for $\Theta(\theta)$ has solutions $\Theta(\theta) = A \sin m\theta + B \cos m\theta$, but the condition $\Theta(0) = 0$ implies $B = 0$. Additionally, the condition $\Theta(\tfrac{\pi}{6}) = 0$ requires $m = 6\ell$ with $\ell \in \mathbb{N}$. Therefore,

$$\boxed{\Theta_\ell(\theta) = \sin 6\ell\theta.}$$

(c) The ODE of the radial part $R(r)$ is an Euler's equation if $k = 0$, and it transforms into a Bessel's equation after rescaling for the case $k^2 > 0$. The case $k = 0$ is discarded. This is because in this situation, we would have that

$$R(r) = \begin{cases} c_1 + c_2 \log r, & k = m = 0, \\ c_1 r^m + c_2 r^{-m} & k = 0, m \neq 0. \end{cases}$$

We know that $m \neq 0$, and therefore, the first possibility is discarded. The second one is also not feasible as the existence of a nontrivial solution requires $\left| \frac{1}{2^m} \frac{1}{2^{-m}} \right| = 0$, which is evidently false. Therefore, it is only possible for $k^2 > 0$, and in this case, the solutions will be of the form

$$\boxed{R(r) = E J_{6\ell}(kr) + F N_{6\ell}(kr),}$$

$E, F \in \mathbb{C}$, in terms of the Bessel functions $J_{6\ell}$ and Neumann functions $N_{6\ell}$, with $\ell \in \mathbb{N}$.

(d) It is necessary to impose the boundary conditions on the radial parts. This leads to the system:

$$E J_{6\ell}(k) + F N_{6\ell}(k) = 0,$$
$$E J_{6\ell}(2k) + F N_{6\ell}(2k) = 0,$$

which has nontrivial solutions $(E, F) \neq (0, 0)$ only if the following transcendental equation is satisfied

$$\boxed{J_{6\ell}(k) N_{6\ell}(2k) = J_{6\ell}(2k) N_{6\ell}(k).}$$

This is, if and only if k is a zero of the function $f_\ell(k) = J_{6\ell}(k) N_{6\ell}(2k) - J_{6\ell}(2k) N_{6\ell}(k)$, which are infinitely many. Let us denote the set of these zeros as $\{k_{\ell,n}\}$, where $k_{\ell,1} < k_{\ell,2} < \cdots$. In conclusion, the energies will be

$$E_{\ell,n} = \frac{\hbar^2}{2M} k_{\ell,n}^2,$$

and the stationary states will be

$$u_{\ell,n}(r,\theta) = \left(N_{6\ell}(k_{\ell,n}) J_{6\ell}(k_{\ell,n} r) - J_{6\ell}(k_{\ell,n}) N_{6\ell}(k_{\ell,n} r) \right) \sin(6\ell\theta),$$

for $\ell, n \in \mathbb{N}$.

§8.4.2. Exercises

(1) Determine the stationary states of a quantum particle enclosed in a cylindrical box of radius a and height h.

(2) The velocity potential of a stationary fluid inside a circular section pipe of radius a is described in cylindrical coordinates by the boundary problem:

$$\begin{cases} \Delta u = 0, \quad 0 \le r < a, \quad 0 \le \theta < 2\pi, \quad z \in \mathbb{R}, \\ \left.\dfrac{\partial u}{\partial n}\right|_{\text{walls}} = 0. \end{cases}$$

Determine the solutions provided by the SVM.

(3) The velocity potential of a stationary fluid inside a h deep circular pond is described in cylindrical coordinates by the boundary value problem:

$$\begin{cases} \Delta u = 0, \quad 0 \le r < a, \quad 0 \le \theta < 2\pi, \quad -h < z < 0, \\ \left.\dfrac{\partial u}{\partial n}\right|_{\text{walls}} = 0. \end{cases}$$

Determine the solutions provided by the SVM.

(4) Under certain conditions certain components of the field confined between two concentric spheres are described by the boundary problem in spherical coordinates:

$$\begin{cases} u_{tt} = c^2 \Delta u, \quad a_1 < r < a_2, \quad 0 \le \theta < \pi, \quad 0 \le \phi < 2\pi, \\ u|_{\text{walls}} = 0. \end{cases}$$

Determine the standing waves of the model.

(5) Prove that in Cartesian coordinates on the unit sphere:

$$Y_{1,0} = \sqrt{\frac{3}{4\pi}}\, z, \qquad Y_{1,1} = -\sqrt{\frac{3}{8\pi}}(x + iy), \qquad Y_{1,-1} = \sqrt{\frac{3}{8\pi}}(x - iy).$$

Determine the development of the functions $u_1 = x$, $u_2 = y$, $u_3 = z$ on the unit sphere in spherical harmonics.

(6) Solve the initial value problem

$$\begin{cases} u_t = \Delta u, \quad 1 < r < 2, \quad t > 0, \\ u|_{t=0} = \dfrac{1}{r}\sin\left(\dfrac{\pi r}{2}\right), \\ \left.\left(u + \dfrac{\partial u}{\partial n}\right)\right|_{r=1} = 0 \\ u|_{r=2} = 0. \end{cases}$$

(7) Solve the initial value problem

$$\begin{cases} u_t = \Delta u + \dfrac{\sin 2r}{r}e^{-t}, \quad t > 0, \\ u|_{t=0} = \dfrac{\sin r}{r} \end{cases}$$

(8) Solve the boundary value problem

$$\begin{cases} \Delta u = \dfrac{C}{r}, & r > R, \\ \begin{cases} u\big|_{r=R} &= z, \\ \dfrac{\partial u}{\partial r}\bigg|_{r=R} &= 0. \end{cases} \end{cases}$$

Bibliography

[1] N. I. Akhiezer & I. M. Glazman, *Theory of Linear Operators in Hilbert Space*, Dover, unabridged republication, 1993.

[2] M. A. Al-Gwaiz, *Sturm-Liouville Theory and its Applications*, Springer, 2008.

[3] G. E. Andrews, R. Askey, and & R. Roy, *Special Functions*, Encyclopedia of Mathematics and its Applications **71**, Cambridge University Press, 1999.

[4] V. I. Arnold, *Lectures in Partial Differential Equations*, Universitext, Springer, 2004.

[5] E. Artin, *The Gamma Function*, Dover, unabridged republication, 2015.

[6] R. Beals & R. Wong, *Special Functions and Orthogonal Polynomials*, Cambridge Studies in Advanced Mathematics **153**, Cambridge University Press, 2016.

[7] W.R. Bennett, Jr. *The Science of Musical Sound*, Springer, 2020.

[8] D. J. Benson, *Music. A Mathematical Offering*, Cambridge University Press, 2007.

[9] M. Bôcher, *Introduction to the Theory of Fourier's Series*, Annals of Mathematics, second series, **7** (3) (1906) 81–152.

[10] D. Borthwick, *Spectral Analysis*, Graduate Texts in Mathematics **284**, Springer, 2020.

[11] U. Bottazzini & J. Gray, *Hidden Geometry—Geometric Fantasies*, Sources and Studies in the History of Mathematics and Physical Sciences, Springer, 2013.

[12] F. Bowman, *Introduction to Bessel Functions*, Dover, unabridged republication, 1958.

[13] B. H. Bransden & C. J. Joachain, *Physics of Atoms and Molecules*, second edition, Prentice Hall, 2003.

[14] D. M. Brink, *Angular Momentum*, third edition, Oxford Science Publications, 1993, corrected 2015.

[15] W. E. Byerly, *Fourier's Series*, Dover, unabridged republication, 1959.

[16] H. Brezis, *Functional Analysis, Sobolev Spaces and Partial Differential Equations*, Springer, 2011.

[17] H. C. Carslaw, *A historical note on Gibbs' phenomenon in Fourier's series and integrals*, Bulletin of the American Mathematical Society **31**(8) (1925) 420–424.

[18] T. S. Chihara, *An Introduction to Orthogonal Polynomials*, Dover, unabridged republication, 2011.

[19] G. Dassios, *Ellipsoidal Harmonics*, Cambridge University Press, 2012.

[20] G. Dassios & A. Fokas, *Electroencephalography and Magnetoencelography. An Analytical-Numerical Approach*, Mathematical and Life Sciences **7**, De Gruyter, 2020.

[21] W. Demtröder, *Atoms, Molecules and Photons*, third edition, Graduate Texts in Physics, Springer, 2018.

[22] S. Dodelson & F. Schmidt, *Modern Cosmology*, second edition, Academic Press, 2021.

[23] D. G. Duffy, *Green's Functions with Applications*, second edition, Advances in Applied Mathematics, CRC Press, Taylor & Francis Group, 2015.

[24] A. R. Edmonds, *Angular Momentum in Quantum Mechanics*, Princeton Landmarks in Physics, fourth printing, 1996.

[25] L. P. Eisenhart, *Separable Systems of Stäckel*, Annals of Mathematics **35** (1934) 284-305.

[26] L. P. Eisenhart, *Stäckel Systems in Conformal Euclidean Space*, Annals of Mathematics **36** (1935) 57-70.

[27] L. P. Eisenhart, *Separable Systems in Euclidean 3-Space*, Physical Review **45**(1934) 427-428.

[28] L. C. Evans, *Partial Differential Equations*, second edition, Graduate Studies in Mathematics **19**, AMS, 2010.

[29] A. R. Forsyth, *Theory of Differential Equations* in six volumes, Dover, unabridged republication, 1959.

[30] I. M. Gel'fand & G. E. Shilov, *Generalized Functions: Volume 1, Properties and Operations*, AMS Chelsea Publications, 2010.

[31] J. W. Gibbs, Fourier series, Nature **59**, April 27, 1899, p. 606.

[32] C. Gordon, D. Webb, and S. Wolpert, *One Cannot Hear the Shape of a Drum*, Bulletin of the American Mathematical Society **27** (1992) 134–138.

[33] J. Gray, *The Real and the Complex: A History of Analysis in the 19th Century*, Springer Undergraduate Mathematics Series, Springer, 2015.

[34] D. J. Griffiths, *Introduction to Electrodynamics*, fourth edition, Cambridge University Press, 2017.

[35] J. Hadamard, *Lectures on the Cauchy's Problem in Linear Differential Equations*, Dover, unabridged republication, 1952.

[36] P. R. Halmos, *Introduction to Hilbert Space and the Theory of Spectral Multiplicity*, Dover, unabridged republication, 2017.

[37] E. Hille, *Ordinary Differential Equations in the Complex Plane*, Dover, unabridged republication, 1997.

[38] L. Hörmander, *The Analysis of Linear Partial Differential Operators* in four volumes, Classics in Mathematics, Springer, 2003.

[39] E. L. Ince, *Ordinary Differential Equations*, Dover, unabridged republication, 1956.

[40] A. Iserles, *A First Course in the Numerical Analysis of Differential Equations*, second edition, Cambridge Texts in Applied Mathematics, Cambridge University Press, 2009.

[41] M. E. H. Ismail, *Classical and Quantum Orthogonal Polynomials*, Encyclopedia of Mathematics and its Applications **98**, Cambridge University Press, 2005.

[42] D. Jackson, *Fourier Series and Orthogonal Polynomials*, Dover, unabridged republication, 2004.

[43] J. D. Jackson, *Classical Electrodynamics*, third edition, Wiley, 1999.

[44] M. Kac, *Can We Hear the Shape of a Drum?*, The American Mathematical Monthly **73** (1966) 1-23.

[45] Y. Katznelson, *An Introduction to Harmonic Analysis*, Dover, unabridged republication, 1976.

[46] T. W. Körner, *Fourier Analysis*, Cambridge University Press, 1989.

[47] B. G. Korenev, *Bessel Functions and their Applications* (Analytical Methods and Special Functions), CRC Press, Taylor & Francis Group, 2002.

[48] M. Mañas & L. Martínez-Alonso, *A hodograph transformation which applies to the Boyer–Finley equation*, Physics Letters A **320** (2004) 383-388. .

[49] W. Miller, *Symmetry and Separation of Variables*, Encyclopedia of Mathematics and its Applications **4**, Cambridge University Press, 1984.

[50] P. Moon & D. E. Spencer, *Field Theory Handbook (Including Coordinate Systems, Differential Equations, and Their Solutions)*, Springer, 1988.

[51] P. M. Morse & H. Feshbach, *Methods of Theoretical Physics, part I*, McGraw-Hill, 1953.

[52] M. A. Naimark, *Linear Differential Operators*, Frederik Ungar Publishing, 1967.

[53] A. F. Nikiforov & V. B. Uvarov, *Special Functions of Mathematical Physics*, Springer, 1988.

[54] F. W. J. Olver, *Introduction to Asymptotics and Special Functions*, Academic Press, 1974.

[55] P. Olver, *Introduction to Partial Differential Equations* (Undergraduate Texts in Mathematics), Springer, 2016.

[56] J. W. S. Rayleigh, *The Theory of Sound*, in two volumes, Dover, unabridged republication, 1945.

[57] M. E. Rose, *Elementary Theory of Angular Momentum*, Dover, unabridged republication, 1995.

[58] M. Reed & B. Simon, *Methods of Modern Mathematical Physics I. Functional Analysis*, revised and enlarged edition, Academic Press, 1980.

[59] M. Reed & B. Simon, *Methods of Modern Mathematical Physics II. Fourier Analysis, Self-Adjointness*, Academic Press, 1975.

[60] R. Remmert, *Theory of Complex Functions*, Springer, Graduate Texts in Mathematics **122**, 1991.

[61] F. Riesz & B. Sz.-Nagy, *Functional Analysis*, Dover, unabridged republication, 1990.

[62] T. Rivière, *Exploring the Unknown: The Work of Louis Nirenberg on Partial Differential Equations*, EMS Surveys in Mathematical Sciences **9** (2022) 1-29.

[63] P. G. S. Stäckel, *Sur une classe de problèmes de dynamique*, Comptes rendus de l'Académie des sciences de Paris **116** (1893) 485-487.

[64] P. G. S. Stäckel, *Ueber die Integration der Hamilton'schen Differentialgleichung mittelst Separation der Variabeln*, Mathematische Annalen **49** (1897) 145-147.

[65] L. Schwartz, *Mathematics for the Physical Sciences*, Dover, unabridged republication, 2008.

[66] L. Schwartz, *Théorie des Distributions*, Hermann, 1966.

[67] B. Simon, *Real Analysis. A Comprehensive Course in Analysis, Part 1*, AMS, 2015.

[68] I. I. Sobelman, *Atomic Spectra and Radiative Transitions*, second edition, Springer Series on Atoms and Plasmas, Springer, 1992.

[69] I. Stakgold & M. Holst, *Green Functions and Boundary Value Problems*, third edition, in Pure and Applied Mathematics Series, Wiley, 2011.

[70] G. Szegő, *Orthogonal Polynomials*, AMS Colloquium Publications **23**, AMS, reprinted with corrections, 2003.

[71] A. Tijonov & A. Samarski, *Equations of Mathematical Physics,* Dover, unabridged republication, 2011.

[72] F. Trèves, *Linear partial Differential Operators with Constant Coefficients*, Gordon & Breach, 1966.

[73] F. Trèves, *Topological Vector Spaces, Distributions and Kernels*, Academic Press, Pure and Applied Mathematics **25**, 1967.

[74] V. S. Vladimirov, *Equations of Mathematical Physics*, Mir, 1984.

[75] G. N. Watson, *A Treatise on the Theory of Bessel Functions*, Cambridge Mathematical Library, Cambridge University Press, 1995 (Second edition first published in 1944).

[76] G. B. Whitham, *Linear and Nolinear Waves*, John Wiley & Sons, 1974.

[77] E. T. Whittaker & G. N. Watson, *A Course of Modern Analysis*, Cambridge University Press, 1927.

[78] H. Wilbraham, *On a certain periodic function*, The Cambridge and Dublin Mathematical Journal **3** (1848) 198–201

Subject Index

Index of Capsule Biographies

Printed in the United States
by Baker & Taylor Publisher Services